KU-627-591

Promises Kept

JOHN F. KENNEDY'S NEW FRONTIER

Irving Bernstein

OXFORD UNIVERSITY PRESS
New York Oxford

Oxford University Press

Oxford New York Toronto
Delhi Bombay Calcutta Madras Karachi
Kuala Lumpur Singapore Hong Kong Tokyo
Nairobi Dar es Salaam Cape Town
Melbourne Auckland Madrid

and associated companies in
Berlin Ibadan

First published in 1991 by Oxford University Press, Inc.,
200 Madison Avenue, New York, New York 10016

First issued as an Oxford University Press paperback, 1993

Oxford is a registered trademark of Oxford University Press

Library of Congress Cataloging-in-Publication Data
Bernstein, Irving, 1916–
Promises kept : John F. Kennedy's new frontier / Irving Bernstein.
p. cm.
Includes bibliographical references.
ISBN 0–19–504641–2
ISBN 0–19–508267–2 (PBK.)
1. United States—Politics and government—1961–1963.
2. United States—Politics and government—1963–1969.
3. United States—Economic policy—1961–1971.
4. United States—Social policy.
5. Kennedy, John F. (John Fitzgerald), 1917–1963.
6. Johnson, Lyndon B. (Lyndon Baines), 1908–1973. I. Title.
E841.B43 1991
973.922—dc20 90–33287

10 9 8 7 6 5 4 3 2 1
Printed in the United States of America

To
Deborah, Jonathan, Judith,
and the memory of David

Preface

Two years, ten months, and two days. That was the brief span of the sparkling and prolific presidency of John F. Kennedy. He planted many seedlings, but lived only long enough to see some flower. Much remained to be done. His accomplishments and those of his successor, Lyndon B. Johnson, are closely linked. This is an important reason for the controversy over the quality of the Kennedy presidency.

Walter Heller was chairman of Kennedy's Council of Economic Advisors and worked with him intimately. In an oral history he made for the Johnson Library in 1971 Heller said, "Had he lived, he would have been a great President. No one is going to disabuse me of that with some of the tinhorn criticisms I see of him in revisionist histories." Who was right, Heller or the revisionists? This book answers that question.

The main focus here is on domestic problems and policies. For these subjects there are large collections of manuscripts in the Kennedy Library, the National Archives, and the Johnson Library, among others. Equally important, there are big secondary literatures on such particular topics as civil rights, tax policy, federal aid for education, and Medicare. Surprisingly, no one in recent years has brought this material together. I have sought to do that here.

The sixties constitute a discrete era in American history. The assassination of John Kennedy on November 22, 1963, ended his life and his Administration, but it barely grazed the issues of the decade. They fell on both sides of that divide. They confronted Johnson as they had Kennedy. This book, therefore, is only half the story. I intend to follow it with another on LBJ's Great Society.

A study of this sort is inherently a collaborative enterprise. An author must acknowledge his large debts to individuals and institutions who extended him essential support. I am particularly grateful to Professor Robert Dallek of UCLA and to Sheldon Meyer, Senior Vice President Editorial, of the Oxford University Press. For reasons that I do not fully understand, perhaps a combination of intellectual sluggishness, myopia, and advancing age, I had a good deal of trouble shaping the structure of this volume. Between them these two devoted bookmen helped me to think through the problems and, it is hoped, resolve them.

I am continually astonished and delighted by the cheerful assistance that librarians and archivists offer me as a matter of course. I doff my cap in particular to those at the Kennedy Library at Columbia Point in Boston and, once again, to those at the UCLA Library system, as well as to those at the Johnson Library in Austin and the National Archives.

The quality of the manuscript has benefited significantly from critical readings by the following individuals: James L. Sundquist, Senior Fellow Emeritus of the Brookings Institution; Willard Wirtz, who was both Kennedy's and Johnson's Secretary of Labor; Lawrence F. O'Brien, who headed congressional relations for both Presidents; James Tobin of Yale, who was a member of Kennedy's Council of Economic Advisors; Walt W. Rostow of the University of Texas, who was an adviser to both Kennedy and Johnson; Francis Keppel of Harvard, who was Kennedy's commissioner of education; William Josephson, who was general counsel of the Peace Corps; and to my UCLA colleagues Professors Benjamin Aaron and Leonard Freedman.

The Academic Senate Research Committee at UCLA has helped me finance travel to the archives mentioned. The Institute of Industrial Relations at UCLA has generously provided funds to allow me to engage Cynthia Taminga to gather bibliography and Rebecca Frazier to prepare the index. I am most grateful.

Sherman Oaks, California Irving Bernstein
April 1990

Contents

Promises Kept

JOHN F. KENNEDY'S NEW FRONTIER

PROLOGUE

The Presidency

DEMOCRATS AND REPUBLICANS: Tweedledum and Tweedledee. Many Americans thought it did not matter, that there was no meaningful difference between the parties. They were not active in politics and many did not even bother to vote. Those who were involved in politics and were concerned about public policy disagreed.

Dean Acheson placed himself among the dissenters. By background, education, position in society, and style he seemed to be a prototypical Republican. But, in fact, he was a dedicated Democrat. In 1956 he wrote an essay entitled *A Democrat Looks at His Party*, which aroused considerable interest. Acheson favored the Democrats over the Republicans for two important and linked reasons.

The first was because, as Jefferson had insisted at the outset, the Democratic party was "the party of the many." Acheson pointed out that the Republican party, by contrast, "centers on the interests deriving from property in its most important form." The party of the many, the Democratic, inevitably became "a party of many interests, including property interests." It brought together in uneasy alliance urban workers, small merchants and bankers, farmers, southern landholders, immigrants, and intellectuals. Republicans, holding to a central interest, tended to be united and orderly. Democrats, having many interests, tended to faction and were unruly.

This brought Acheson to his second reason for favoring the Democrats. It was no great challenge for a Republican leader to hold his party together; for a Democratic leader, on the other hand, this was a major burden. As a result,

Republican Presidents tended to be ineffective, Democratic Presidents forceful. Since the nation often faced crises, foreign and domestic, Acheson much preferred the party of "strong and vibrant personalities" who "redirected the party to meet the emerging problems of each new era."

The historical record supported Acheson's analysis. In 1962 the elder Arthur M. Schlesinger, himself a distinguished American historian, took a poll of seventy-five of his expert colleagues to rank the Presidents as great, near great, average, below average, or failures, excluding the first Harrison and Garfield, who served too short a term to allow for judgment. The numerical ranking is shown in parentheses.

There were five great Presidents, of whom three were Democrats: Franklin Roosevelt (3), Wilson (4), and Jefferson (5); one was a Republican, Lincoln (1); and one had no party, Washington (2). There were six near greats, of whom four were Democrats: Jackson (6), Polk (8), Truman (8), and Cleveland (10); one Republican, Theodore Roosevelt (7); and one Federalist, John Adams (9).

There were two failures, both Republicans, Grant (28) and Harding (29). Six Presidents came out below average: three Whigs, Taylor (22), Tyler (23), and Fillmore (24); two Democrats, Pierce (26) and Buchanan (27); and one Republican, Coolidge (25). It is significant that the second-rate Democrats served just prior to the Civil War when their party was torn apart by the slavery controversy. There were 12 average Presidents. This is the category with the highest concentration of Republicans, seven in all. They were Hayes (13), McKinley (14), Taft (15), Hoover (18), Harrison (19), Arthur (20), and Eisenhower (20). There were four average Democrats: Madison (11), J. Q. Adams (12), Van Buren (16), and Monroe (17). There was one Unionist, Andrew Johnson (21).

The general conclusion is clear: Democratic Presidents tended to come out at the higher end in the ratings, Republicans at the lower. This was dramatically the Democratic case in the twentieth century prior to 1960, that is, during most of the lifetime of Dean Acheson. The three Democratic Presidents were either ranked great—Wilson and FDR—or near great—Truman. The reason for Democratic dominance, James MacGregor Burns has pointed out, was that historians applied essentially the same criterion: "one crucial characteristic—strength in the White House. The great Presidents were the strong Presidents." They were the masters of events, they influenced history, they shaped the nation's destiny, and they brought in talented administrators and advisers.

On January 14, 1960, John F. Kennedy, as he launched his campaign for the White House, addressed the National Press Club on his image of the presidency. He agreed with Dean Acheson. Kennedy excoriated President

Eisenhower's vision of his office. While hopes had been "eloquently stated," he had not followed through. Eisenhower had failed "to override objections from within his own party, in the Congress or even in his Cabinet." Perhaps "this detached limited concept of the Presidency" was appropriate in the fifties, when the nation needed "a time to draw breath." But now the country was entering a new decade, "the challenging and revolutionary sixties," which will demand more than "ringing manifestoes issued from the rear of the battle." It will be vital for the President to "place himself in the thick of the fight, that he care passionately."

> He must above all be the Chief Executive in every sense of the word. He must be prepared to exercise the fullest powers of his Office—all that are specified and some that are not. He must master complex problems as well as receive one-page memorandums. He must originate action as well as study groups. He must reopen the channels of communication between the world of thought and the seat of power. . . .
>
> We will need . . . what the Constitution envisaged: a Chief Executive who is the vital center of action in our whole scheme of Government. . . .
>
> It is the President alone who must make the major decisions of our foreign policy.
>
> That is what the Constitution wisely commands. And, even domestically, the President must initiate policies and devise laws to meet the needs of the Nation. And he must be prepared to use all the resources of his office to insure the enactment of that legislation—even when conflict is the result.

Was this empty campaign rhetoric or was Kennedy genuinely committed to the Democratic tradition of the strong President? More important, did he prove himself an activist and effective leader when he occupied the office? The historiography since his death on November 22, 1963, has gone through two contradictory phases, which is typical of a very important national figure.

The shock of the Kennedy assassination and the two early biographies—Arthur M. Schlesinger, Jr.'s *A Thousand Days* (1965) and Theodore C. Sorensen's *Kennedy* (1965)—temporarily shut off debate. Both authors were insiders who worked for Kennedy, knew him intimately, and held him in the highest esteem. Both were also extremely good writers and their books solidified the favorable image of the assassinated President that prevailed in the late sixties.

But in the seventies a revisionist assessment of the Kennedy performance emerged. Those who wrote from this viewpoint were hardly a "school" because they shared little else in common. But they did agree that Kennedy's legislative program had been a failure, that he had been unable to move a reluctant Congress to accept the policies he advocated.

In *The Promise and the Performance: The Leadership of John F. Kennedy* (1975),

Lewis J. Paper wrote of Kennedy's relationship with Congress that "he was a skeptic who doubted his ability to take more than that first step of a thousand-mile journey." Grant McConnell in *The Modern Presidency* (1976) found that Kennedy was "perhaps the most cautious of the Democratic presidents" who left "a record of few achievements." In *Decade of Disillusionment: The Kennedy-Johnson Years* (1975), Jim F. Heath wrote that JFK "achieved few concrete results" and that he was "just a mediocre President." Henry Fairlee in *The Kennedy Promise: The Politics of Expectation* (1973), claimed that Kennedy was a man of words, not of deeds, that his "achievements were less significant than those of James K. Polk."

Alan Shank in *Presidential Policy Leadership: Kennedy and Social Welfare* (1980) studied housing, education, Medicare, and civil rights and concluded that "his overall performance in domestic policy can only be rated as disappointing. . . . Kennedy's caution with Congress indicated an understanding of the prevailing opposition coupled with a discomfort of the legislative process." He did not "take advantage of all presidential powers in functioning as a legislative leader."

In *JFK: The Presidency of John F. Kennedy* (1983), Herbert S. Parmet declared that Kennedy

> capitulated to Congress. He followed a domestic course that precluded battling for the fulfillment of the economic and social welfare needs of the Democratic Party's postdepression constituency. His effectiveness on Capitol Hill was limited, and he even appeared submissive. He had vowed to "get America moving again" but failed to deliver in key ways.
>
> He found that the much-advertised "corridors of power" were really Byzantine labyrinths. His assessment argued for caution, for harnessing resources to fight the real battles some other day.

Gary Orfield's *Congressional Power: Congress and Social Change* (1975) was a special case. His basic argument for the era following World War II was that "Congress is more important and less conservative than is generally believed." Orfield contended that, excepting Lyndon Johnson's 1964 landslide, there had been "a long-term condition of close partisan division in the country." Truman in 1948, Kennedy in 1960, and Nixon in 1968 won by "razor-thin electoral margins." Even Eisenhower, who won handsomely in 1952 and 1956, was unable to break the stalemate with Congress. According to Orfield, Kennedy was trapped in this logjam and was unable to move a recalcitrant Congress.

In *Pragmatic Illusions: The Presidential Leadership of John F. Kennedy* (1976), Bruce Miroff was a critic on the left. An avowed revisionist, he sought to reinterpret the New Frontier as an attempt to "rationalize the economy in a manner favored by the most influential of the corporate elite." Corporate power in America, according to Miroff, was so formidable that every liberal President—T.R., Wilson, FDR, Truman, JFK—became its hostage. "Once we set aside the assumption that the twentieth-century Presidency has been a progressive institution, it is possible to see that Kennedy's record is not untypical."

Kennedy's performance, Miroff argued, "cannot . . . sustain his reputation as a progressive" because of his "service to established power and established values." His Keynesian economic program was designed as a partnership with business to lift the whole economy. But the main beneficiary "in terms of power, ideological sanction, and money—was an ungrateful corporate community." With civil rights Kennedy was afforded the opportunity to establish "a truly creative relationship with the black movement." But he missed it because he was "rooted in elite values and assumptions [and] could not come to terms with a politics of mass participation."

This book, based on manuscript sources and the large secondary literature published since 1963, is an assessment of John F. Kennedy's performance on domestic issues. The naked conclusion is that he was a very successful President, that the revisionists were dead wrong.[1]

1

America in 1960

THE COLD WAR set the main stage on which post–World War II presidents acted out their presidencies, starting with Truman and Eisenhower. The conflict between the superpowers dominated American foreign policy and made wide inroads into domestic life in the U.S. It had its own dynamic: for the most part the Soviet Union acted and the United States reacted. Thus, depending upon the Kremlin leadership, the Cold War was sometimes "colder" than at other times.

It was never more frigid than at the end of World War II. Stalin asserted Soviet dominance over all of eastern Europe—Finland, the Baltic states, Poland, East Germany, Czechoslovakia, Hungary, Rumania, Bulgaria, and Yugoslavia. The U.S.S.R. annexed the Japanese islands in the North Pacific and occupied Manchuria and northern Korea. In 1949, Mao Tse-tung's Communists took over China. Also in that year the Soviet Union built its first atomic bomb and in 1953 added the hydrogen bomb.

Jolted, the U.S. and its European allies moved to contain this juggernaut. In 1947 the Truman Doctrine, by providing aid to Greece and Turkey, established a policy "to support free peoples who are resisting attempted subjugation." That year, as well, the Marshall Plan was launched to rebuild war-torn western Europe. In 1948, when the Soviets blockaded Berlin, the U.S. countered with a massive airlift. NATO was established in 1949 for the common defense of the West. In 1950, after North Korea invaded, American forces landed to defend South Korea. The presidency of Harry Truman was devoted primarily to waging the Cold War.

But Stalin died on March 5, 1953, and the conflict warmed to a "thaw."

The collective successors—Malenkov, Molotov, Khrushchev, and Beria (the feared head of the secret police, who was soon executed)—relaxed Stalinist repression in both the Soviet Union and the eastern European satellites and sought "peaceful coexistence" with the West. An armistice ended the Korean War in 1953 and the peninsula was divided at the thirty-eighth parallel. By 1955, Soviet forces had withdrawn from Austria, Finland, and Port Arthur, the U.S.S.R. had established normal relations with West Germany and Japan, and the Soviet Union had recognized Tito's independent path in Yugoslavia. At the 20th Party Congress in 1956, Khrushchev dramatically denounced Stalin and Stalinism.

The captive nations of eastern Europe, savoring their gains, demanded more. There were great demonstrations in 1953 in East Berlin and in 1956 in Poland which were resolved without bloodshed. But the Hungarian uprising in 1956 spun out of control, and the Soviet Union used tanks and artillery to crush the rebellion.

Dwight Eisenhower became President on January 20, 1953, and his first term coincided almost exactly with the thaw in the Cold War. Thus, he became the "peace" President. He took domestic credit for the Korean armistice and improved relations with the Soviet Union. In 1956 the voters returned him for a second term by a large majority.

But in 1957 the Cold War entered a new and much more dangerous phase. Khrushchev won supreme power in the Kremlin and retired the other members of the collective. While shunning Stalin's ruthlessness and cruelty, he was determined that the Soviet Union should surpass the U.S. militarily and economically. In 1957 the Soviets launched two *Sputniks* in space and tested a very powerful hydrogen bomb. These were dramatic achievements in themselves and also demonstrated that the U.S.S.R. could build intercontinental missiles capable of striking the U.S. with atomic weapons. Another shocking fact was that the Soviet Union had leapfrogged the U.S. technologically. In 1958–59 there was a second Berlin crisis when Khrushchev demanded that the western powers evacuate the city. But the West held firm and he did not make good his veiled threat to resort to atomic warfare. In 1959, Castro's revolution triumphed in Cuba and, shortly, a Soviet client stood only ninety miles from Florida.

Starting in 1956, the U.S. had flown U-2 reconnaissance planes at very high altitudes over the U.S.S.R. to photograph military installations. A summit conference to deal with the festering Berlin question was scheduled to open in Paris on May 16, 1960. But on May 5, Khrushchev announced that a U-2 flight from Pakistan to Norway had been shot down 1250 miles inside Soviet territory; its espionage equipment had been recovered and its pilot, Gary Powers, had been captured. At Paris, Khrushchev made impossible demands upon an embarrassed Eisenhower and, when they were rejected, aborted the summit.

Thus, Eisenhower's second term was the reverse of his first. He was in continual confrontation with Khrushchev and, if he was not losing the Cold War, he was certainly having trouble keeping up. On the domestic side there was a similar pattern of sharp difference between Eisenhower's first and second terms.

Later many Americans, particularly conservatives, looked back at the decade of the fifties as an era of good feelings, a time often described as "placid," "quiet," "contented," "complacent." The country, it was said with satisfaction, enjoyed both peace and prosperity. President Eisenhower was felt to be a comforting symbol of consensus, confidence, and calm with his quintessentially American avuncular style. Eisenhower, Richard H. Rovere wrote, "hates nothing so much as a scene and shuns controversy on any plausible excuse." This analysis was only partly true.

The quiet period lasted not eight years, but four. It began with the Korean armistice in 1953 and collapsed in 1957 with violent racial conflict in Little Rock and the worst recession since World War II, to say nothing of *Sputnik*. For Republicans, 1958 was, if anything, worse. They suffered a devastating defeat in the congressional and state elections in November.

Eisenhower, in fact, immersed himself almost totally in the issues of war and peace. He seemed uninterested in, bored by, and uninformed about domestic problems and policies. Rovere wrote: "Once Eisenhower has found a first-class automobile dealer, cotton broker, or razor manufacturer to head a department, he has acted as if the public interest has been satisfied and his own responsibility discharged." Occasionally, and to his annoyance, a domestic crisis compelled his attention, as when he had to send the Army to Little Rock to assure compliance with a court order desegregating the schools. But this event and others were no more than momentary diversions. Eisenhower, his administration, and his party seemed to have neither sense of nor interest in the profound changes under way in American society in the fifties.[1]

In the decade and a half between the end of World War II in 1945 and the election of John F. Kennedy as President in 1960, American society was transformed. New forces compelled modifications of virtually every institution and activity. The greatest engine of change was the family. In an atomic age it seemed as though the tremendous energy unleashed by the shattering of the atomic nucleus had flowed into the reproductive process. The result, of course, was babies . . . in enormous numbers. In the hackneyed terms of the time, this was the "population explosion," "the baby boom."

During the Great Depression of the thirties, young people tended either to remain single or to defer marriage. A dramatic change began as war loomed in 1940. In the census decade 1940–50, the percentage of persons never married and the median age at first marriage dropped by as much as they had during the preceding half-century. The trends toward marrying and doing so at an earlier age continued during the 1950–60 decade. Thus, the number of married couples in the United States skyrocketed from 28.2 million in 1945 to 38.9 in 1960, a gain of 10.7 million. Further, those women who married and some who did not bore many more children than their predecessors. The birth rate—that is, the number of children born annually per 1000 women—rose from about 19 during the depression to around 21 during the war and to about 25 during the fifties.

The consequence was an enormous annual increase in births. During the thirties the number of babies born yearly was usually fewer than 2.5 million. In the war period this figure surpassed 3 million only in 1943. The big leap upward began immediately after the war, from 2.9 million in 1945 to 3.4 million in 1946. This was followed by a steady climb, passing 4 million in 1954 and remaining above that mark through 1960. The total number of babies born between 1946 and 1960 was a staggering 59.4 million. By 1960, the number of persons under age 15 was 55.8 million. Both the number of births between 1946 and 1960 and the number under 15 in 1960 much exceeded the total population of the United Kingdom, France, or Italy.

This enormous increase in births had two critical demographic consequences. The first was a sharp rise in the size of the American population. It leapt upward from 141 million in 1946 to 180.6 million in 1960, a gain of 39.6 million people. Immigration, of course, contributed to this growth, but, because of restrictions, only minimally. Population also grew because of a decline in the death rate. The result was a marked increase in the number of old people. Life expectancy advanced from 62.9 years in 1940 to 69.7 in 1960, a gain of almost 7 years. There were 9 million people 65 and over in 1940 and 16.7 million in 1960, a rise of 7.7 million. Further, it was certain that the number of the aged would continue to increase rapidly thereafter.

The second significant result of the expansion in the number of births was a new pattern of population distribution. Most of the postwar growth flowed into the suburbs. The number of people living on farms continued in secular decline, from 30.5 million in 1940 to 15 million in 1960. These people migrated into the towns. But large cities did not grow. Of the sixteen with a population over 600,000 in 1960, eleven declined in size over the preceding decade. Of the five that grew, three were in the booming Southwest—Los Angeles, Houston, and Dallas—but Los Angeles was hardly a typical concentric city and the Texas cities increased in large part by annexation of fringe areas. Thus, the great movement of population was to the urban fringe, to the suburbs, and to nonfarm rural areas, part of which were suburban. According to Landon Y. Jones, "No less than 83 percent of the total population growth in the United States during the 1950s was in the suburbs, which were growing fifteen times faster than any other segment of the country."

A different redistribution took place among the black population. While the white birth rate was extremely high between 1945 and 1960, the black rate was even greater. During the fifties, Negro women averaged nearly four children per mother, the highest fertility rate of any group since the Civil War. The black population grew from 12.9 million in 1940 to 18.9 in 1960, a gain of 6 million. The rural South, where Negroes had historically concentrated, could not support this great increase in population. Although black migration to urban areas had begun earlier, it accelerated dramatically during World War II and continued at a high rate after the war. By 1960, therefore, a very large number of rural southern Negroes had settled in cities, particularly in the North. In the latter year two-thirds of the black population lived in metropolitan areas, half the total in central cities. Here, where whites with rising incomes were moving to the suburbs, Negroes lived in vast ghettos in towns such as Washington, Philadelphia, New York, Cleveland, Detroit, and Chicago, along with the California cities. They were joined by large numbers of Puerto Ricans in New York and Mexicans in the southwestern cities such as Los Angeles, San Antonio, and Houston.

As the experience of Negroes demonstrated, in the fifteen years following the war, population redistributed itself regionally, with a marked tilt for whites to the West and the South. The largest percentage increase from interregional migration took place in the mountain states, particularly Arizona, Nevada, and New Mexico. The greatest number flowed into California. Texas and Florida were also large gainers. The East, the Midwest, and much of the South lost population from emigration. Interregional migration for whites, unlike blacks, was between cities. Typically, a white family moved from a Chicago suburb to a Los Angeles suburb. The boom towns grew phenomenally. Between 1940 and 1960, Los Angeles increased in population from 1.5 to 2.5 million, San Diego from 203,000 to 573,000, Houston from 385,000 to 938,000, Seattle from 368,000 to 557,000, Phoenix from 65,000 to 439,000.[2]

In the postwar era the nation became, as Max Lerner put it, "Suburbia, U.S.A." Millions of Americans, especially young couples and their offspring, spilled into the rims of the cities. The exact number is not known. This is because no one, including the Bureau of the Census, could define a suburb precisely. Metropolitan areas differed markedly one from another. In many cases it was impossible to draw the line between either the central city and the suburban ring or between the latter and the rural sector beyond. The Census offered two different concepts: Standard Metropolitan Areas and Urbanized Areas. Louis H. Masotti proposed a not unreasonable, though hardly precise, definition of suburbia: "Politically independent residential communities outside the corporate limits of a large central city, but within reasonable commuting distance of it, and culturally and economically dependent upon it." The uncertainty over definition was evident in a fussing over names. Most writers accepted "suburb," derived from the Latin *suburbium*, by which the Romans meant near the city. Others wrote, often with greater imprecision, of "the urban ring," "the urban fringe," "the city's rim," or "the cluster city." One writer even distinguished between "the Urban Fringe" (caps) and "the urban fringe" (l.c.).

No matter how defined, the suburbs received many millions of Americans between 1945 and 1960. One illustration suggests the magnitudes. The original Levittown arose on potato fields in New York's Nassau County on Long Island. The first families moved into their new rented houses in 1947. When Levitt started to sell homes on a Monday evening in March 1949, the firm made 1400 contracts in three and one-half hours. By 1960, there were 15,741 units in Levittown, housing 65,276 people. This pattern spread across the nation, though less markedly in the South. Both presidential candidates in 1960 had been born and raised in suburbs, Kennedy in Boston's Brookline and New York's Bronxville, Nixon in Yorba Linda and Whittier, appendages of the sprawling Los Angeles metropolis. Prior to World War II suburban living had been atypical; after the war it became the dominant American way of life. The change, in Lerner's view, was "revolutionary."

The reasons for the great move to the suburbs were patent. Many white families left because they wanted a better education for their children than the deteriorating schools in the central cities provided. There was a severe shortage of housing, estimated in 1946–47 at 3 to 6 million units. Very few homes had been built in the preceding decade and a half and now there was a population boom. Young adults held steady jobs and enjoyed rising incomes. They owned automobiles for commuting to work. Government at all levels built roads, culminating in 1956 in the launching of the Interstate Highway System, for starters $37 billion for 41,000 miles of expressways. The suburbanites were eager to own single-family homes and could afford them if the terms were favorable. The FHA and the VA guaranteed mortgages. In a risk-free market lending institutions made loans with long-term paybacks at low interest rates. Land was cheap in the outskirts of the cities. Developers rushed in to buy this land and build the houses to meet the great demand. Some of the developments were enormous, like the Levittowns outside New York, Philadelphia, and Trenton, Park Forest near Chicago, Park Merced in San Francisco, and Drexelbrook in Philadelphia. Many more were smaller. Few were planned and most developers squeezed every house they could onto the available land, leaving the construction, if any, of shopping centers, schools, churches, and parks to others.

There was a sameness to these developments and a stereotypical image emerged. There was row upon row of new small houses of the same basic designs but with cosmetic external differences—Cape Cod, ranch-style, split-level. The streets curved, the lawns were neatly mowed, two cars nestled in virtually every garage, young trees reached for the sky, and bicycles and tricycles remained where they had been left on sidewalks. The kitchens were fitted with labor-saving utilities and, of course, there was a television in every living room. Nearby, often, was a new school, a modern shopping center, and a rakish church.

The occupants, in significant part, were cut from the same pattern. At Park Forest, according to William H. Whyte, Jr., "the modal man . . . is a twenty-five to thirty-five-year-old white-collar organization man with a wife, a salary between $6,000 and $7,000, one child, and another on the way." Likely as not, he was an engineer, in middle management, a lawyer, a salesman, an insurance agent, a teacher, a bureaucrat, a junior officer in the military—all middle-class. Both Mr. and Mrs. Modal had some college. Mr. commuted to work, usually in the central city. During the day the suburb became a matriarchy devoted to child-watching and child-rearing. The women held daily kaffeeklatschs at which they gossiped or, in serious moments, discussed Spock's *Infant Care* or Gesell's *The First Five Years of Life*.

Suburbia was consumerland, in *Fortune*'s words, "the most important single market in the country." Here were millions of families with relatively high incomes that allowed for discretionary buying. Manufacturers, marketers, and advertising agencies took direct aim at them, relying on what *Fortune* called "the science of consumption." The sellers hustled houses, automobiles, white goods, furniture, TVs, diapers, bikes, barbecues, casual clothes, and "the fabulous market for food." But the editors of *Fortune* were writing in 1955 and that was only the beginning. The prospects for the future, they glowed, were "exhilarating."

The rise of suburbia induced a flood of newspaper and magazine articles, novels, movies, TV shows, and scholarly analyses. Everyone wanted to know what was happening out there. The answers often produced culture shock. The critics attacked suburbia on three grounds.

The first was that suburban living imposed conformity on the residents.

Lewis Mumford wrote of "a multitude of uniform, unidentifiable houses, lined up inflexibly, at uniform distances, on uniform roads, on a treeless communal waste, inhabited by people of the same class, the same income, the same age group. . . ." David Riesman deplored the lack of "diversity, complexity, and texture." Where, he asked, were the old people, the servants, the teenagers? The women were chained to their homes by their children and the TV soap operas. Whyte posited that Americans were being transformed into automatons, organization men, by the large bureaucratized institutions for which they worked and by "the communal way of life" in the "packaged" suburbs. They were dominated by the conformity in lifestyle that governed the suburbs. This even extended to politics. "People from big, urban Democratic wards," Whyte wrote, "tend to become Republican and, if anything, more conservative than those whose outlook they are unconsciously adopting."

The second criticism was that life in the suburbs was barren, in Mumford's words, "a low-grade uniform environment from which escape is impossible." Unless the residents returned to the city from which they had fled, they found no esthetic or intellectual stimulation. The suburb was not only child-centered; "it was based on a childish view of the world." Riesman titled his essay "The Suburban Sadness." He found "aimlessness, a pervasive low-keyed unpleasure." Suburban families had lost control over their lives. For those who had taken "the vows of our organization life," Whyte wrote, the dominant values had become security, "love that system," "the fight against genius."

Finally, Mumford insisted, the divorcement of suburb from city defied history and reason. Archaeologists had dug up the suburbs of Ur in Mesopotamia. "All through history, those who owned or rented land outside the city's walls valued having a place in the country." They sought escape from the city's disease, filth, and congestion, but they kept the link to the city's bustle and civilization. Only now did suburbanites try to sever that cord. The modern suburb, he claimed, was "an asylum for the preservation of illusion." In fact, it was indissolubly tied to the city, economically, politically, culturally. A. C. Spectorsky playfully proposed that an escapee from New York should leapfrog over suburbia into what he called exurbia, where the people were richer and the houses more elegant.

Several intensive studies revealed that these criticisms were based on an image of the suburbs which was only partly true, that suburbs differed in kind and changed over time. Class, these writers concluded, was more important than where people lived. Not all suburbs were middle-class; some were working-class. Suburban living did not much change the values the residents had brought with them.

Ford in 1955 closed its assembly plant in grimy Richmond on the east side of San Francisco Bay and opened a new factory in semi-rural Milpitas at the south end of the Bay. Virtually the entire work force transferred and became homeowners in a new suburb near Milpitas. Would suburbanization and home ownership convert these workers into Whyte's organization men? After careful study, Bennett M. Berger concluded that they had not changed. While the auto workers preferred living in the suburbs and were proud of their homes, their styles and values remained working class. Insofar as politics went, four-fifths said they were Democrats and Eisenhower received only 18 percent of their votes in 1956.

Levittown on Long Island, the oldest of the mass suburbs, allowed for a comparison between 1950 and 1960. William M. Dobriner, again, found that class rather than the suburb was the decisive force. In 1950 no one had been rich and none had been poor. The residents were young, homogeneous, "middle class America." Over the decade, those whose incomes had risen bought more expensive homes and sold their Levittown houses to craftsmen, foremen, skilled and semi-skilled workers, who were heavily Catholic. The community was now bipolarized between the values of the middle class and the working class. A conflict over education arose. Middle-class parents, mainly Protestant and Jewish, wanted academically oriented schools that prepared their children for college and were willing to pay the price. Working-class parents, mainly Catholic, were little interested in curriculum and opposed higher taxes. Moreover, as Mumford argued, Nassau County was growing and industrializing rapidly. Levittown was no longer a suburban island. It was becoming increasingly urban, part of both a local city and the vast urban mass of Greater New York.

Levitt and Sons announced in 1955 that it had purchased virtually all of rural Willingboro Township, New Jersey, seventeen miles from Philadelphia, for a third Levittown. Herbert J. Gans, a sociologist, was among the first to move into the community in 1958. He lived there for two years and returned for two more of interviewing.

Most of the residents, Gans found, liked their homes and the outdoor living, found compatible friends, and were not bored. Certain groups suffered social isolation—older couples, the well educated, the poorly educated, women who came from a working-class or strongly ethnic or extended family background, and adolescents. People brought their lifestyles and values with them; the suburb produced little change. Rather than being homogeneous, Levittown was "conflict-ridden." There were disputes between the dominant lower-middle-class and the smaller working- and upper-middle-class groups over education, taxation, political power, teenagers, and legally mandated desegregation.

Did people vote Republican because they had become homeowners and suburbanites (conversion) or did they bring their politics with them from the city (transplantation)? Politicians and political analysts struggled with this question. Robert C. Wood, weighing the conflicting evidence, concluded that both factors were at work in the fifties with the probability that transplantation was more important. But there were qualifiers. In the suburbs, as in the nation as a whole, party allegiance was eroding. Voters increasingly split their tickets in presidential, congressional, gubernatorial, and state legislative elections. Suburbanites tended to be apathetic toward party identification in these broader canvasses. They were, however, deeply concerned about local issues—schools, zoning, streets—but on a nonpartisan basis. The evidence, Wood wrote, "suggests that the distinguishing mark of suburban politics is not a difference in partisan attitudes between suburbanites and other Americans. If any conclusion is possible . . . , it is that the suburbs, as such, have no significant influence on the fortunes of the two parties."

The great sleeping issue in suburbia, as in the nation as a whole, was race. The suburbs, of course, were almost exclusively white. While there was a handful of segregated black developments and a few middle-class Negro families bought homes in the white areas, they were the exceptions. There seem to have been no blacks at all in Park Forest. The New York Levittown had 220 nonwhite residents in 1960, an insignificant proportion of its population of 65,276. Levittown, New Jersey, grudgingly agreed to admit a few black families when required to do so by state law. The reasons for black exclusion were clear. Poor rural Negroes who migrated north could not afford suburban homes. Most would probably have been uncomfortable living in a white middle-class community. Many, though not all, whites did not want black neighbors. Some whites were committed segregationists and had fled the increasingly black central cities in order to escape racial integration. Developers, concerned about property values, had no interest in social engineering in so sensitive an area. Towns used zoning to screen out the poor, which, of course, included most blacks.[3]

The word "ghetto" seems to be derived from the Italian *gietto*, the name for the Venetian cannon foundry near which the walled Jewish quarter was established in 1516. Jews had lived in segregated settlements in European and North African cities long before the name was invented, some by law, others by choice. With time, the name was applied to other Jewish neighborhoods in Europe and throughout the world, including the United States. It also came into use as the name for urban quarters of other ethnic groups, particularly blacks.

American blacks had lived in ghettos prior to World War II, notably Harlem and the black Belt on Chicago's South Side. But the numbers were relatively small. They swelled enormously in the fifteen years following the war, in substantial part by natural increase, but also by the huge migration from the rural South. There were four streams. Many moved from the farms to the cities of the South, particularly into Atlanta, Birmingham, New Orleans, and Houston. Another group migrated up the eastern seaboard from Georgia, the Carolinas, and Virginia into Washington, Philadelphia, Newark, and New York. A third, especially from Mississippi and Alabama, took the route of the Illinois Central Railroad north to Chicago and then into other industrial cities of the Middle West. Finally, blacks from the Southwest moved to Los Angeles and the San Francisco Bay area, especially Oakland.

Between 1940 and 1950 there had been a net exodus of 1,244,700 Negroes from the South to the North. In the decade ending in 1960 the total was 1,457,000. In 1910, 73 percent of blacks had lived in rural areas; by 1960, 73 percent resided in cities. In the latter year, in fact, Negroes were more urbanized than whites. Moreover, in 1960, almost 80 percent lived in the central cities of metropolitan areas, that is, mainly in ghettos. The black population of the five biggest cities was: New York, 1.1 million, 14 percent of the city total; Chicago, 890,000, 24 percent; Philadelphia, 529,000, 26 percent; Detroit, nearly half a million, 29 percent; and Los Angeles, 464,000, 14 percent. Several southern and border cities had even higher concentrations: Baltimore, 35 percent; Houston, 23; Washington, 54; St. Louis, 29; New Orleans, 37; Memphis, 37; and Atlanta, 38 percent. There were, however, important cities in the northern tier and the West in which the Negro population was spare: Boston, Milwaukee, Minneapolis, Seattle, Denver, San Antonio, Phoenix, and San Diego.

By 1960 American cities were almost totally segregated. Karl E. and Alma F. Taeuber devised an index of segregation by city block: if the residents were in the same proportion as the white-black ratio in the whole city, the index was 0; if the block was all white or all black, it was 100. In 207 cities in 1960 (virtually all those over 50,000 in population and a number of smaller towns), none fell below 60.4 on the index. Only eight were under 70; only 31 below 79. Half the cities had values above 87.8 and a fourth over 91.7. Segregation was the pattern in all regions, in all kinds of cities, and without regard to the proportion of Negroes or to desegregation laws or policies. A study of Detroit concluded that ghettoization led to "the creation of two cities bearing a single name."

In the twelve largest cities the black ghettos had the highest concentration of aged buildings, the largest percentage of substandard dwellings, and the greatest frequency of crowding. Segregation was imposed by exclusion (by property owners, real estate boards, mortgage lenders, intimidation, violence, and restrictive covenants until held unconstitutional in 1948) as well as by white flight. Housing segregation led inevitably to low-quality schools, hospitals, libraries, parks, water supplies, garbage removal, and police protection. Vice and crime flourished. Cramped quarters made meaningful family life almost impossible.

The overwhelming majority of ghetto residents were poor, both those with jobs and those without them. In 1960, 44 percent of Negro families had annual incomes below $3000. The unemployment rate for blacks stabilized at roughly double that for whites. Negro men suffered higher jobless rates than black women. The teenage rate was enormous and it was hardly surprising that teenage gangs throve in the ghettos. Negro employment concentrated in what would later be called the secondary labor market, characterized by lack of skills, little or no unionization, low wages, and frequent layoffs.

The Negro came to the North with a weak family structure, and ghetto life eroded the little he had. In New York City in 1960, for example, a quarter of the black families were headed by women, compared with one-tenth for whites. The rate of black illegitimacy was fourteen or fifteen times as high as that for whites. A large proportion of the female-headed black families was on welfare under the Aid for Dependent Children program.

Ghetto life, particularly poverty, inadequate or unaffordable medical and hospital care, lack of sanitation, and poor garbage collection exacted a toll on Negro health. Maternal mortality rates were four times those for whites; about three times as many black babies died before they were one year old; life expectancy was significantly higher for whites; mental hospitals admitted many more Negroes than whites. Rat bites were common ghetto experiences.

The problems of educating black children seemed almost insoluble. They flooded into the ghetto elementary schools which became primarily or entirely black. Severely disadvantaged when they entered, they then faced inferior teachers, deteriorating buildings, and low academic standards. A very large proportion, including some who received high school diplomas, never learned to read or write. Of those blacks who took the Armed Forces Mental Test in 1962, 56 percent failed, compared with 15 percent of the whites.

Crime was endemic in the ghetto and law-abiding citizens lived in perpetual fear. Morton Grodzins asked, "How does a mother keep her teen-age son off the streets if an entire family must eat, sleep, and live in a single room?" Blacks, in fact, committed the great majority of criminal acts against members of their own race. In Chicago the reported crime rate against persons was thirty-five times higher in a poor ghetto district than it was in a high-income white area. If unreported crimes had been included, the probability is that the ratio would have been significantly greater. In 1960, Negroes constituted 37 percent of the inmates of federal and state prisons. In state penitentiaries in that year 56 percent of the homicide and 57 percent of the assault offenders were black.

Lee Rainwater studied the 10,000 residents of Pruitt-Igoe, the all-black public housing project in St. Louis. "Living in the heart of the ghetto, then, *black* comes to mean not only deprivation and frustration, but also membership in a community of persons who think poorly of each other, who attack and manipulate each other, who give each other small comfort in a desperate world." The residents tried to live ordinary lives in an environment of "deprivation and social danger." People complained about their "nerves," and thought that they were aging rapidly. Marriages were unstable, matriarchy dominated, males were marginal to the family, and mothers were little involved with their children. The boys took to the streets, where they quickly learned about sex, drugs, and weapons. Rainwater concluded:

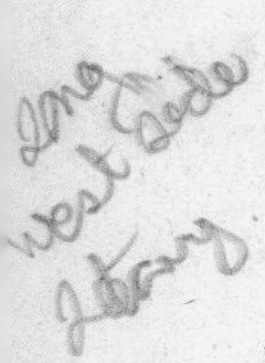

> In Pruitt-Igoe the lower-class world is defined by two tough facts of life as it is experienced from day to day and from birth to death: deprivation and exclusion. The lower class is deprived because it is excluded from the ordinary course of American working- and middle-class life, and it is excluded because it is deprived of the resources necessary to function in the institutions of the mainstream of that life. The lower-class Negro is doubly deprived and doubly excluded.[4]

During the 1950s many economists were preoccupied with economic growth. They were impressed with the great postwar boom in North America, Europe, and Japan and explored the means for extending it to the underdeveloped world. The economic historian W. W. Rostow, offered a bold analysis in *The Stages of Economic Growth*, published in 1960.

Rostow distinguished five stages in the economic development life cycle: (1) the traditional society—limited production based on a pre-Newtonian science and technology with a heavy concentration of resources in agriculture; (2) the preconditions for take-off—the emergence of modern science and technology, education, entrepreneurship, capital accumulation, and small-scale industry; (3) the take-off—the watershed stage in which the new forces accelerate dramatically and growth bolstered by compound interest becomes the norm; (4) the drive to maturity—a long stage in which the forces fostering growth permeate most of the system; and (5) the age of high mass-consumption—the culminating phase in which the problems of production are largely solved and "a large number of persons gained a command over consumption which transcended basic food, shelter and clothing." For the United States, Rostow identified stage 5 as having emerged in the 1920s, as being restrained by the Great Depression and World War II, and as reaching its apex in the postwar era.

Between 1946 and 1960 this stage 5 nation enjoyed an enormous expansion in the consumption of goods and services. To be sure, there were recessions in 1948–49, 1953–54, and 1957–58. But the two earlier dips were shallow and brief. While the decline in the late fifties was deeper and more persistent, it did not stop the expansion. Gross national product rose from $210.7 billion in 1946 to $502.6 in 1960, an advance of almost two and one-half times. When inflation is removed, GNP rose 36 percent and personal consumption expenditures went up even more—42 percent. These aggregate gains, of course, were reflected in the incomes of families and unrelated individuals. While the number of these units increased sharply from 43.3 million in 1946 to 56.1 in 1960, a rise of almost 23 percent, their average incomes grew even more rapidly, from $3940 in 1946 to $6900 in 1960, 43 percent. After removing inflation, the real advance was 16 percent.

These record sums of money available for consumption transformed the American standard of living. As George Katona wrote,

> Today in this country minimum standards of nutrition, housing, and clothing are assured, not for all, but for the majority. Beyond these minimum needs, such former luxuries as homeownership, durable goods, travel, recreation, and entertainment are no longer restricted to a few. The broad masses participate in enjoying all these things and generate most of the demand for them.

Between 1946 and 1960 more than 21 million housing units were built and by the latter year 52 percent of consumer units in metropolitan areas owned their homes. The passenger car population of the United States shot up from 25.8 million in 1945 to 61.6 million in 1960. By 1960, 72 percent of consumer units owned an automobile. In 1957, of all wired homes 81 percent had a television, 96 a refrigerator, 18 a freezer, 67 a vacuum cleaner, 87 an electric washer, 12 an electric or gas dryer, and 8 air conditioning.

There were similar gains in paid leisure time for workers. The forty-hour work-week norm established by the Fair Labor Standards Act in covered industries became the actual schedule in most workplaces by 1960, almost always accompanied by the eight-hour day. The self-employed and farm workers, who were not covered, worked substantially longer hours, though less than they had formerly. In a few unionized industries and in offices, schedules below forty hours prevailed. Before World War II few blue-collar workers received paid vacations. By 1957, 91 percent of the workers covered by major collective bargaining agreements got paid vacations, usually to a maximum of three weeks annually. Office workers fared even better. Prior to the war few industries paid their hourly rated employees for holidays even when the companies shut down. By the early sixties virtually all industries paid for holidays and most did so for seven days a year, many for nine or more. Since most Americans now enjoyed a significant amount of paid leisure time, the industries that catered to leisure activities mushroomed.

The postwar era also witnessed a marked improvement in insurance for workers and their dependents against the risks of old age, illness, and death. Excepting farm and domestic workers, virtually the entire labor force was covered by a public old-age pension system—Social Security, railroad retirement, or a federal, state, or local government program. In addition, a Bureau of Labor Statistics survey of 188 metropolitan areas in 1959–60 showed that about three-fourths of the office workers and two-thirds of the plant workers were also provided with supplemental private pension plans. Disability insurance was available under Social Security. The BLS study revealed that 83 percent of office workers and 86 percent of plant workers were eligible for hospital insurance. The figures for medical care were 61 percent for office and 59 percent for plant workers. Life insurance was available to 92 percent of office and 89 percent of plant employees.

The rise in the standard of living, along with an increase in many prices, compelled BLS to revise fundamentally its City Workers' Family Budget. The original version had been developed in 1946–47 for a family of four—an employed husband of thirty-eight, a wife/homemaker, a boy of thirteen, and a girl of eight, who lived in a rental dwelling in a large city or its suburbs. The budget was intended to show the cost of a "modest but adequate" standard of living for this family in order to meet current levels of health, efficiency, the nurture of children, and participation in community activities. BLS first priced it in 1946 and abandoned it in 1951 because, even then, the budget no longer reflected rising standards. The Bureau completely revised "the market basket" in the late fifties and priced it late in 1959 in twenty large cities. The total cost was approximately 40 percent higher than it had been in the same cities in 1951. More than half the increase was due to a rise in living standards.

Strong economic growth, particularly in mid-decade, generated optimism about the future. Corporations and individuals increased their investments in capital goods. Many Americans, at least implicitly, came to assume that prosperity was now perpetual, that some divine force had repealed the business cycle. In his highly successful campaign for re-election in 1956, Eisenhower both took credit for and encouraged the national euphoria over good times. The sharp recession of 1957–58, therefore, came as a painful blow.

Though severe, the contraction itself was mercifully brief, lasting only nine months from the peak in July 1957 to the trough in April 1958. In the face of shrinking markets, manufacturers of automobiles and consumer durables cut back sharply on output. Investment in plant and equipment in manufacturing fell by one-third. Unemployment seasonally adjusted, which had clustered at 4 to 4.5 percent (actually 3.9 in March and April 1957), shot up to 7.5 percent in July and August 1958, the highest level since the Great Depression.

Equally worrisome was the fact that the upswing that began in the spring of 1958 was the weakest and shortest of the postwar recoveries. Capacity utilization in manufacturing averaged only 80 percent in 1959–60. The unemployment rate never got under 5 percent and the number of long-term jobless stuck at a high level. The country slid into another recession in 1960.

The performance of the economy between 1957 and 1960 suggested to many that the postwar consumer goods boom had come to an end, that the nation had entered a new secular phase with limited demand and underutilization of resources, including labor. Moses Abramovitz, the Stanford economist, explained what had taken place:

> The last few years represent the culmination of a long boom in which demand was heavily supported by needs and desires suppressed and postponed during the War and even during the preceding Great Depression. As those backlogged needs were made good, the growth of demand slowed up, recoveries became weaker, excess capacity accumulated, unemployment tended to become unduly large, and inflationary pressures in commodity markets disappeared.[5]

The American political system in the fifties hung in uneasy balance between the Republicans and Democrats. It was as though two gladiators had locked in mortal combat with neither able to put the other down. Eisenhower was President, he was the leader of the Republican party, and these seemed to be "the Eisenhower years." They were, in part. V. O. Key, Jr., described the period with a musical image—"Republican Interlude." He noted that, excepting 1952 and 1956, the Democrats had won every presidential election between 1932 and 1960. Key died in 1963. If he had lived another year, he would have added 1964 to the Democratic skein. Samuel Lubell favored an image of the solar system—a dominant Democratic sun and a subordinate Republican moon. "Neither party," he wrote in mid-decade, "seems quite capable of commanding the trust of an effective majority of the country. Although party allegiances have weakened sufficiently to rob the established majority of vigor and unity, the forces of realignment have not progressed sufficiently for a new majority to emerge."

The game of domestic politics in the fifties was played out on a board which rested on two solid feet—the New Deal legislation and the voting coalition Roosevelt had forged in 1936—labor, Catholics, ethnics, including the Irish, Italians, Jews, and blacks, the liberal establishment, and the South.

Eisenhower, who was almost exclusively interested in foreign affairs, carefully avoided attacks on the New Deal programs and, in fact, supported modest improvements. While FDR's voting tapestry had frayed since 1936—Eisenhower himself had been an important unraveler, particularly with Catholics, with suburbanites, and in the South—the aging coalition proved remarkably durable.

The Survey Research Center of the University of Michigan took party identification polls in every year between 1952 and 1958, except in 1956, when two were taken. Participants were asked to identify themselves as strong, weak, or independent Democrats or Republicans, as independents, or as apolitical. Omitting the independents and apoliticals, the Democrats received heavy majorities of the two-party identifiers in every year. Excepting the two polls in 1956, the year of Eisenhower's re-election triumph, the Democrats got between 60 and 63 percent of the vote. In 1956 their margins dropped to 56 and 58 percent. Significantly, they bounced back to 63 percent in 1957 and 62 in 1958.

The continuing strength of the New Deal coalition was evident in the congressional elections. In 1950 the Democrats won handsomely in the House and narrowly in the Senate. In 1952 the Republicans, riding on Eisenhower's coattails, captured both chambers by close margins—221 to 211 in the House and 48 to 47 in the Senate with one independent. In 1954 the Democrats reciprocated, winning in both—232 to 203 in the lower chamber and 48 to 47 in the upper with one independent. In 1956, despite the Eisenhower landslide, the Democrats increased their margins slightly—233 to 200 and 49 to 47. The decisive congressional elections took place in 1958 with a Democratic sweep—283 to 153 in the lower body and 64 to 34 in the upper. The Eisenhower magic was now gone and the basic majority status of the Democratic party reasserted itself.

"It is within the majority party," Lubell wrote, "that the issues of any particular period are fought out." In the fifties and particularly the latter part of the decade the Democratic party provided the arena in which the domestic agenda of the sixties took shape. According to James MacGregor Burns, the Democratic party, like the Republican, was really two parties. One was presidential, with its eye on 1960—northern, urban, progressive, activist. The other was congressional, rooted in the congressional seniority system—southern, rural, conservative, suspicious of government initiative. Two Texas professionals, Speaker Sam Rayburn in the House and Majority Leader Lyndon B. Johnson in the Senate, presided over their balky team, trying to keep both horses in tandem with what Burns called "coalition politics."

The presidential Democrats in both houses pushed strongly for a liberal program, introducing proposals, holding hearings, moving to get their bills out of committee and onto the floor, and urging extra-congressional program committees and study groups. Rayburn and Johnson insisted on working within the congressional machinery, opposed statements of policy that would divide the party, and allowed the activists consideration on the floor only if they could get a bill through committee, usually controlled by southern conservatives. Following the 1958 congressional elections, the presidential party took command programmatically and, in effect, shaped the 1960 Democratic platform and the domestic agenda for the sixties. The main policy proposals were: civil rights, a Keynesian economic policy to provide jobs, area redevelopment to deal with structural unemployment, concern for the poor, aid to education, and Medicare for the aged.

A number of talented Democrats became prominent during the decade. They were mainly men who had gone to school with Franklin Roosevelt and several who had worked in his administrations. They had also been touched by Adlai Stevenson's eloquence. A number were governors, the historic recruiting ground for presidential candidates. There were also notable Democrats in the House. But it was in the Senate that the Democratic luminaries shone brightest. While there were individual differences, of course, as a group they were unusually intelligent, well-educated, youthful, vigorous, articulate, ambitious, and politically sophisticated. Among the stars were John F. Kennedy of Massachusetts, Hubert H. Humphrey of Minnesota, Joseph S. Clark of Pennsylvania, Paul H. Douglas of Illinois, Edmund S. Muskie of Maine, Stuart Symington of Missouri, Wayne Morse of Oregon, Mike Mansfield of Montana, Henry M. Jackson of Washington, Frank Church of Idaho, Estes Kefauver of Tennessee, J. W. Fulbright of Arkansas, and, of course, Lyndon Johnson. Several, particularly Kennedy, Humphrey, Johnson, and Symington, were eager to make the run for the presidency and were convinced that the time was ripe for the Democratic party to break with tradition and pick its presidential candidate from the Senate.

These Democrats were aware of the profound changes that had taken place in American society since the end of the war. They recognized that the new problems could no longer be glossed over. Kennedy, Theodore C. Sorensen wrote, "believed four more years of Republican rule would be ruinous," and this view was generally held by the others.

American political history, Arthur M. Schlesinger had pointed out in 1939, moved with the ebb and flow of tides.

> A period of concern for the rights of the few has been followed by one of concern for the wrongs of the many. Emphasis on the welfare of property has given way to emphasis on human welfare. An era of reaction—for stability usually suits the

> purposes of the conservatives—has usually been succeeded by one of rapid movement.

He defined these alternations, likening them to a spiral rather than a pendulum because of constant upward movement, for the long period 1765 to 1931. Though Schlesinger did not note them, in the twentieth century at the federal level the activist phases had come in fairly short bursts followed by longer periods of digestion and reaction. A full turn of the spiral seemed to take a generation, for example, from Woodrow Wilson's New Freedom just before World War I to FDR's New Deal of the thirties. If the tide continued to flow, and the 1958 elections seemed a portent, there might be another era of "concern for the wrongs of the many" in the sixties. The Democratic nomination for the presidency in 1960, therefore, could become a great prize.[6]

During 1960 the country continued to move moderately to the left politically on domestic issues. This was evident in the platforms both the Democratic and Republican parties adopted. The Democratic platform had an odd and significant history.

In October 1959 Senator Kennedy had asked Chester Bowles to become his adviser on international affairs. Bowles had been head of the Office of Price Administration during the war, governor of Connecticut, and ambassador to India, and was now a congressman. He was a respected authority on world affairs and had an impeccable liberal record on foreign and domestic issues. Bowles accepted and Kennedy announced the appointment in February 1960.

At the same time Paul M. Butler, chairman of the Democratic National Committee, invited Bowles to become chairman of the platform committee for the Los Angeles convention scheduled for early July. Bowles demurred, concerned that his association with Kennedy would invite criticism from other candidates, particularly Lyndon Johnson and Stuart Symington. Butler said that he had discussed the appointment with the party leaders and, despite the Kennedy attachment, they all wanted him. Bowles accepted the assignment.

He read the old platforms of both parties and compared them with the victors' later performances. He found this a depressing exercise. He concluded that the time had come to try something new. Bowles decided on a 3000-word platform that he would write himself and, because of its brevity, would be largely confined to civil rights, certain aspects of foreign policy with emphasis on the Third World, and a fresh statement of Franklin Roosevelt's "Four Freedoms" in the context of 1960.

At the time of the Bowles appointment, Butler asked Charles S. Murphy, former counsel to President Truman, to act as his liaison with Bowles, and Murphy, in turn, invited James L. Sundquist to become secretary of the platform committee. He was Senator Clark's administrative assistant, and had worked with Murphy in the Truman White House and in drafting statements for the Democratic Advisory Council, the coordinating body of the northern liberals, particularly the Council's comprehensive policy position which appeared early in 1960 and foreshadowed the party platform. Bowles did not think he needed additional help, saying that he planned to write the 3000 words himself, but agreed that the committee would need a secretary and accepted Sundquist. Murphy and Sundquist, along with Thomas Hughes, Bowles's legislative assistant, and Abram Chayes of the Harvard Law School, protested the 3000-word limit, contending that the party's key support groups expected their traditional planks. Bowles did not budge. Sundquist, Hughes, and Chayes then met and decided that Sundquist must assemble a complete platform so that, as Sundquist later put it, "the planks would be ready when Bowles would be forced to change his mind." Further, they wanted the measures sponsored in Congress by the northern liberals "to get a ringing endorsement by the convention."

Thus, there were two draft platforms that, in Sundquist's words, came to be known as Part I and Part II. Bowles and Harris Wofford wrote the first, Bowles on the Four Freedoms and foreign affairs, and Wofford on civil rights. The last, obviously, would be an extremely important and potentially divisive plank in the platform, and Bowles handled it with care. He wanted a strong civil rights position because this was the view of the northern liberals who would dominate the convention and also, as party hearings across the nation had shown, of the Democratic rank and file. On the other hand, Bowles did not want to rile the southerners needlessly, which might provoke a disruptive fight or, worse, a southern walkout. He instructed Wofford to draft two planks, expressing maximum and minimum positions. The former, which asked for everything the imaginative Wofford could think of, became the version submitted to the platform committee. The latter, considerably shorter, contained those positions on which supporters of civil rights would accept no compromise. Bowles and Wofford anticipated a bitter fight within the committee and expected to wind up with the minimum plank.

Meanwhile Sundquist took an office in Washington and with the help of several other senatorial assistants wrote the much longer Part II of the platform. While his work was in progress, he informed Bowles of what he was doing and said that it was necessary to have other planks for the interest groups. Bowles reluctantly agreed on condition that only Part I would be read to the convention. But Sundquist, supported by Hughes, prepared Part II to dovetail with Part I. Several weeks before the convention opened

Sundquist moved into a secure suite at the Biltmore Hotel in Los Angeles to put the second part into final shape.

The drafting subcommittee of the platform committee met before the convention opened and had both parts on the table. One member said that he assumed that Part II would be adopted by the convention and would have the same force as Part I. "There was a general nodding of heads," Sundquist wrote later, "and Bowles had to go along." The decision made, Sundquist then put the parts together and had them printed as one document, though the former Part I appeared in boldface. Sundquist met Bowles in the elevator with the printed platforms on the way to the press conference. Bowles seemed shocked, but said, "I guess it is too late now to change it, isn't it?" Sundquist readily agreed. "It was the most insubordinate and manipulative episode of my life," he wrote afterwards, "but it had to be done."

Bowles handled the civil rights plank masterfully and, to his and Wofford's amazement, the maximum position sailed serenely through both the platform committee and the convention with only *pro forma* opposition from the South. The northern liberals had a solid majority and were in no mood for compromise. It was said that Lyndon Johnson did not want a bruising battle over civil rights and Bowles, by his courtesy and consideration, made the southern defeat more palatable.

While the inclusion of Part II in the platform appears to have been a minor event, it was, in fact, of historic significance. As Murphy, Sundquist, Hughes, and the other senatorial aides intended, the 1960 Democratic platform became the bridge between the agenda shaped by the liberal northern Democrats in Congress in the late fifties and Kennedy's New Frontier of the early sixties. While Kennedy does not seem to have been involved in shaping the platform, he would base his legislative program on that document and the earlier congressional proposals from which it arose. Further, since Bowles, Wofford, and Sundquist were very good writers, this platform had some literary distinction, even eloquence, and was unusually specific. It also had a theme that, perhaps for the first time in platform history, was expressed in the title—Bowle's resonant phrase "The Rights of Man."

The civil rights plank called for decisive federal intervention to eliminate discrimination in American society based on color, race, religion, or national origin. This would demand support from the Congress and the executive agencies, but, "above all, it will require the strong, active, persuasive, and inventive leadership of the President of the United States." The platform hailed the current peaceful lunch-counter demonstrations in the South as "a signal to all of us to make good at long last the guarantees of the Constitution."

"The right to vote," the Democrats asserted, "is the first principle of self-government." Literacy tests and poll tax requirements for voting must be eliminated. The Civil Rights Acts of 1957 and 1960 should be enforced to secure the franchise. If they proved inadequate, new legislation would be needed.

The Democrats would use the full power of the federal government, "legal and moral," to enforce *Brown v. Board of Education* so that racial discrimination shall be ended in public education. "We believe that every school district affected by the Supreme Court's desegregation decision should submit a plan providing for at least first-step compliance by 1963, the 100th anniversary of the Emancipation Proclamation."

A new law should be enacted to establish a fair employment practices commission "to secure effectively for everyone the right to equal opportunity for employment." The executive power must be used to abolish racial discrimination in the federal service and on government contracts. The same policy should extend to federal housing programs.

The U.S. Commission on Civil Rights should become permanent and its scope and powers broadened. It should assist communities, industries, and individuals to obtain constitutional rights in education, housing, employment, transportation, and the administration of justice.

Beyond civil rights, the platform declared that the Democratic party "reaffirms its support of full employment as a paramount objective of national policy." Recent levels of joblessness were intolerable. The growth rate of the American economy must be almost doubled to 5 percent annually. This was necessary to provide jobs for the anticipated massive entry of young people into the labor force in the 1960s and for older workers displaced by technological change.

The platform called for enactment of the depressed areas bill to assist localities suffering from high and persistent unemployment. The Democratic Congress had twice passed the Douglas bill only to have it vetoed by Eisenhower. The legislation would offer low-interest loans to private enterprises to create new jobs, assist communities with the construction of needed public facilities, and retrain workers who had been displaced.

The minimum wage should be raised from $1.00 to $1.25 per hour and coverage under the Fair Labor Standards Act should be extended to several million workers presently unprotected. Farm workers, particularly migrants, should receive decent wages, adequate health and housing conditions, Social Security, education for their children, and welfare services.

The Democrats offered a comprehensive program to improve the nation's health. The most important proposal was for Medicare, health insurance for the aged under Social Security. They urged federal support for hospital construction, for expanding and modernizing schools of medicine, dentistry, nursing, and public health, and for improving mental health programs. Finally, they proposed coordinated research on major diseases—cancer, heart ailments, arthritis, and mental illness.

The platform stressed "the financial crisis" in education. "The tremendous increase in the number of children of school and college age has far outrun the available supply of educational facilities and qualified teachers." States and localities were straining under the load. The federal government should help finance schools and provide loans and grants to college students. The Democrats also recommended a youth conservation corps for underprivileged young people.

Finally, the Democrats called attention to the problems of the decaying central cities and their mushrooming suburbs. They asked for a ten-year coordinated program to eliminate slums and blight, to assist in metropolitan area planning, to develop urban transportation, to combat air and water pollution, and to expand recreational facilities.

Kennedy accepted his party's nomination before a large crowd in the Los Angeles Memorial Coliseum on July 15, 1960. He expressed gratitude for "such an eloquent statement of our party's platform. Pledges which are made so eloquently are made to be kept." Standing "facing west on what was once the last frontier," he gave his and the Democratic Party's program a name, the New Frontier. Here he identified himself with the programs of two of his illustrious Democratic predecessors, Woodrow Wilson's New Freedom and Franklin Roosevelt's New Deal. But the New Frontier "is not a set of promises—it is a set of challenges. . . . It holds out the promise of more sacrifice instead of more security."

Republicans had traditionally treated their platform as a necessary inconvenience, to be dispatched expeditiously in bland language to which almost no one could dissent. This was not to be at the Chicago convention in late July 1960.

Eisenhower, jolted by the Democratic electoral sweep in 1958, had called for a re-examination of "the Republican philosophy." In 1959 he established the committee on program and policy under Nixon. He, in turn, chose Charles H. Percy, a young and successful Chicago businessman, to chair the committee. His assignment was to draft what would become the 1960 convention platform in order to satisfy both the conservative and progressive wings of the party. Percy's main problem was with the latter, dominated by Governor Nelson A. Rockefeller of New York.

Aside from his wealth, energy, organizational skills, and magnetism, Rockefeller, almost alone among Republicans, had won a stunning electoral victory in 1958 over Averell Harriman, and his performance as governor was much admired. He intended to become President and in the fall of 1959 he took systematic soundings of the Republican regulars. They could not have been more dismal. Rockefeller was much too progressive for the Republican establishment and Nixon already had a firm lock on the nomination. At the end of December Rockefeller announced that he was not a candidate.

Though the curtain had fallen, Rockefeller refused to leave the stage. During the spring of 1960 he spoke out critically of Eisenhower administration policies on defense and foreign affairs. On June 8 he announced a nine-point program, each supported by a study paper, on foreign policy, defense, civil rights, economic growth, health care for the aged, and education, among others. This was, Theodore H. White wrote, "open warfare with the leadership of his own Party." Eisenhower was outraged; Nixon was deeply concerned.

In early July, Percy came to New York to show Rockefeller the bland platform he had drafted. The language was intended to gain the assent, if not the approval, of Eisenhower, Nixon, and Rockefeller. The governor expressed no interest in compromise; he insisted that the party adopt his platform; and he set up a command post in Chicago to monitor his convention operation. Now, in defiance of Republican tradition, the platform had become a central issue that could split the party, a prospect that appalled Nixon.

Through an intermediary, Nixon asked Rockefeller for a secret meeting. The governor, whose distaste for the vice president was hardly masked, responded that he would meet Nixon only if certain humiliating conditions were fulfilled: Nixon must phone Rockefeller personally to ask for the meeting; they must meet at Rockefeller's apartment at 810 Fifth Avenue; the meeting must be secret until Rockefeller issued the press release; the release must say that Nixon had asked for the meeting; and the statement of policies must be inclusive and detailed. To Rockefeller's surprise, Nixon immediately accepted all these terms.

"The Compact of Fifth Avenue," as it came to be known, was negotiated between 7:30 p.m. and 3:20 a.m., July 22-23, 1960. At the outset Nixon offered Rockefeller the vice presidential nomination, which the latter rejected. Rockefeller then insisted that the Republican platform endorse his nine points. Nixon bargained as best he could to tame the language in order to pacify Eisenhower and the Republican regulars. Rockefeller issued the press release within two hours of Nixon's departure from his apartment.

On important domestic issues, the statement read, "the Vice President and I also reached agreement on the following specific positions": on civil rights the Republicans "must assure aggressive action to remove the remaining vestiges of segregation or discrimination in all areas of national life—voting and housing, schools and jobs." The statement supported "the objectives" of both black lunch-counter demonstrators and the businessmen who agreed to serve Negro customers.

The rate of economic growth, the compact declared, must be accelerated "by policies and programs stimulating our free enterprise system." Instead of calling for a 5 percent growth rate, the statement referred to a speech Nixon had made in which he noted that such a rate would sharply increase tax revenues.

On health insurance for the aged, Rockefeller and Nixon called for a contributory system on "a sound fiscal basis" and for the option for the elderly to purchase private health insurance.

In order to meet educational needs, the statement recommended federal grants-in-aid with matching state funds for school construction, classrooms, and laboratories, a loan program for dormitories, student loans and graduate fellowships, and federal scholarships for gifted undergraduates.

Rockefeller himself concluded that, if the Republicans endorsed these points and his foreign and defense policies, "they will constitute a platform that I can support with pride and vigor."

Few Republicans, aside from Nixon, were interested in his support. The compact, in fact, caused explosions of anger. Eisenhower considered the foreign and defense policies a repudiation of his presidency and was furious. Senator Barry Goldwater of Arizona, the leader of the right wing, called the compact a "surrender," "the Munich of the Republican Party," and a guarantee of defeat in November. The platform committee, made up of regulars, expressed, according to White, "pain, outrage, and fury." In a single night, they said, two men meeting in New York had wiped out all their work.

Nixon arrived in Chicago to find the convention in disarray. He needed to unite the party by appealing to both Eisenhower and Rockefeller and by establishing control over the delegates. He decided to give Rockefeller most of the civil rights plank. Unlike the compact, the Percy draft said nothing about either federal intervention to secure black equality in the job market or support for the lunch-counter demonstrations. This posed a strategic gamble. Many Republicans, notably Goldwater, were convinced that the Percy platform guaranteed a sweep of the South and victory in November. Others thought this would present the industrial states as a gift to Kennedy by forfeiting the Negro vote in the North. Nixon opted generally for the Rockefeller position and insisted that the platform committee change the plank. They agreed to do so by votes of 50 to 35 with many abstentions. There were also language changes in the national defense plank that won grudging approval from both Eisenhower and Rockefeller. Thus, the Republicans entered the campaign more or less united.

In the platform the civil rights plank declared that racial discrimination was unconstitutional, immoral, and unjust. It called for equal opportunity without regard to race, religion, color, or national origin in housing, education, and employment. Discrimination, the platform stated, was a problem in the North as well as the South and was not limited to blacks.

Specifically, the Republicans pledged enforcement of the civil rights laws to guarantee the right to vote and of the Supreme Court decision to desegregate the public schools. They backed a law to create a commission on equal job opportunity "to make permanent and to extend the excellent work being performed by the President's Committee on Government Contracts" (headed by Nixon). They called for a law forbidding labor unions to discriminate in admission to membership. They urged the end of discrimination in federally subsidized housing and segregation in public transportation. The Republicans said they would try to change Senate Rule 22, which required a two-thirds' vote to close debate and prevent a filibuster. "We reaffirm the constitutional right to peaceable assembly to protest discrimination in private business establishments. We applaud the action of businessmen who have abandoned discriminatory practices in retail establishments, and we urge others to follow their example."

The Republicans favored "vigorous economic growth," but expected it to come from the private sector. They rejected "artificial growth forced by massive new federal spending." But they favored "federal-local action" to aid areas suffering from chronic unemployment.

The Republicans called for "upward revision" of the minimum wage and extension of coverage to several million workers. While they did not ask for legislation, they urged "action" to improve manpower training and to aid migratory farm workers.

The platform declared that primary responsibility for education must remain with the states and localities. But it asked for federal aid for school construction, college housing, and student loans and graduate fellowships.

The Republicans came out for a contributory system of health care for the aged provided that the option of private health insurance was protected.[7]

A presidential election is hardly a proper forum for candidates to debate the issues and 1960 was no exception. Theodore H. White wrote of 1960: "Rarely in American history has there been a political campaign that discussed the issues less or clarified them less." The stress, probably more than usual, was on personalities, particularly on Kennedy for his youth and alleged inexperience, his good looks, his religion, his wealth, his family. Nixon, too, was an issue—his below-the-belt style, whether he was now the "new" (high road) or "old" (low road) Nixon. The focus on the candidates was reinforced by the extraordinary importance of campaigning on television, particularly in the so-called "great debates." Viewers were far more interested in how the candidates looked and who won than in what they said. In his campaign, Nixon used a set speech which he repeated with few changes everywhere. While Kennedy varied his remarks, he, too, made key points again and again. Finally, both Kennedy and Nixon were young, extremely energetic, and exceptionally determined to expose themselves to the maximum number of voters. Thus, their schedules left no time for deliberation and barely enough for survival.

A fundamental factor shaping the campaign was the disparity in strength between the parties. The Democrats had overwhelming majorities in both houses of Congress, held 36 governorships, and controlled the great majority of state legislatures. A Gallup poll early in 1960 showed that 47 percent of the voters were Democrats, 30 percent Republicans, and 23 percent independents. "To win in 1960," Nixon wrote, "the Republican candidate would have to get practically all the Republican votes, more than half of the independents—and, in addition, the votes of between five and six million Democrats." Thus, Nixon avoided identifying himself as a Republican, though he did trot out Eisenhower (who was "above politics") in the closing weeks of the campaign. By contrast, Kennedy ran as a Democrat, invoked the traditions and heroes of his party, associated himself with its candidates, and cited its platform . . . selectively. The campaign strategists, White wrote, insisted that "Kennedy must make clear that the two parties were wholly different in goals and pin the Republican label on Nixon as tightly as possible, hammering him as the spiritual descendent of McKinley, Harding, Hoover, Landon, and Dewey."

Nixon, therefore, referred to the Republican platform on domestic issues only rarely, and then, when he had no choice. The first television debate, which had 70 million viewers and was supposed to deal with domestic issues, actually became a personality contest (a big question was whether Nixon had five o'clock shadow). Several reporters asked questions that were not germane to the topic and Nixon mouthed platitudes on such issues as taxes and schools.

Kennedy did more. His campaign had a theme, which he expressed this way in his election eve speech at the Boston Garden: "This race is a contest between the comfortable and the concerned, between those who believe that we should rest and lie at anchor and drift, and between those who want to move this country forward in the 1960s."

In his speeches, invariably brief, Kennedy laid stress on unemployment in depressed areas, on low wages and the minimum wage, on federal aid for education, on medical care for the aged, and on the eradication of urban blight. He campaigned especially vigorously in the suburbs of the big cities in the large industrial states, endorsing aid to education to the parents of baby-boomers who were now reaching college age. He often added humor.

Kennedy, concerned about alienating the white South, avoided civil rights. But he became the beneficiary of an extraordinary incident that nailed down the black vote. Harris Wofford gave this account: On October 19, Martin Luther King, Jr., led a group of Negro demonstrators demanding service at the Magnolia Room restaurant in Rich's department store in Atlanta. Many, including King, were arrested. He refused to put up bail, saying that he would sit in jail until Rich's desegregated. The demonstrations spread. After several days Atlanta's moderate mayor, William Hartsfield, intervened. Wofford, who was working in Kennedy's civil rights section, was told that the mayor had made an agreement with the black leaders that would lead to the release of King and the others who had been jailed "in response to Senator Kennedy's personal intervention." Kennedy had not intervened, knew nothing about the incident, and was being urged by his white Georgia supporters not to get involved with King for fear of losing the state. Hartsfield told Wofford: "You tell your Senator that he and I are out on a limb together, so don't saw it off. I'm giving him the election on a silver platter."

Wofford got through to Kennedy in a motorcade in Kansas City and his staff issued a noncommittal statement that, at least, did not call Hartsfield a liar.

But King was not released from jail. The previous spring, while driving in Atlanta with a white woman, the novelist Lillian Smith, he had been arrested for operating a vehicle without a Georgia license (he had one from Alabama), had been fined $25, and had been placed on probation. The De Kalb County judge now decided that his arrest for the sit-in violated the probation and reinstated the earlier conviction. The judge went further, sentencing King to six months at hard labor in state prison, an unreasonably harsh penalty for a minor traffic offense. Negroes and many white Americans were shocked and outraged and King's mistreatment became a world event.

Wofford prepared a strong public statement for Kennedy denouncing the sentence and demanding King's release, which Kennedy at first accepted. But his Georgia supporters, headed by Governor Ernest Vandiver, said it would give Georgia and several other southern states to Nixon. Vandiver agreed to get King out of jail if Kennedy issued no statement. The senator agreed.

But three days after the other prisoners had been released King was still in jail. Worse, he had been put into handcuffs and leg chains during the night and had been taken to Reidsville State Prison in "cracker country." Coretta King, who was pregnant and was convinced that her husband would be lynched, called Wofford in desperation. At the time he could give her little reassurance. Later that day he got the idea of having Kennedy do something "direct and personal, like picking up the telephone and calling Coretta." Sargent Shriver, the candidate's brother-in-law and Wofford's boss, liked the idea and proposed it to Kennedy in Chicago. He said, "Why not? . . . Get her on the phone." He spoke to her, as Wofford put it, "warmly, seriously, reassuringly."

Robert Kennedy was furious. He told the civil rights people: "Do you know that three Southern governors told us that if Jack supported Jimmy Hoffa, Nikita Khrushchev, or Martin Luther King, they would throw their states to Nixon?" The election, he said, could be "razor close" and "you have probably lost it for us." Upon reflection, he turned his moral rage upon the judge and phoned him to ask for King's release. He went free shortly thereafter.

The calls came ten days before the election and caused a sensation in the black community. The Negro vote swung sharply to Kennedy. Martin Luther King, Sr., was the most notable switcher. Earlier he had joined a number of other southern Negro Baptist preachers who had opposed Kennedy because he was a Catholic. Now he announced his support for the senator because of the call to Coretta. Kennedy said to Wofford: "Imagine Martin Luther King having a bigot for a father!" He grinned and added, "Well, we all have fathers, don't we?"

The election campaign did give Kennedy the impetus to announce a new proposal—the Peace Corps. The idea was in the air. Several years earlier Senator Jacob Javits, Republican of New York, had urged the enlistment of a million young Americans to serve overseas in an "army of peace." Representative Henry Reuss, the Wisconsin Democrat, had met four young American school teachers setting up elementary schools for UNESCO in Cambodia in 1957 and was much impressed with their work. He returned to propose a Point Four Youth Corps and in 1960 got Congress to support a study of the idea by Colorado State University. Senator Humphrey had urged the proposal in his primary campaign against Kennedy. Milton Shapp, a Philadelphia businessman, had suggested it to Sorensen. At the suggestion of Bill Moyers, Lyndon Johnson came out for a "volunteers for peace" program in campaigning in September. General James Gavin gave a speech with the same idea that fall. Kennedy was aware of these proposals and was sympathetic. He discussed them with Humphrey and Bowles and asked several academic people to prepare position papers.

Like the King phone call, Kennedy's endorsement of the Peace Corps was accidental. As Wofford reconstructed the event, Nixon probably provided the stimulus. In the third televised debate on October 14 he charged that in the past half-century three Democratic and no Republican Presidents had led the nation into war. Kennedy had peace on his mind when he addressed 10,000 people on the steps of the Michigan Student Union in Ann Arbor later that night (actually 2:00 a.m., October 15). He asked the students whether they were willing to perform service overseas and received an ovation. Professor Samuel Hayes of Michigan was doing a background paper for Kennedy on the Peace Corps and, with his help and a follow-up address by Bowles, the students quickly organized and drafted a petition in which almost a thousand signers agreed to volunteer.

Much impressed with the response, Kennedy gave a major speech at the Cow Palace in San Francisco on November 2 in which he promised that, if elected, he would establish a Peace Corps to "work modern miracles for peace in dozens of underdeveloped nations." He hammered the point in the last days of the campaign. Representatives of the Michigan students, who met him in Toledo, asked whether he was serious. "Until Tuesday," Kennedy responded gaily, "we'll worry about this country. After Tuesday—the world!"[8]

The 1960 presidential election was, according to the classification of the University of Michigan Survey Research Center, a "reinstating" as distinguished from a "deviating" election. That is, the majority party, after being denied the presidency for two terms, recaptured that prize. But there were significant peculiarities about the victory.

For one, Kennedy's margin was extremely narrow. Of the 68.8 million ballots cast, he won by only 119,450 votes—49.7 percent to Nixon's 49.6, with 0.7 percent going to minor parties. Kennedy ran ahead more comfortably in the electoral college—303 to 219, not counting 15 votes cast for Senator Harry F. Byrd of Virginia, 6 in Alabama and 8 in Mississippi, which hardly boded well for the victor.

The other complicating factor was Kennedy's Catholicism. Religion had not been a force in any presidential election since 1928. In 1960 it seems to have been, excepting party loyalty, the most important issue to voters. It, of course, cut both ways. Many Catholics who had deserted the Democratic party to vote for Eisenhower now came home. This was evident in the heavily Catholic cities of the East and the Midwest. In Boston's predominantly Irish Ward 16 (Dorchester), Stevenson's 1956 margin of 53.6 percent leapt to Kennedy's 82.4 percent of the vote in 1960. In four low-income Polish wards in Buffalo the gain was from 56.2 to 78.6 percent. In a German Catholic upper-middle income ward in Cincinnati the increase was from 30.5 to 56.5 percent. The five Italian wards in Providence gave Kennedy 78.5 percent of the vote.

The Survey Research Center concluded that "American Protestants were remarkably preoccupied by the fact that Kennedy was a Catholic." This was evident in the voting of heavily Protestant districts. Eleven fundamentalist counties in Tennessee went unanimously for Nixon, most by heavy margins. All had been strongly Democratic since the Civil War, though Eisenhower had carried two in 1952 and four in 1956. Kennedy lost three traditionally Democratic fundamentalist counties in southern Illinois. He suffered stinging defeats in five of six Protestant counties in southern Ohio.

Since the country was overwhelmingly Protestant, religion exacted more from Kennedy than he gained. While no one knows the exact figure, one informed estimate was that 4.5 million Protestant Democrats switched from Stevenson in 1956 to Nixon in 1960. According to Sorensen, this was "far more than any Catholic vote gains could offset." If Kennedy had been a Protestant, he would have run moderately less strongly in the East and the Midwest and would have performed substantially better in the South, the Border, the lower Midwest, and the West. His popular vote victory margin would have been handsome (the Survey Research Center estimated a 54 to 46 split) and he would have piled up a landslide in the electoral college.

This conclusion is reinforced by the results in the congressional and gubernatorial elections. In the downstate Illinois counties the Democratic candidate for governor ran about five points ahead of Kennedy. In the southern Ohio counties the Democratic candidates for senator and governor in 1958 beat Kennedy in 1960 by about ten points. In ten Lutheran counties in Minnesota, Governor Orville Freeman and Senator Humphrey easily outdistanced Kennedy. In six Catholic counties, however, Kennedy ran even with Humphrey and well ahead of Freeman.

The congressional election results, like the presidential, exhibited crosscurrents. Democratic candidates, demonstrating greater strength than the head of their ticket, captured 55 percent of the vote nationally compared with 45 percent for the Republicans. But the Democrats, while retaining strong control in both houses—64 to 36 in the Senate and 262 to 175 in the House—lost two seats in the upper chamber and twenty-one in the lower. A substantial majority of the newcomers were more conservative than the new President. Thus, the historic alliance of Republicans and southern Democrats would control the next Congress. Kennedy could anticipate strong opposition in getting the country to move forward. David Halberstam wrote:

> It was easier to stir the new America by media than it was to tackle institutions which reflected vested interests and existing compromises of the old order. In a new, modern, industrial, demographically young society, this was symbolized by nothing so much as congressional control by very old men from small southern towns, many of them already deeply committed, personally and financially, to existing interests; to a large degree they were the enemies of the very people who had elected John F. Kennedy. He was caught in that particular bind.

In mid-December 1960 Kennedy lunched at his Georgetown house with Senator Joseph S. Clark of Pennsylvania. He wanted to know the odds on a "Kennedy 'must' legislative program in the Senate." A few days later Clark sent him a careful review, and it was bleak.

Medicare would have to clear the Finance Committee, of which Senator Byrd was chairman. He considered it "socialized medicine" that would bankrupt the Social Security trust fund. In a test vote in August that committee had voted it down 11 to 5. A vacant seat would have to be filled by a favorable senator and three Democrats would have to be persuaded to switch, something most unlikely to occur. Clark considered the prospect "poor."

A rise in the minimum wage to $1.25 and significant increase in coverage would go to Labor and Public Welfare. Prolonged hearings and "dilatory tactics" by Republicans Goldwater and Dirksen would cause extended delay. Nevertheless, Clark thought the prospects "fair."

A substantial majority of the Senate favored federal aid for education. The school construction bill would pass the Labor and Public Welfare Committee easily. The teacher-salary supplement would have rougher sailing, but, with some breaks, the prospect was "good."

The depressed areas bill would clear the Banking and Currency Committee readily. The big question was how much money would be pumped into the program. Though the majority had been niggardly in the past, some might be turned around. Clark thought the prospect "fair."

There were two important civil rights bills—school desegregation and fair employment practices—which were before Labor and Public Welfare. There was a northern Democratic–liberal Republican majority on the committee "in favor of a desegregation measure." But Chairman Lister Hill of Alabama "will clearly use all dilatory tactics at his command." If he failed, which seemed unlikely, "a filibuster is assured." There had been nine attempts on civil rights bills since 1917 to close debate, all unsuccessful. "There is no reason to believe that the South will not be able to muster 34 votes needed to defeat cloture." Clark thought the prospects for civil rights legislation "bad."

If anyone in the Kennedy entourage, and the new President was certainly not among them, thought that he would get what he wanted from a pliant Congress, as FDR had during the Hundred Days, the Clark memorandum laid that hope to rest.[9]

For President-elect Kennedy the 72 days of interregnum between November 8, 1960, and January 20, 1961, were extremely busy. He had three main tasks: to effect a transition of power from the Eisenhower administration; to shape a substantive program; and to staff his administration. As to the first, Eisenhower and Kennedy got along very well and this facilitated an exceptionally smooth transition.

Even before his nomination, Kennedy had outlined his intention with regard to program when he became President: "To determine what the unfinished business was, what our agenda was, and set it before the American people in the early months of 1961." During the summer of 1960 he had asked experts for reports on important policy questions. After the election he established "task forces" to deal with a broad array of issues. One count reached 29 prior to the inauguration and about half a dozen in the next few weeks. They took many shapes—from Senator Douglas's 23-member force on depressed areas, which broke into subcommittees and held hearings in West Virginia, to a confidential report on civil rights by Harris Wofford. There were five important reports that are relevant for present purposes: Wofford's on civil rights; Paul Samuelson's on economic policy; Douglas's on area redevelopment; Frederick L. Hovde's on federal aid for education; and Wilbur Cohen's on health and Social Security with particular reference to Medicare. They will be discussed in detail in later chapters.

When the President-elect met with his most trusted advisers in a planning session at his brother's house on Cape Cod on November 10, 1960, the problem of staffing the administration was addressed—perhaps seventy-five key cabinet and policy positions and about 600 lesser, but important, posts. They agreed on a massive talent search, combing the universities and professions, the civil rights movement, business, labor, foundations, wherever. Kennedy asked Sargent Shriver, his brother-in-law, to head the hunt. He, in turn, recruited Wofford, Adam Yarmolinsky, and Louis Martin, who was black. Shriver afterwards recalled that Kennedy told him on November 10 to find "the brightest and best people possible." Later, according to Wofford, they learned of an 1811 hymn—"Brightest and best of the sons of the morning/Dawn on our darkness, and lend us thine aid." Lyndon Johnson would use the phrase "brightest and best" to describe the men he inherited from Kennedy as he slid into the Vietnam quagmire. David Halberstam in his incisive analysis of Vietnam inverted the phrase and used it ironically as the title of his book, *The Best and the Brightest*.

Shriver's group worked long hours under great pressure and literally scoured the country for talent. According to a form Yarmolinsky worked up, the standards they applied were: "judgment, integrity, ability to work with others, industry, devotion to Kennedy's program, and toughness." The President-elect did not want "faint-hearted" people in his administration. The word got out and some jobseekers phoned to say, "I'm tough." Overall the results were extraordinary, perhaps unmatched by any other new administration. If the people chosen were not the brightest and the best the nation had to offer, they must have come close. Kennedy's primary interest was in international affairs. As Halberstam put it, "The real action was in determining the role America played in the world. . . . It was where the real excitement was." Thus, the priority was to man the White House, State, Defense, and Treasury. But the staffing of the domestic agencies was not slighted.

As Wofford had stressed in his report, the Department of Justice would be the critical agency in the emerging crisis over civil rights. There were two key positions—Attorney General and assistant attorney general for civil rights. Picking an Attorney General proved to be extremely complicated.

At the outset it seemed easy. Abraham Ribicoff, the popular and attractive governor of Connecticut, who had been an early Kennedy supporter, was offered Justice. He turned it down because he thought it would be unseemly for a Catholic President and a Jewish Attorney General to put black children into schools in the Protestant South. Robert Kennedy also thought that Ribicoff hoped to be named to the Supreme Court when Felix Frankfurter left and Ribicoff did not want to face the wrath of southern senators. Ribicoff went to Health, Education, and Welfare. Justice also got involved in the extremely complicated problem of placing Adlai Stevenson. After the Ribicoff rebuff, Kennedy decided to name Stevenson Attorney General. A sounding revealed that he was not interested. Stevenson went to the United Nations.

Joseph P. Kennedy, the President-elect's father, had urged his brother for Attorney General. But Robert did not want the job. He was worried about the charge of nepotism. Also, "I had been chasing bad men for three years and I didn't want to spend the rest of my life doing that." Finally, he did not relish the prospect of the "Kennedy brothers" taking the heat on civil rights. He really wanted to be the number two man at Defense responsible for closing the so-called "missile gap." But Robert S. McNamara, the new secretary, thought it would be awkward having the President's brother taking orders from him. John, therefore, reluctantly concluded that Robert should go to Justice. He wanted someone he trusted near at hand and he was certain that Robert would make a fine Attorney General. Both expected sharp criticism. When asked how he would make the announcement, John said, "I'll open the front door of the Georgetown house some morning about 2:00 a.m., look up and down the street, and, if there's no one there, I'll whisper, 'It's Bobby.'" The appointment got a bad press. John tried to laugh it off: He wanted to give his brother "a little legal experience before he goes out to practice law."

Robert Kennedy had just turned thirty-five. At Harvard he had performed with something less than distinction in both his studies and football. He barely made it into the University of Virginia Law School and he was not among that institution's notable scholars. He worked briefly in the criminal division of the Justice Department and, in 1952, managed his brother's campaign for the Senate impressively. His father got him a job with Senator McCarthy's red-baiting committee. He left after six months, upset by McCarthy's tactics and appalled by the senator's assistants, Roy Cohn and David Schine.

In the late fifties he was counsel to Senator John L. McClellan's Select Committee on Improper Activities in the Labor or Management Field, which concentrated on exposing corruption in the Teamsters union. In the hearings Kennedy was the righteous grand inquisitor and received great publicity. He destroyed Dave Beck, the union's president, and went relentlessly after his successor, Jimmy Hoffa. He became a celebrity and wrote (with help) a book on organized corruption, *The Enemy Within*. In late 1959 Robert took over the management of his brother's campaign for the presidency and performed brilliantly.

When Robert Kennedy became Attorney General in 1961, he knew that civil rights would be a big problem. But he had not known many black people, knew little about segregation, and had not considered the federal role in promoting desegregation. As he noted later, "I won't say I stayed awake nights worrying about civil rights before I became Attorney General." He thought that Negroes would assimilate rapidly like other ethnic groups. In fact, he expected the United States would have a black President by the year 2000. When his grandfather had immigrated to Boston, he declared, "the Irish were not wanted there. Now an Irish Catholic is President of the United States. There is no question about it. In the next forty years, a Negro can achieve the same position my brother has."

The choice for assistant attorney general for civil rights seemed obvious: Harris Wofford. His great-grandfather had been a colonel in the Confederate army and he had been born in Tennessee and had family there and in Arkansas. After World War II, Wofford and his wife went to India, were captivated by Gandhi's message of nonviolent direct action, and wrote *India Afire*. In 1950 he was the first white man to enroll in the Howard University Law School because he wanted to practice civil rights law and felt that he had to live and work with Negroes. At the dean's urging, he finished at both Yale and Howard. He practiced law with the noted Washington firm Covington and Burling. In the mid-fifties Wofford and Martin Luther King, Jr., were drawn together by their interest in nonviolence. He arranged for King to visit India and became his adviser. He worked for passage of the Civil Rights Act of 1957 and then joined the staff of the new Civil Rights Commission.

Father Theodore Hesburgh, the president of Notre Dame, was a member and persuaded Wofford to teach at the law school in South Bend. In May 1960, Robert Kennedy called him in to work with Shriver as coordinator in the civil rights section of the campaign organization. It was hard to imagine anyone better qualified to head the civil rights division of the Department of Justice. The problem with appointing Wofford was that he must clear the Senate Judiciary Committee, of which James O. Eastland of Mississippi was chairman and Richard Russell and Herman Talmadge of Georgia as well as Allen Ellender of Louisiana were members. The new President would not need that kind of fight within the Democratic party. While Robert Kennedy felt that Wofford was the obvious choice, "I was reluctant to appoint him because he was so committed on civil rights emotionally." John Kennedy made Wofford his assistant on civil rights in the White House, charged with coordinating the federal government's programs.

Wofford, when he had been on the search committee, had glowingly recommended Burke Marshall as assistant attorney general. They had worked together at Covington and Burling, had jointly taught the corporations course at Howard Law School, and had read Toynbee's *A Study of History* together in a small seminar. Wofford urged Marshall because he "exuded fairness, thoughtfulness, and responsibility, and instilled a sense of confidence and readiness to reason in others." Robert Kennedy said, "I wanted a tough lawyer who could look at things objectively and give advice—and handle things properly. And that's why I settled on Burke Marshall." Byron R. White, the new deputy attorney general, who had known Marshall at the Yale Law School, joined in supporting him.

Marshall's career had moved impressively in a straight line, but hardly in the direction it now took. He had been schooled at Exeter, Yale, and the Yale Law School. He had then gone to Covington and Burling and was now a partner. He was a specialist in antitrust law, an esoteric and complex branch of corporate law that happened to be extremely lucrative. He was slightly older than his boss, thirty-eight, and probably knew even less about civil rights than Kennedy, and that was mainly learned from Wofford. Articles written later about him called him "quiet fighter" and "quiet gun." Like many others who did not make a great deal of noise, he was to prove exceptionally courageous.[10]

2

Civil Rights: Confrontation

THE ORIGIN OF THE TERM "Jim Crow," C. Vann Woodward wrote, "is lost in obscurity." Thomas C. Rice composed a song and dance called "Jim Crow" in 1832, but no one seems to know how it got its later meaning, which attained broad usage by the 1890s. It then came to describe the system of both law and practice in the South by which black people were locked into an inferior status by discrimination and segregation.

Between the abandonment of Reconstruction in 1877 and the early years of the twentieth century, Jim Crow, with the blessing of the Supreme Court, strapped the Negro population of the South into second-class citizenship. Blacks were denied the right to vote by state constitutional barriers based on property and literacy (with loopholes for whites), by the poll tax, and by the all-white primary. In Louisiana in 1896 there had been 130,334 Negro voters; by 1904, when the restrictions were in place, there were 1,342. Railways, trolleys, and steamboats, along with stations and waiting rooms, were segregated by law. Soon the "Whites Only" and "Colored" signs, without legal requirement, invaded theaters, ticket windows, boarding houses, toilets, and water fountains.

Educational systems from kindergarten through the university were separated for white and black young people. Work places were segregated by race and many unions denied blacks membership. There was a similar situation in hospitals, mental institutions, jails, homes for the aged, orphanages, and so on. Jim Crow ruled in public parks, circuses, recreations, fraternal orders, and sports. Urban residential segregation hardly needed to be imposed because blacks were already ghettoized by poverty in the "Darktown" city slums.

Some white racists went to extremes. Mobile had a 10:00 p.m. curfew for Negroes. In Oklahoma the telephone company installed segregated booths. North Carolina and Florida used separate textbooks in the public schools. In Atlanta's courts white and black witnesses swore on different Bibles. In New Orleans prostitutes were segregated.

Georgia as of 1950 will serve as an illustration of legalized Jim Crow, many provisions of which appeared in the state's constitution. "Persons of color" were defined as "all Negroes, mulattoes, mestizos, and their descendents, having any trace of either Negro or African, West Indian, or Asiatic Indian blood in their veins." A "white person" was of "the white or Caucasian race" with no trace of the blood of a person of color. All individuals must be registered by race with the State Board of Health. "It shall be unlawful for a white person to marry anyone except a white person." Miscegenation was a felony. A birth certificate showing the birth of a legitimate child to mixed parents opened them to criminal proceedings. "Any charge or intimation against a white female of having sexual intercourse with a person of color is slanderous without proof of special damage."

All voters must be registered. In order to register, an individual must answer orally and correctly ten or thirty questions asked of him by the registrar. They dealt with federal, state, and county governments and were set forth in the law. Each voter's race was listed.

In Georgia in 1950 the following institutions were segregated by race: public schools and the University of Georgia system, training schools for juvenile delinquents, the state hospital for mental patients, prisons, common carriers furnishing passenger transportation (conductors and other employees of railways, streetcars, and buses were vested with police powers to require seating segregation by race; but colored nurses and servants "in attendance on their employers" were exempted), and billiard and pool halls.

Blacks, of course, detested Jim Crow, but for several generations were without power or will to mount a significant challenge. World War II was the turning point, for now a large part of the black population was in the North, was urbanized, was better educated, and had the suffrage. After the war, Truman by executive order desegregated the armed forces and prohibited discrimination in federal employment. More important, largely as the result of suits brought by the National Association for the Advancement of Colored People, the Supreme Court held segregation laws unconstitutional for interstate carriers, in all-white primaries, in jury service, in admission to universities, particularly graduate and professional schools, and struck down the enforcement of restrictive housing covenants. The culminating event, of course, was the Court's momentous and unanimous decision in *Brown v. Board of Education,* which overruled *Plessy v. Ferguson* and held that racially segregated schools were inherently unequal under the Fourteenth Amendment. The decisiion, Woodward wrote, "marked the beginning of the end of Jim Crow. But the end was to be agonizingly slow in coming."

The nation held its breath as it asked, "What will the white South do?" Mississippi immediately proclaimed its defiance by launching Citizens' Councils to wage war for segregation. But elsewhere many voices of moderation spoke out. The federal courts nullified segregation statutes in a number of states. The District of Columbia, Tennessee, Arkansas, Texas, and Florida, among other states, started to open their schools to both races. Negroes began to demand equality of treatment in other areas, notably in the Montgomery, Alabama, bus boycott in 1955.

But in 1956, Woodward wrote, "resistance hardened up and down the line, and in places stiffened into bristling defiance." The Citizens' Council movement spread from Mississippi to other states. Those who favored compliance with the Court's decision suffered economic pressure. Violence at the University of Alabama blocked the admission of Autherine Lucy, a black graduate student. Virginia, the historic leader of the South, took the hard line. Senator Byrd called for "massive resistance." The pre–Civil War doctrine of interposition was trotted out, and Virginia declared that it would make its public school system private in order to prevent desegregation. A number of other states imposed legal barriers to compliance with the Court's decision.

The white South was now encased in a thick blanket of segregationist conformity. The Southern Manifesto, signed by the great majority of southern senators and representatives, proclaimed that the Supreme Court had abused its powers and called for resistance against integration by all lawful means. In the late fifties, Woodward noted, "the lights of reason and tolerance and moderation began to go out." Politicians inclined to a middle position learned that political survival demanded adherence to the segregationist line. Senator J. William Fulbright of Arkansas signed the Manifesto. Governor Orval Faubus of that state called out the national guard to defy a federal court order to prevent nine black students from entering Central High School in Little Rock.

The year 1960, Woodward observed, was the "turning point," "the year of massive awakening for the Negroes of the South—indeed Negro Americans generally." The key event was the college student lunch-counter sit-in in Greensboro, North Carolina, which started the bonfire. It became, Woodward wrote, "one of the great uprisings of oppressed people in the twentieth century." It was not just blacks against whites; it was also "an uprising of youth against the older generation."

Thus, the Kennedy administration came to power in January 1961 with the weight of Jim Crow on its back. The opposing forces, both formidable, were locked in mortal combat. While the immediate battlefield was the South, it was a certainty that the struggle would engulf the entire nation in its greatest domestic crisis, certainly since the Great Depresssion, and perhaps since the Civil War.[1]

During the interregnum the President-elect asked Harris Wofford to prepare a policy report on civil rights. In effect, he would be a one-man task force. Wofford submitted his memorandum on December 30, 1960. It reflected his own mastery over and commitment to civil rights along with his political sophistication. Highly sensitive, it was not released to the press.

Victory in the election, Wofford argued, arose from an "artful balancing" of "a strong civil rights platform and campaign, and Lyndon Johnson and southern support." For the years of power he urged Kennedy to continue "by strongly supporting and building on *both* parts of 1960's winning combination." Johnson was the "essential ingredient." "The more of the load of civil rights that Johnson will carry, the better, and the sooner, the better."

The federal government "has been flying on only one of its three engines—the Judiciary—while the Congress and the Executive have been stalling." The latter two, or at least the executive branch, must bring their full weight to bear on the elimination of racial discrimination.

Wofford divided the nation into three regions. In the Deep South the immediate issue was the right to vote. "This requires action in some 100 rural counties which are the stronghold of disfranchisement." There should be "token" school desegregation to break "the solid front of massive resistance" to the Court's decision in *Brown v. Board of Education.* This should include the admission of Negroes to higher educational institutions and the introduction of "Nashville-type, first-grade, grade-a-year public school desegregation," at least in major cities. The government should insist upon the elimination of discrimination in jobs in the federal service, among government contractors, and on federal grant-in-aid projects. He expected little progress towards integrated housing.

In the Upper South, Southwest, and Border states the battle for voting rights had been substantially won and the remaining pockets of resistance should be mopped up. The present token spots of school desegregation should be pushed to full compliance with the Court's decision. Equality of job opportunity in areas of federal control was needed. One could expect only tokenism in housing.

The rest of the country was a tinderbox. "The northern vicious circle of Negro slums taking over the central cities, with white suburban nooses tightening around an expanding, demoralized Negro population, is as explosive as any southern school crisis. The present rising tensions could lead to serious race riots." Wofford was especially concerned about "growing *de facto* school segregation." In a general way, he urged the opening of jobs and integrated housing.

Wofford thought it politically critical not to single out the South on the question of race. "The pressure would be national, not focussed on the South alone. This is as it should be, for the problem is in fact national."

In the short run Wofford called for "a minimum of civil rights legislation and a maximum of executive action." While it was "heresy" in the civil rights movement to give up on immediate legislation, "political necessities" dictated this policy. There were a few minor legislative exceptions—extension of the life of the Civil Rights Commission, which expired in November 1961, and the "symbolic" anti–poll tax constitutional amendment.

The central legislative issue, Wofford argued, was a change in Senate Rule 22, which required two-thirds of the senators voting to close debate and prevent a filibuster. The fair rule on cloture was a simple majority, but "the votes for this do not now exist, and the southern fight against this would inflict a serious wound to your other legislative prospects even if the votes did exist." He thought a 60 percent cloture rule attainable. There were 41 solid votes, 6 probables, and at least three from a group of 10 mainly Border and western Democratic senators. "A call from you or Johnson would swing more than three of these." Civil rights leaders would regard a three-fifths rule as a victory, the groundwork for future legislation.

Wofford then turned to executive action. He urged voting suits brought by the Department of Justice, particularly to outlaw literacy tests or to accept minimal educational attainment as an alternative to the tests.

On jobs, Wofford looked forward to energizing the committee on government contracts. It was good that Johnson would be chairman and that Arthur Goldberg, the new Secretary of Labor, would be vice-chairman. Jurisdiction should be extended to the nongovernment work of firms partly on federal contract.

Kennedy, Wofford advised, should instruct the Justice Department to bring "well-chosen suits" to protect constitutional rights. He had memoranda from distinguished academic authorities and two prominent Washington law firms that the Supreme Court would probably hold this within the powers of the President.

School desegregation was the toughest issue because there was a clear need for legislation. But the Clark-Celler bill, which would have enacted the Democratic platform plank (ironically Wofford was its author), would send this session of Congress "up in flames." At a minimum, there would, by 1962, probably be a need to provide federal financial aid to districts that desegregated (New Orleans) or to provide education for black children in districts that abandoned their public schools (Prince Edward County, Virginia).

Wofford urged shifting "the spotlight" of publicity from lunch-counter sit-ins and school desegregation to voting rights. Aside from the merits, this was "the point on which there would be the greatest area of agreement and the greatest progress could be made." Southern Democratic leaders had learned from the election results that the Negro vote was vital to their survival. The Democratic party, rather than the NAACP and Martin Luther King, Jr., should launch a massive registration drive with southern leaders prominently on display. The Justice Department should follow up by enforcing the 1957 and 1960 Civil Rights Acts on voting rights in rural counties.

Kennedy, Wofford recommended, should issue three executive orders promptly after taking office: establishing the general principle of nondiscrimination in all federal operations and requiring agencies to report on compliance within ninety days; forbidding discrimination in all federally assisted housing programs; and conditioning federal aid to institutions of higher education on nondiscrimination.

Wofford then turned to staffing the critical federal agencies. Most important was the civil rights division of the Department of Justice. The assistant attorney general in charge of this division was "the key appointment in the civil rights field." He must be "vigorous and imaginative." Lyndon Johnson would need "a strong and effective staff director" for the government contracts committee. The Civil Rights Commission required new blood and more black faces both as members and as staff.

Wofford proposed a list of names of qualified Negroes to appoint to subcabinet level positions, to the regulatory agencies, and to federal judgeships. He suggested that Judge William Hastie of the Third Circuit should receive "most serious consideration" for an opening on the Supreme Court.

Wofford urged "symbolic action," such as refusal by the President to use segregated facilities or to take part in segregated meetings, the invitation of leading Negroes to the White House, and presidential acknowledgement of "the courage and Americanism" of parents and children for attending desegregated schools in the South.

Finally, Wofford thought it indispensable that these policies should be coordinated. While one might consider the assistant attorney general for this task, other experts recommended "a White House coordinator," the job that Wofford got.[2]

The Kennedy policy on civil rights in 1961 called for a minimum of legislation and a maximum of executive action. This stemmed from a realistic appraisal of the lineup in the Congress, particularly the Senate. As Wofford and Clark had pointed out, there was no chance of passing a basic civil rights bill dealing with voting, public accommodations, and fair employment. The votes for cloture were simply not there. This policy was not announced publicly and the civil rights leaders were not officially told that the administration would not seek legislation. This fooled no one. Roy Wilkins of the NAACP was "floored" by the decision. He thought it was like telling the other football team that you were not going to use the forward pass.

But there was a secondary legislative matter that had to be faced—extension of the life of the Commission on Civil Rights, which otherwise would expire in the fall of 1961. The Civil Rights Acts of 1957 and 1960 had created the commission and had given it two functions—a vague oversight over voting rights and the authority to study the status of civil rights in the nation. Eisenhower had balanced the six members of the commission carefully—three from the North (Chairman John A. Hannah, the president of Michigan State, Father Theodore A. Hesburgh, the president of Notre Dame, and J. Ernest Wilkins, a black and Republican assistant secretary of labor, who was soon succeeded by another black Republican, George M. Johnson, former dean of the Howard Law School) and three from the South (John S. Battle, former governor of Virginia, Doyle E. Carleton, who had been governor of Florida, and Robert G. Storey, dean of the Southern Methodist Law School). Hannah and Hesburgh were firm integrationists, Battle and Carleton were convinced segregationists, and Johnson and Storey were moderates. In 1959, Battle, frustrated as the commission moved away from his views, resigned and was succeeded by Robert S. Rankin, a Duke political scientist, who was, like Battle, a Democrat and a southerner, but was more in favor of civil rights. The commission had held and published hearings on voting, education, housing, and civil rights in general. It had also issued reports on these and other topics, which had a considerable educational impact.

In January 1961 the members offered their resignations to allow the new President to name his own commission. Hannah met with him on February 7 and Kennedy asked the members to continue to serve. Hannah urged the President to take three executive actions—a proclamation of nondiscrimination in federal employment, investigation of the possibility of withholding federal funds from universities that practiced segregation, and an executive order forbidding discrimination in housing financed by the government. Hannah thought Kennedy reacted favorably to these proposals.

In March, Carleton and Johnson resigned and the President replaced them with two law deans, Erwin N. Griswold of Harvard and Spottswood W. Robinson III of Howard. While bipartisanship and an equal North-South split were maintained, the changes gave integrationists command. Griswold sided with Hannah and Hesburgh, and Robinson had formerly represented the Virginia NAACP. Further, on March 15, at the commission's urging, Kennedy named Berl I. Bernhard as staff director. Aside from being an excellent administrator, Bernhard strongly supported civil rights. These appointments generated southern opposition in the Judiciary Committee, particularly over Robinson, but the Senate approved all three.

While Kennedy seemed to support the commission, he stalled in asking Congress for an extension of its life. By July 1961, therefore, the agency started to cut back. Northern Democrats and liberal Republicans in the Senate introduced a bill to allow it to continue for two years. Hannah wrote Senator Clark that the commission "could operate far more effectively if given a longer lease on life, not less than four years." Now the President and the Attorney General finally spoke up, the latter urging permanent status or at least four to six years. After a bitter attack by southerners, the Senate approved a two-year extension and the House concurred. Kennedy quietly signed the bill on September 21, 1961.

By contrast, Kennedy, with Wofford nipping his heels, moved decisively by executive action with one notable exception. That was the executive order forbidding discrimination in federally assisted housing. The commission had asked Eisenhower to issue such an order in 1959, advice he ignored. During the presidential campaign the next year Kennedy had promised that he would remedy this with, in a phrase Wofford supplied, "one stroke of the pen." Hannah, after the February 7, 1961, meeting, concluded that Kennedy would soon do so. But time dragged on and nothing happened. Pens arrived at the White House marked "One stroke of the pen." The President, annoyed, instructed the mail room: "Send them to Wofford." He told his brother that he could not understand how he had ever made the promise. He would mutter: "Who put those words in my mouth?"

Over the Thanksgiving weekend in 1961 the President met at Hyannis Port with the Attorney General, Burke Marshall, Ted Sorensen, Kenny O'Donnell, and Larry O'Brien to consider a number of issues. At the end of a long session the President said plaintively, "Nothing but tough ones today. Why doesn't someone bring me an easy one?"

According to Wofford, the order was ready for signature. But Marshall, who had been working on it, was less sanguine. He thought there were problems over its scope and enforcement. Any governmental intrusion of race into housing was certain to arouse deep emotions. The housing order, furthermore, was entwined with two other Kennedy objectives—the creation of a new department of urban affairs and the appointment of Robert C. Weaver, who was black, as secretary. If the housing order was issued first, it would probably be impossible to get the other two through Congress. The

President, therefore, made the political decision to go for the department first, the secretary second, and the order last. Since this could not be stated publicly, he would have to take the heat for a year.

During 1962 the Civil Rights Commission added another complication: the executive order should direct the Federal Deposit Insurance Corporation (FDIC) to forbid its member banks to lend to builders who refused to agree not to discriminate. Weaver, the civil rights movement, and many builders supported the proposal. Marshall opposed it for the following reasons: the order dealt with federally "assisted" housing (FHA and VA) and the FDIC did not assist housing. While some in the Justice Department felt that the President had the authority to act by executive order, Marshall thought it probably unconstitutional. The FDIC wanted nothing to do with a civil rights issue, and, as an independent agency, it might decline to comply with the order. Since the issue would arouse passions, it made sense to move by steps. Finally, there was a question of whether Congress would appropriate funds to support a new function of the FDIC which it had no hand in establishing. Marshall persuaded the President.

In May 1962, just before he left for a Peace Corps mission in Africa, Wofford had a final meeting with Kennedy. He noted the unfinished business—"the big steps which I believe you should take." The housing order was first on the list. Kennedy assured him that, "with time," he would deal with all these matters.

In late November 1962, when Wofford was in Ethiopia, he learned that Kennedy had finally signed the order. Not much more than "symbolic" to start with, it had been further watered down. The proposal for an urban affairs department had failed and Weaver could not be secretary. The 1962 elections were now safely history. The signing of the order, therefore, was hardly a triumphant ceremony. Sorensen wrote:

> [Kennedy's] desire was to make a low-key announcement that would be as little divisive as possible. He found the lowest-key time possible on the evening of November 20, 1962. It was the night before he and country closed shop for the long Thanksgiving weekend. The announcement was deliberately sandwiched in between a long, dramatic and widely hailed statement on Soviet bombers leaving Cuba and another major statement on the Indian border conflict with China.

For the rest, the Kennedy executive performance on civil rights was impressive. At the inauguration parade he saw no black faces in the Coast Guard honor guard. That night he phoned Treasury Secretary Dillon, who had jurisdiction over the Coast Guard, to ask that something be done. In fact, there had never been a Negro student in the eighty-six years of the Coast Guard Academy. The first entered in 1962.

At the first cabinet meeting Kennedy discussed this incident and asked each member to review the racial balance in his department. The results were a jolt. At State there were only 15 blacks among 3,674 foreign service officers; at Justice but 10 of 950 lawyers in Washington and 9 of 742 in the field offices were black (all 56 messengers, however, were black). So it went through the departments. Although 13 percent of federal employees were Negroes, very few were in the higher ranks—for example, 15 of 6900 in grades GS-12 to GS-18 in Agriculture. The departments were told to remedy this imbalance and were required to report on their progress. Kennedy was especially pleased with his high-level appointments: Weaver at Housing; Andrew Hatcher, associate press secretary at the White House; five Negroes to the federal bench, including the NAACP's counsel, Thurgood Marshall, to the court of appeals, and the first woman, Marjorie Lawson, to the district court in Washington; George Weaver as assistant secretary of labor; Clifton R. Wharton as ambassador to Norway. "It got to be a kind of sub rosa joke around Washington even among Negroes," Wilkins noted, "that Kennedy was so hot on the Department heads, the Cabinet officers, and Agency heads that everyone was scrambling around trying to find himself a Negro in order to keep the President off his neck."

Kennedy set up two interagency groups to push desegregation by administrative action. One, headed by Wofford, met weekly in his office with his legal assistant, William Taylor, who had been assigned to him from the staff of the Civil Rights Commission. The regulars were Marshall, Bernhard, Louis Martin of the Democratic National Committee, and John Feild, director of the Committee on Equal Employment Opportunity. They reviewed civil rights in the whole federal service. The other was launched by Fred Dutton, the secretary to the cabinet, and met monthly in the Cabinet Room. It consisted of representatives of the major agencies at the assistant secretary level, along with Wofford. While there was some muttering about turf (not by Wofford, who welcomed support from any source), Dutton soon turned the subcabinet group over to him. It offered an opportunity to push important agencies that otherwise might have been missed—Defense, Health, Education and Welfare (HEW), Commerce, Atomic Energy, the Federal Aviation Administration, and General Services. Dutton and Wofford were surprised and pleased by how much these agencies accomplished quietly.

There were a number of what might be called "symbolic" actions. Kennedy announced that neither he nor members of his administration would attend functions at segregated facilities. There was a great to-do about the Metropolitan Club in Washington, which declined to admit blacks. Robert Kennedy headed a list of administration luminaries who resigned. General Ulysses S. Grant III, head of the Civil War Centennial Commission, planned a meeting at segregated facilities in Charleston, South Carolina. Kennedy, furious, ordered the commission to meet at an integrated naval station. (Washington query: Did Grant surrender to Lee at Appomattox?) In presenting his credentials, the black ambassador from the new African nation of Chad told Kennedy that he had been "thrown on my rear end as a result of entering the Bonnie Brae restaurant over on Route 40." Kennedy complained to Angier Biddle Duke, chief of protocol in the State Department. He and Wofford launched a restaurant desegregation campaign along Route 40 in Maryland. Wofford would send or sometimes read himself a message he had written "from the President" to the owners and got half of them to agree to serve blacks.[3]

The power of the South in Congress foreclosed any legislation to end discrimination in employment. Thus, Kennedy could act only by executive order. This meant that wholly private employers were beyond reach. The federal government under executive order could exert control only over its own agencies and its contractors. Nor did the southerners make even this limited authority easy. Senator Richard Russell of Georgia had attached a rider to a 1945 appropriations bill that restricted an agency created by executive order from dealing with discrimination in employment for more than one year without legislative authorization. The effect of the Russell rider was to limit the structure, the budget, and the staff of any such agency. After Kennedy issued the executive order and the rules and regulations were published, Senator Lister Hill of Alabama, chairman of the Committee on Labor and Public Welfare, took the next step. He notified the President's Committee on Equal Employment Opportunity that "the assumed authority of the Committee constitutes an unconstitutional usurpation of the legislative powers of the Congress" as well as "an unauthorized and unwise extension of Federal interference with and control of the Nation's private business."

Kennedy's policy was in the mainstream of his predecessors—FDR's Committee on Fair Employment Practice, Truman's Fair Employment Board and Committee on Government Contract Compliance, and Eisenhower's Committees on Government Employment Policy and Government Contracts. All had been flawed, particularly in enforcement, and had accomplished little. Kennedy insisted on a more vigorous policy.

Since Lyndon Johnson would head the program, during the interregnum the President-elect had asked him to draft the executive order. His staff, Bill Moyers and George Reedy, worked on it, and Gerald W. Siegel, who had assisted Johnson in the Senate and was presently at the Harvard Business School, drafted the order. Several experts offered their views: Abe Fortas,

Johnson's close friend and a notable Washington attorney, Theodore W. Kheel, the grand vizier of New York City's labor relations and former president of the Urban League, and Joseph A. Jenkins, a member of the National Labor Relations Board. The Vice President did not dally. Siegel had a working draft of the order by January 3, 1961. It was then reviewed by the Labor Department and the Department of Justice.

Johnson submitted his report along with the proposed order, evidently in late February. While the Eisenhower committee had done some good work, he wrote, its "achievements have fallen far short of the goal." It was vital to strengthen its authority. "The Government of the United States cannot remain in a position where equal opportunity for employment and advancement is denied to citizens employed on the Government's own work, because of their race, creed, color, or national origin. This is an intolerable paradox in a democratic society." It was not enough only to act negatively, that is, merely to forbid discrimination. "It is necessary that affirmative action be taken to make equal opportunity available to all who, directly or indirectly, are employed by the nation."

While Johnson hoped that "much can be accomplished by [the] voluntary cooperative effort of management, labor and citizens of good will," more was needed. For the first time in the history of the program, therefore, there would be "sanctions and penalties." They comprised publication of the names of violators, the cancellation of contracts, and referral to the Attorney General.

Although the labor movement as a whole supported equal employment, some unions insisted on discriminating by exclusion from membership and other means. There were no "completely satisfactory corrective measures" to solve this problem. Johnson suggested a study. A similar difficulty arose over federal grants to state and local agencies which discriminated. Again, he proposed further study.

President Kennedy accepted Johnson's proposals and on March 6, 1961, issued Executive Order 10925. Unlike the Truman and Eisenhower programs, the order combined the federal employment and government contractor agencies into a unified Committee on Equal Employment Opportunity (CEEO). As Eisenhower had picked Nixon as chairman, Kennedy named Johnson. Secretary of Labor Arthur Goldberg, whose department would be administratively responsible for the committee, was executive vice-chairman. The other members were from three groups—cabinet officers, including the Attorney General, the heads of agencies which financed CEEO, and prominent citizens drawn from industry, labor, the churches, and public life, several of whom were black. Because of the Russell rider, the budget and staff were ridiculously inadequate. As of October 12, 1961,

$425,000 was available ($140,000 from Defense, $50,000 each from Atomic Energy, National Aeronautics and Space Administration, and General Services, $30,000 each from the Tennessee Valley Authority and the Federal Aviation Administration, and so on) and these agencies had supplied 23 employees.

The order forbade federal agencies to discriminate in employment because of race, creed, color, or national origin and required each agency to develop "positive measures for the elimination of any discrimination, direct or indirect, which now exists." Government contracts, with limited exceptions, must contain the following provision:

> The contractor will not discriminate against any employee or applicant for employment because of race, creed, color, or national origin. The contractor will take affirmative action to ensure that applicants are employed, and that employees are treated during employment without regard to their race, creed, color, or national origin. Such action shall include, but not be limited to, the following: employment, upgrading, demotion, or transfer; recruitment or recruitment advertising; layoff or termination; rates of pay or other forms of compensation; and selection for training, including apprenticeship.

This was the first use of the affirmative action concept in federal antidiscrimination policy, language borrowed from Section 10(c) of the National Labor Relations Act of 1935.

The Kennedy order, like its predecessors, provided for conciliation. But its enforcement teeth went far beyond the others in allowing for the publication of the names of violators, suits by the Department of Justice, cancellation of the contract, and a bar against future contracts. This program, while suffering from shortcomings which will be noted, was, as Michael I. Sovern wrote, "the powerful offspring of feeble progenitors." The jurisdiction, even the budget and staff, the enforcement authority, and, as Sovern put it, the "energy and élan" were much expanded.

One of CEEO's tasks was to increase black employment in the 2 million civilian jobs in the federal service. To establish the facts, the Civil Service Commission started to gather and publish the statistics of minority group employment by agency and by state. Federal employment was highly concentrated, six agencies accounting for four-fifths—40 percent in Defense, 25 in the Post Office, 5 in the Veterans Administration, and 10 percent in HEW, Agriculture, and Treasury combined.

Defense was so enormous and decentralized that it was hard to find the handle. Further, Secretary Robert McNamara's efforts to increase efficiency had led to the elimination of many blue-collar jobs in which blacks concentrated. But he strongly supported the elimination of discrimination and established an equal opportunity office in the department. Significant gains were made in the secretary's office. Among the services, the Navy was notorious for practicing segregation, and important changes were introduced. Between 1961 and 1965 black employment in Defense in civil service grades GS-5 to GS-18 advanced from 1.4 to 5 percent. In the secretary's office under all pay plans the 1965 figure was 19 percent.

The Post Office, which was highly decentralized and provided many low-skilled jobs, had always offered a good deal of employment to blacks, especially as letter carriers, even in the South. There was a sharp increase during the early sixties. By 1965, 16 percent of postal employees were black.

Prior to World War II the Veterans Administration had discriminated against Negroes, but reversed its policy after the war and came to be known as "the government's most integrated agency." Under the Kennedy order this policy was pushed aggressively at all levels. By 1965, 25 percent of VA employees were black.

HEW also made gains, particularly in the Social Security Administration, and was up to 20 percent by 1965. Agriculture, with big operations in the South and beholden for appropriations to congressional committees dominated by southerners, was called "the last bastion of the Confederacy." It responded with massive inertia to the severe criticism it received. There were painfully won modest gains. Treasury, with two-thirds of its employment in Internal Revenue, was more responsive. Justice was a special problem. While Robert Kennedy insisted on hiring more blacks, the gains were modest and virtually nonexistent in the FBI. Director J. Edgar Hoover was a racist and flatly refused to employ blacks as agents. Kennedy was told that Hoover believed that Negroes were less intelligent than whites because their brains were 20 percent smaller. Despite repeated requests, he even declined to tell the Attorney General how many of his agents were black. He finally admitted that there were only five of 5000. Kennedy was furious and demanded that the FBI hire more. Yet, by the end of 1962, there were only ten on the payroll. State, with strong support from Secretary Dean Rusk, a Georgian, and Under Secretary Chester Bowles, was a model of compliance and Negroes made notable gains, particularly in the higher ranks.

Since tens of thousands of business firms with about 20 million employees held government contracts, this was much the larger part of CEEO's program. In fact, contracts had contained nondiscrimination clauses for almost twenty years, but they had not been enforced. The Kennedy order offered the opportunity to close this gap. But Johnson had little stomach for using CEEO's powers, much preferring conciliation and voluntary compliance. This had a certain legal legitimacy, as Senator Hill's letter demonstrated. Sovern pointed to the probability that "a contractor disciplined for his breach of his promise not to discriminate would seek an adjudication that

Executive Order 10925 was an unconstitutional usurpation of legislative power." No one knew what the court would hold.

When CEEO opened its doors for business on April 7, 1961, Herbert Hill, national labor director for the NAACP, entered to file a complaint against Lockheed Aircraft Corporation. One of the nation's largest defense contractors, the company had just received a billion-dollar order for military aircraft to be manufactured at its huge complex in Marietta, Georgia, near Atlanta. Lockheed had long enjoyed a reputation for progressive labor relations in its Burbank, California, operations. Like other national corporations that had migrated to the South, Lockheed had conformed to "local custom" by imposing segregation. Despite NAACP complaints, the Nixon committee had found the Marietta operation in compliance with the government's antidiscrimination policy.

But, NAACP charged, Negroes were hired only for unskilled or semiskilled jobs, were excluded from the aprenticeship training program, were forced to use segregated facilities, and could join only an all-black local of the Machinists union. After investigation, the CEEO staff found that these complaints were "substantially correct." The Machinists quickly merged the segregated locals. Lockheed removed "White" and "Colored" signs from restrooms and drinking fountains, and open lunch counters replaced segregated facilities.

On May 25, 1961, Johnson and Courtland Gross, the president of Lockheed, signed the Plan for Equal Job Opportunity at Lockheed Aircraft Corporation. It provided for the elimination of discrimination, a policy the company pledged to disseminate widely. Lockheed would "aggressively seek out" qualified minority candidates for jobs, including requests to college placement officers. It would review the capabilities of existing black employees to learn whether they were eligible for upgrading. The apprenticeship training program would be opened to Negroes.

Lockheed, in fact, made a good-faith effort to fulfill these promises. There were formidable obstacles. Few southern blacks had the education or skills for the higher level jobs, and even fewer with the competence in the North wanted to move to Georgia. Only a handful of high school students applied for the apprenticeship program. White workers on layoff, of course, held seniority over prospective new black hires. Despite these difficulties, the company moved ahead. In late 1962 a Southern Regional Council study concluded that "Lockheed is carrying out its pledge." The Marietta operation was on its way to desegregation.

With the Lockheed announcement in May 1961, CEEO with Johnsonian ballyhoo launched Plans for Progress. The idea was to persuade large employers to commit themselves to equal employment opportunity on the dubious premise that they would actually provide it. Nevertheless, the highly publicized program proved attractive to the corporate giants and by August 1964, 268 firms with over 8 million employees had signed on. Most of the plans followed the Lockheed model with a fairly detailed promise to eliminate discrimination, though many were merely letters of intent. A high official of the company would come to Washington where he and Johnson would sign the agreement, often at the White House.

The results of Plan for Progress were uneven. A January 1963 Southern Regional Council study of twenty-four signers in the Atlanta area showed that only three were in full compliance—Lockheed, Western Electric, and Goodyear. Four others gave evidence of some progress. But the other seventeen revealed either ignorance or indifference. Another defect was that many firms thought the program was limited to blue-collar jobs. A nationwide study for the period May 1961 to June 1963 of 103 corporations showed a Negro gain from 28,940 to 42,738 salaried and from 171,021 to 198,161 hourly paid jobs. This was a combined increase of 40,938 new positions for blacks.

In addition, CEEO disposed of a large number of complaints, almost three and one-half times as many in two and one-half years as the Eisenhower committee had dealt with in seven years. A number eliminated racial discrimination. For example; Socony Mobil at its Beaumont, Texas, refinery had maintained separate seniority rosters based on race with blacks confined to service jobs and provided segregated entrances, time clocks, locker rooms, cafeterias, and drinking fountains. As the result of CEEO intervention, the seniority lines were merged and the facilities were integrated. In a number of cases the CEEO used a specific complaint to gain a "broad pattern-changing 'breakthrough'" for an entire corporation, community, or industry. In 1962 the committee claimed forty-eight such agreements.

In a few cases of obdurate noncompliance CEEO threatened to invoke its authority to bar the firm from future government contracts. One involved Comet Rice Mills, with plants in Arkansas, Louisiana, and Texas, which employed blacks only in low-skilled jobs. Another was Danly Machine Specialties of Cicero, Illinois, which hired no Negroes. The threatened loss of contracts brought both firms into quick compliance, each pledging to promote equal employment opportunity and to submit progress reports. The cases received considerable publicity which, doubtless, persuaded other firms to come into line voluntarily.

Since many labor unions, particularly the crafts, historically discriminated against blacks by prohibiting membership, establishing separate locals based on race, and denying apprenticeship training, CEEO established a parallel program for labor organizations. Both the Industrial Union and the Civil Rights Department of AFL-CIO cooperated.

On November 16, 1962, Johnson and officials of 116 national unions along with President George Meany for 300 locals chartered directly by AFL-CIO signed Programs for Fair Practices, covering about 11 million members. The unions promised to work for the elimination of discrimination based on race, creed, color, or national origin in employment, membership, and apprenticeship training. They also pledged to seek nondiscrimination clauses in collective bargaining agreements. While gains were made in this program, some unions did not sign on and a few, particularly in the building trades, did no more than affix their signatures.

There was a good deal of dispute over the effectiveness of Lyndon Johnson's CEEO. Later, Robert Kennedy was sharply critical, charging Johnson with failure to give the program direction and to follow up on Plans for Progress. He thought the Vice President milked the agency for his own public relations advantage. This judgment seems tainted by Kennedy's intense dislike for Johnson. While Marshall found some merit in the Attorney General's complaints, he thought the program far superior to Nixon's and, on the whole, good. Wofford, who attended only the committee's early meetings, found Kennedy "not respectful" of the Vice President and was quite favorably impressed with Johnson's performance. Roy Wilkins thought that Johnson took "a very personal concern" in CEEO and in finding jobs for Negroes. Kheel made a comprehensive review of the agency after fifteen months, which, while suggesting a number of changes to improve efficiency, reached the general conclusion that CEEO had made "a record in which all associated with the Committee may take just pride. I am compelled to add that this is undoubtedly due in large measure to the enthusiasm and interest which the Chairman and Vice Chairman have, to my personal knowledge, consistently demonstrated."

Sovern concluded that the accomplishments of CEEO were both significant and trivial. They were important in three ways. First, CEEO was a pioneer among federal agencies in seriously promoting racial integration, and its experience demonstrated that gains could be made. Second, as a consequence of this intervention, a small number of employers and unions fully accepted desegregation in the work place and a much larger number took their first halting steps in that direction. Finally, the program and the hoopla that went with Plans for Progress educated management, labor, and the public at large to accept the concept of racial desegregation in employment.

The accomplishments were trivial in the sense that only a small fraction of the nation's labor markets were pried open for Negroes. The black unemployment rate remained much higher than the white and the great majority of employed blacks who worked for federal agencies or for federal contractors gained nothing. In April 1963 the Bureau of Labor Statistics completed an analysis of CEEO compliance reports from 7,887 contractors covering about 3.5 million workers, 7.4 percent of private nonagricultural employment. These firms employed only 224,000 blacks, 6.5 percent of the total, of whom 40,800 were women. "Negro employment was concentrated in lower paying jobs which require little or no skill." Only 15,700 blacks had white-collar jobs, 1 percent of the total, exceeding that percentage only in New York, Philadelphia, and Chicago. There was a long way to go.

At a cabinet meeting in June 1963, Kennedy asked about the effect of motivation and education on black employment. Secretary of Labor Willard Wirtz replied off the top of his head and promptly thereafter consulted the Bureau of Labor Statistics experts and their data. On June 21 he sent the following conclusions to the White House:

> A review of the immediately available information confirms the tentative view I expressed that the reasons for this unemployment are, *in this order:* (1) lack of job opportunities in the economy; (2) inadequate education, training, and motivation; (3) *present* prejudices and discriminatory practices.
>
> It is an important part of this summary analysis that the best evidence we have supports the conviction that the "motivation" factors are *all* the *result* of previous discrimination and lack of educational opportunity (with lack of motivation also representing, in a different sense, part of the reason for a failure by Negroes to use their educational opportunities to the fullest possible extent).

This was much like the pessimistic conclusion Burke Marshall reached over voting rights when he learned that 70 percent of the blacks in Mississippi were functionally illiterate. The Kennedy administration was learning that the integration of the Negro into American society was to be extraordinarily complex and difficult.

The other "gain" which seemed trivial related to those covered by the executive order who suffered discrimination for reasons other than race. CEEO seems to have dealt only with blacks. Persons who might have had claims based on religion, color (other than black), or national origin either did not file or were overlooked. And those who were the victims of prejudice because of their sex or age were outside the system.[4]

"For Southern blacks," Catherine A. Barnes wrote, "segregated transportation was long one of the most despised forms of discrimination. The Jim Crow railroad car, the seats at the back of the bus, the 'colored' waiting room in a rail or bus depot or an airport were all blatant indicators of blacks' second-class status." The accommodations and service were invariably inferior and Negroes often had to take abuse from white employees, passengers, or policemen. "A color line in public transit was a humiliating personal affront."

The NAACP began the legal challenge to Jim Crow in transportation in the thirties and stepped up the campaign during and after World War II. The advances came with painful slowness.

On July 16, 1944, Irene Morgan, a black woman, was traveling from Baltimore to Gloucester County, Virginia, by Greyhound bus. In Saluda, Virginia, a white couple got on and the driver asked her to give up her seat. She refused to stand. Morgan was arrested, convicted of violating the state Jim Crow law, and fined $10 and costs. NAACP brought suit and the state supreme court sustained the law, holding, as it must in this case, that it applied to an interstate carrier. In *Morgan v. Virginia* in 1946 the U.S. Supreme Court set aside the state law as a burden on interstate commerce.

The Interstate Commerce Commission administered both the Interstate Commerce Act for railways and the Motor Carrier Act for buses, both of which banned discrimination, a prohibition not observed in the South. In 1953 NAACP filed a comprehensive complaint with the ICC against twelve railways that operated in that region. On August 1, 1952, Sarah Keys, a black woman who was a member of the Women's Army Corps, had received a two-week leave from Fort Dix, New Jersey, and bought a bus ticket for travel from Trenton to her family's home in Washington, North Carolina. She sat in the front of the bus. A new driver took over for the Carolina Coach Company in Roanoke Rapids, North Carolina, and ordered her to sit in the Jim Crow section. She refused. He moved the other passengers to another bus and would not let her board. She argued and he called the police. They arrested her, put her in jail overnight, and she was later convicted of disorderly conduct. The NAACP joined the two cases.

The Supreme Court's decision of May 17, 1954, in *Brown v. Board of Education* challenged Jim Crow in areas beyond education and influenced the ICC. The commission handed down its decisions in *NAACP v. St. Louis–San Francisco Railway Co.* and *Keys v. Carolina Coach Co.* on November 7, 1955. They held that state laws which imposed segregation on interstate passengers on trains and buses and in waiting rooms violated federal law. In effect, the ICC rulings set aside the "separate but equal" doctrine enunciated in *Plessy v. Ferguson* for interstate carriers.

During the Montgomery bus boycott in 1955 the black leaders formed the Montgomery Improvement Association to carry on the protest. MIA, with the blessing of NAACP, filed suit on behalf of five black women in *Browder v. Gayle,* asking the federal courts to hold the Alabama and Montgomery Jim Crow laws for local transit unconstitutional. The Supreme Court did so on December 20, 1956, ruling that these statutes violated the equal protection clause of the Fourteenth Amendment, thereby driving another nail into the coffin of *Plessy.*

On December 20, 1958, Bruce Boynton, a Howard law student, was traveling from Washington to his home in Selma, Alabama, by Trailways bus. In Richmond he sat at the white lunch counter and was refused service. He declined to move to the colored section. A policeman arrested him for violating the Virginia anti-trespass law. The NAACP took *Boynton v. Virginia* to the Supreme Court. It ruled that the terminal, including its eating facility, was an integral part of the interstate bus operation and that segregation at the lunch counter violated the Motor Carrier Act.

Thus, by 1961 there was a substantial body of law holding Jim Crow illegal or unconstitutional in both interstate and intrastate transportation. Many carriers, particularly in the upper South, had come into compliance. But there remained a hard core of defiance, especially in the Deep South. The Congress of Racial Equality (CORE) determined to assault residual Jim Crow with a nonviolent demonstration that would compel the federal government to intervene. As James Farmer, its director, said later,

> We planned the Freedom Ride with the specific intention of creating a crisis. We were counting on the bigots of the South to do our work for us. We figured that the government would have to respond if we created a situation that was headline news all over the world, and affected the nation's image abroad. An international crisis, that was our strategy.

Farmer had been born in Texas and held a divinity degree from Howard. He had worked for the Fellowship of Reconciliation, the Upholsterers Union, the League for Industrial Democracy, the State, County and Municipal Workers, and the NAACP. He was handsome, urbane, imposing, eloquent, and courageous.

The *Boynton* decision came down on December 5, 1960, and provided the springboard for CORE's Freedom Ride. Scouts moved into the South to plan the route and volunteers were trained in nonviolence. In late April 1961 Farmer sent mimeographed releases explaining the Ride to the President, the Attorney General, the ICC, and the heads of the Greyhound and Trailways bus systems. The trip would start in Washington and end in New Orleans on May 17, the anniversary of *Brown*. "We intend to challenge . . . every form of segregation met by the bus passenger." The White House seems to have lost the release. In the Justice Department it was routed to Marshall, who, evidently, did not receive it. The Attorney General may have been informed of the Freedom Ride earlier, but did not remember it until a bus was burned near Anniston, Alabama.

Farmer led six other blacks and six whites onto two buses, one Greyhound and the other Trailways, in Washington on May 4. As they moved south, they tested segregation on the buses, at stations, and at rest stops. It took them 10 days to reach Atlanta. While there were minor incidents in Charlotte, North Carolina, and Rock Hill and Winnsboro, South Carolina, they met no serious resistance. Nor did they attract much attention. They met with Martin Luther King, Jr., in Atlanta on May 13.

They knew that trouble lay ahead: Alabama, Mississippi, and Louisiana. On May 14, Mother's Day, they boarded the buses in Atlanta and set out for Birmingham. When the Greyhound pulled into Anniston, sixty miles from Birmingham, a Ku Klux Klan mob attacked with steel bars and clubs. They dented the sides, smashed windows, and punctured tires. A plainclothesman from the Alabama Public Safety Department blocked entry at the stairwell. The police arrived, cleared a route, and the bus left. A few miles out of Anniston a tire went flat and the driver pulled over. The mob, which had followed in cars, smashed the bus and set it afire. Several Freedom Riders were beaten as they got off. The police arrived, dispersed the mob, and took the injured to the hospital.

When the driver of the Trailways bus learned in Anniston of what had happened, he told his passengers that it was too dangerous to proceed with an integrated bus. He was certainly right. A group of thugs entered and attacked the Freedom Riders. James Peck required 53 stitches in the head. Walter Bergman, a 61-year-old retired school administrator from Detroit, suffered permanent brain damage.

Nevertheless, the buses staggered on into Birmingham. Police Commissioner Eugene (Bull) Connor, perhaps the nation's most notorious racist, had promised the Klan 15 minutes to assault the riders before his police arrived. He later said they were late because he was shorthanded; many were paying their respects to their mothers. Gary Thomas Rowe, Jr., an FBI informant inside the KKK, had tipped off his superiors in advance, but they had failed to inform the Attorney General. The mob attacked the riders with baseball bats, lead pipes, and bicycle chains. Horribly beaten and bloodied, they sat in the terminal, where John Seigenthaler, Kennedy's assistant, found them. Since they were in no condition to complete the ride, he took them to New Orleans by air. But a group of black students from Fisk University in Nashville, determined to keep the ride alive, drove to Birmingham to replace the wounded.

Now the Freedom Ride truly was an international incident, the shocking events in headlines and on TV screens all over the world. The President had recently suffered the humiliation of the Bay of Pigs and was about to depart for Europe for high-level talks with the British and French and, more important, his first meeting with Khrushchev in Vienna. CORE could not have come at him at a worse time and he was furious. "Tell them to call it off! Stop them!" he said to Wofford. Wofford replied that nobody could stop them.

The Attorney General, Marshall, and Byron White went to the White House for an early breakfast. The President was still in his pajamas. There was no question over strategy, whether the government should act. As Marshall recounted later,

> I just had the feeling that within the course of, I don't know, fifteen or twenty minutes, he understood the whole thing, all its implications; that he was prepared to take the necessary action; that he realized that there were going to be all sorts of consequences in the future; and that he saw this and accepted it and digested it and that was it.

Thus, the questions were tactical.

Marshall stated that he would be asking "a very strong judge," Frank Johnson, for an injunction against interference with interstate bus movement and against the police for failing to protect interstate travelers. Assuming its issuance, the department could enforce the court order without a presidential proclamation, or, if law and order broke down, the President could act independent of the injunction. Shortly, Judge Johnson issued the order.

Although the Army had been notified and troops were on the alert at Ft. Benning, Georgia, everyone at the breakfast agreed that calling out soldiers was the absolute last resort. Little Rock was the awful example, and military occupation of a southern city was to be avoided at almost any cost. White had begun to improvise a force at Maxwell Air Force Base—federal marshals, guards from the Bureau of Prisons, alcohol and tobacco tax agents from the Treasury, and border patrol officers from the Immigration Service. Only the border patrol officers were experienced and White doubted that his force would be effective. They agreed that the President would call Alabama Governor John Patterson, a friend of the Kennedys, to urge him to protect the riders.

The Attorney General had already spoken with Patterson, who had agreed to provide protection. But the governor, a confirmed segregationist who had been elected with Klan support, quickly changed his mind. He issued a statement advising the "agitators" to leave Alabama at once and asserting that he would not be responsible for their safety. He then "disappeared" for four days so that neither of the Kennedys could reach him.

The new Freedom Riders from Fisk insisted on proceeding with the next leg of the trip to Montgomery. Seigenthaler, back in Alabama, "found" Patterson. He persuaded the governor to provide protection, and Floyd Mann, the chief of the Alabama Highway Patrol and a reliable professional, gave his assurance. But Greyhound said it could not find a driver willing to risk the 90-mile run. Robert Kennedy acidly ordered the bus superintendent in Birmingham that "Mr. Greyhound" locate a driver. One was found. The bus, now carrying 21 riders, 18 of them black, set out for Montgomery, protected by state troopers. But Mann peeled off his men at the city line and the bus pulled into the terminal unprotected.

Seigenthaler, John Doar of the civil rights division, and William Orrick of the civil division were waiting for the riders, but so was a mob armed with chains and axe handles. There were no police in sight, probably a reprise of the Birmingham arrangement with the Klan. Orrick was an eyewitness:

> At the Montgomery bus depot a mob—which could have been readily controlled by proper police work—rioted and attacked the group of riders, beating them with pipes, sticks, clubs and their fists. At least four out-of-town reporters and photographers were beaten and an ambulance which arrived on the scene was chased away. People with no apparent connection to the trip were beaten, a boy's leg was broken and another boy had inflammable liquid poured over him and set on fire. Mr. Seigenthaler, who attempted to help a girl escape from the mob, was struck from behind and lay on the sidewalk for 25 minutes before police took him to a hospital.

On learning of this event, Robert Kennedy phoned to tell Seigenthaler that he had done the right thing. "Let me give you some advice," Seigenthaler said, "never run for governor of Alabama. You couldn't get elected."

Kennedy concluded that civil authority had broken down in Montgomery. On May 21 the Attorney General, acting under Judge Johnson's order, directed the force of 500 men under White to move into the city from Maxwell Air Force Base.

That same day Martin Luther King, Jr., arrived in Montgomery to preach that evening for the Freedom Riders at Ralph Abernathy's First Baptist Church. He would be a natural target of white hatred, and Kennedy and Marshall feared for his safety. Fifty marshals escorted him from the airport to the church. More than a thousand people jammed the building to cheer the Freedom Riders and to call for a massive attack on segregation and for federal intervention. A hostile mob gathered outside and became extremely threatening. White rounded up about a hundred marshals to hold them off. There was a nasty skirmish and fear that the mob would set fire to the church. White considered the situation "very touch-and-go." Patterson, now alarmed, called out the Alabama National Guard.

King phoned Kennedy and Marshall from the church about 10:00 p.m. Marshall thought he was "panicky," "very scared." They told him guardsmen were on the way to reinforce the marshals. He was not reassured so Kennedy tried a joke: "As long as you are in church, you might say a prayer for us." King was not amused. The guardsmen arrived shortly after midnight and the mob gradually drifted away. In the early morning hours the guardsmen escorted the parishioners home. Kennedy pulled out the marshals. They had been in Montgomery for less than twenty-four hours.

The Nashville contingent of Freedom Riders said that they wanted to continue the ride with Farmer to Jackson, Mississippi. They told King privately that he was morally obligated to come with them. White strongly urged them to abandon the dangerous journey. King, citing legal problems in Georgia, said he could not risk more trouble. The young riders accepted neither the caution nor the explanation.

Robert Kennedy had hoped to stop the Freedom Ride in Montgomery to avoid having anyone killed. He asked for a "cooling-off" period. Farmer replied, "We had been cooling off for 100 years. If we got any cooler, we'd be in the deep freeze." Kennedy did not think they would reach Jackson alive. Former Mississippi governor James P. Coleman warned Marshall that his successor, Ross Barnett, could not be trusted and that the riders would be murdered before they got to Jackson.

Kennedy turned to Senator Eastland. Though a bizarre combination, they understood each other. As Kennedy said later, "He always kept his word and he always was available, and he always told me exactly where he stood and what he could do and what he couldn't do." Eastland promised that there would be no violence in Mississippi. He said the riders would be arrested when they arrived in Jackson. Kennedy said his primary concern was to avoid having them beaten or killed, implicitly accepting Eastland's proposal to have them arrested.

Two buses with a total of twenty-seven Freedom Riders left Montgomery on May 24. They were escorted by the Alabama Highway Patrol and National Guard to the state line, where the Mississippi National Guard took over. There were airplanes and helicopters overhead. The Jackson terminal was cordoned off when the buses arrived. The first group of riders was arrested for breach of the peace when they tried to use the white restrooms, the second as they went to the white cafeteria. All were convicted and received 60-day sentences and $200 fines. Twenty-two decided to work off their fines in jail; at $3 a day, they faced 63 days of incarceration.

In fact, the Freedom Ride now became a movement. CORE, the Student Nonviolent Coordinating Committee, and the Southern Christian Leadership Conference (SCLC) set up a Freedom Ride Coordinating Committee to call for volunteers and to arrange trips. During the summer of 1961 over 300 riders came to Jackson, were arrested, and were stuffed into Mississippi's bulging jails. Governor Ross Barnett later described the process: "When the jails were all filled and the mayor's chicken coops down on the fairground were all filled, there were thirty-two of them left, and it was my happy privilege to send all of them to the State Penitentiary at Parchman and put them in maximum security cells." CORE also arranged Freedom Rides from Washington to Florida, from St. Louis to New Orleans, from Virginia to Arkansas, among others. There were rides independent of CORE, notably one led by William Sloane Coffin, Jr., the Yale chaplain, which ended in the Montgomery jail.

Earlier King had told Kennedy that the ICC should end segregation at the terminals. The Attorney General pointed out that the commission was an independent agency with a notorious reputation for delay. But, after Jackson, T. Robert Saloschin and Nathan Siegel, attorneys in the office of legal counsel, proposed a petition to the ICC. On May 29, Kennedy asked the commission to issue regulations prohibiting segregation in interstate bus terminals, and Marshall pushed the commissioners. The ICC issued the regulations on September 22, 1961. Over the next year or two the Department of Justice, with some difficulties in Mississippi, quietly persuaded local officials to remove the "White" and "Colored" signs, first in bus and later in railroad and airline terminals. By the end of 1962 James Farmer's Freedom Riders and Robert Kennedy's lawyers had abolished Jim Crow in interstate transportation.[5]

"Freedom to vote," the Commission on Civil Rights reported in 1961, "is the cornerstone of democracy." Yet many counties in eight southern states denied Negroes the franchise. In sixty-nine counties 38 percent of the voting-age population was black, but only 6.2 percent was registered. In the summer of 1960, John Doar, a new attorney in the civil rights division, went down to Haywood County, Tennessee, to see for himself. He met with Negro sharecroppers in a packed country church who had tried to register. He asked whether any had received notices of eviction from their homes. Nearly all the hands shot up. In thirteen counties with large Negro populations no blacks were registered. Mississippi and Alabama were the only states in the nation which continued to require payment of the poll tax as a condition for casting a ballot. The Mississippi constitution had been amended in 1954 to impose a grandfather clause to make it harder for Negroes to vote. Of 500,000 voting-age blacks in that state in that year, only 22,000, 4.4 percent, were registered.

"Ending discrimination in voting," Wofford wrote President-elect Kennedy on December 30, 1960, "is the point of which there would be the greatest area of agreement and the greatest progress could be made." Both Kennedys and Burke Marshall agreed with this assessment. The Fifteenth Amendment, after all, read, "The right of citizens of the United States to vote shall not be denied or abridged by the United States or by any State on account of race, color, or previous condition of servitude." Thus, denial of the franchise to blacks because of their race was unconstitutional on its face. Further, the right to vote was the door that opened onto other civil rights. As Marshall put it:

> Other federal rights cannot successfully be asserted where the right to vote is not protected. Only political power—not court orders or other federal law—will insure the election of fair men as sheriffs, school board members, police chiefs, mayors, county commissioners, and state officials. It is they who control the institutions which grant or deny federally guaranteed rights. . . . Any elected official represents not the people in his district, but the people in his district who vote. When it is true, as it is only in the case of Negroes, that a large class of people are not permitted to vote, the elected officials represent only those who are opposed to Negro rights.

Thus, the President, the Attorney General, and the head of the civil rights division repeatedly urged the leaders of the civil rights movement to concentrate their energies upon gaining the franchise. In principle the mainline organizations strongly supported action to increase black voter registration in the Deep South—Wilkins of NAACP, Young of the Urban League, Farmer of CORE, and King of SCLC. In fact, they had long taken this view. In 1957, for example, King had declared at the Lincoln Memorial: "Give us the ballot. Give us the ballot and we will no longer have to worry the federal government about our basic rights." He had then launched the unsuccessful Crusade for Citizenship to register 5 million southern Negroes. SNCC, however, was divided on this issue. Part of the organization went along with the others. But the militant wing regarded the pressure from "the Kennedys" to concentrate on voting as a guileful stratagem to divert the civil rights movement from direct action, like the Freedom Rides, into an area where the administration would not be embarrassed in its dealings with the white South.

But even if the government and the civil rights organizations agreed on the principle of a voting rights campaign, almost everything about it would prove to be knotty and intractable. A fundamental problem was that denial of the ballot was geographically particular rather than, as in the case of transportation, general. That is, while the framework of disfranchisement was set forth in state law, the actual decision to deny registration to a prospective voter was made by the county registrar. This compelled pressure, negotiation, and litigation county by county, which required an enormous commitment of time and manpower. The attorneys representing the civil rights division, as John Doar and Dorothy Landsberg put it, were "not prepared to take the terrible risk of losing a single case because of lack of proof." The instructions to the FBI for the investigation of registration infractions in Choctaw County, Alabama, went on for 174 pages. The Montgomery case became legendary, requiring the analysis of 36,000 voter registration applications. A team of one lawyer, two law clerks, and two secretaries spent three months on the preparatory work; four and sometimes five lawyers prepared for the trial. The brief ran to 293 pages and there were 69 exhibits, one of which filled five filing cabinets. There were 160 witnesses at the trial. Three and at times four lawyers spent two weeks on the appeal.

The outcome of these suits depended critically upon the federal judges who heard them. But, as Marshall pointed out, "the bench reflects the customs and attitudes of the community." Both senators from the state in which the judge would serve must approve him before the Senate would do so. "Cases enforcing the civil rights of Negroes penetrate more deeply than any other into the society in which a judge lives and is personally a part. It is inevitable that most district judges want to do as little as possible to disturb the patterns of life and politics in their state and community."

This problem, with a few exceptions, was troublesome enough in the case of federal judges in the Deep South appointed prior to 1961. But in the first two years of the Kennedy administration it actually got worse. The President named five new district court judges in the Fifth Circuit—Florida, Texas, Georgia, Alabama, Louisiana, and Mississippi. While, under the Constitution, the President appointed judges with the advice and consent of the Senate, according to a contemporary quip, southern senators appointed them with the advice and consent of the President. In fact, Byron White made the investigations and Robert Kennedy made the decisions. The notorious five who cleared this process were E. Gordon West of Louisiana, Robert Elliott of Georgia, Clarence Allgood and Walter Gewin of Alabama, and William Harold Cox of Mississippi, every one, Victor S. Navasky wrote, "anti–civil rights, racist, segregationist, and/or obstructionist."

West denounced *Brown* from the bench. But his specialty was obfuscation and delay. In defiance of orders from both the Fifth Circuit and the Supreme Court, he declined to issue a decision in the St. Helena Parish school desegregation case, which had been in litigation for eleven years. Elliott refused to apply *Brown* and held against Negroes in 90 percent of the cases that came before him. Allgood and Gewin were cut from the same cloth. But Cox was the prize-winner.

Kennedy had named Thurgood Marshall to the Second Circuit in 1961. Perhaps the most distinguished black member of the bar in the nation, he had been director-general of the NAACP Legal Defense Fund and had argued many of the leading civil rights cases. Senator Eastland, chairman of the Judiciary Committee, held up his appointment for a year. One story had it that Eastland spotted Robert Kennedy in a corridor of the Capitol and said, "Tell your brother that if he will give me Harold Cox I will give him the nigger." Another view was that the Department of Justice was romantic, believing that "every redneck who came down the pike was another Hugo Black."

Harold Cox was no Hugo Black. From the bench he called Negroes "niggers" and said some of them were acting like "a bunch of chimpanzees." He systematically ignored the desegregation decisions of the higher courts. He wrote John Doar that "I am not favorably impressed with you or your tactics in undertaking to push one of your cases before me. I spend most of my time in fooling with lousy cases brought before me by your Department in the civil rights field." He accused Doar of stupidity, impudence, and harassment.

How in the world did John F. Kennedy come to appoint William Harold Cox to the federal bench? The fact that he had been Eastland's college roommate should have been a red flag. Roy Wilkins wired the President that "for 986,000 Negro Mississippians Judge Cox will be another strand in the barbed wire fence." But the American Bar Association gave him its highest rating—"exceptionally well qualified"—and the FBI found nothing against him. Marshall was more perceptive; Cox made him "nervous." Robert Kennedy called Cox in and asked him whether he would enforce the law, including the Supreme Court's interpretations of the Constitution. Cox said he would. Kennedy believed him. He later concluded that Cox had lied.

Another problem in winning voting rights, along with other civil rights, was the system of federalism imposed by the Constitution as interpreted by the Kennedy Justice Department. Marshall propounded the problem this way:

> A typical instance, which has happened often, might involve a student leader. He informs the Justice Department that the next day he is going to lead a group of Negroes down to register to vote in a small town in the Deep South where no Negro has attempted to register for decades. He asks for federal protection, or at least for a show of "federal presence," in the terminology of civil rights groups. When he is told that there is no national police force, that federal marshals are only process servers working for the courts, that the protection of citizens is a matter for the local police, and that there is nothing to do until and unless something happens, the gap between his vision of government and the reality, between the expectations set forth in the Constitution and Supreme Court decisions and the hope of their fulfillment becomes too great. In the mind of the student, he has looked into the eye of federal authority, has asked for the help of the federal government in exercising and realizing federal rights, and has been turned away.

Marshall granted that this system was grossly inadequate for providing the suffrage to blacks. The gains made in registration in the Deep South came painfully and with agonizing slowness. Further, Marshall was concerned because "civil rights issues cut into the fabric of federalism" and "cut most deeply where police power is involved." He was disturbed because this was causing "the loss of faith in law."

Were there no alternatives? In theory there were two, though neither was viable. The first was to create a federal police force to enforce the constitutional rights of blacks in the South. Aside from the practical problems of creating such an agency, it was a political mirage. No one wanted it, not even the civil rights movement and certainly not the FBI. Robert Kennedy said, "I wouldn't want that much authority, much more authority, in the hands of either the Federal Bureau of Investigation or the Department of Justice or the President of the United States." The other alternative was a new voting rights law that would establish federal oversight of the registration process. In the early sixties, for the reasons already noted, this option was foreclosed politically. The realistic choice, therefore, was to file and prosecute lawsuits. The President instructed the Justice Department: "Keep pushing the cases." "You've just got to keep going back," John Doar said. "You've got to come *here* [Mississippi]. . . . You'll think it's taking forever. . . . So you do the only thing you can do."

Another sticky problem was jurisdictional conflict between the Department of Justice and the Civil Rights Commission over voting rights cases. "When bureaucratic warfare between Burke Marshall and commission staff director Berl Bernhard first broke out," Wofford wrote, "I couldn't believe it; it couldn't happen between friends of such obvious goodwill and common dedication." But it did.

In part, Wofford thought, it was just another battle between federal agencies over turf. But it was more. Robert Kennedy told the commission: "You're second-guessers. I am the one who has to get the job done." The commissioners pointed out that the right to vote was the one substantive area over which they had jurisdiction. It was also, Wofford noted, a conflict between two very strong men—Kennedy and Father Hesburgh. The Attorney General did not want "a runaway grand jury" making political trouble in the South. Father Hesburgh saw the commission as "the conscience of the nation," not as a "burr under the saddle of the administration." Relations soured and the two agencies battled over commission access to FBI files, its right to hold hearings in Louisiana and Mississippi, and the publication of commission reports.

Even the President failed to resolve the jurisdictional dispute. He was surprised to learn that the commission was unanimous and was galled to learn that this included Dean Griswold of the Harvard Law School. "Who the hell appointed Griswold?" Kennedy asked. "You did," Bernhard replied. "Probably on the recommendation of Harris Wofford," the President said.

A final problem, though one that was resolved, was money. Voter registration drives, which required house-to-house canvassing mainly in rural areas, were extremely expensive. The civil rights organizations were perpetually broke and competed with each other for financial support. The government, obviously, could not be the source of the aid. But Marshall and Wofford were instrumental in arranging for assistance. Harold C. Fleming, the outgoing executive director of the Southern Regional Council, and Stephen R. Currier, head of the Taconic Foundation and the Potomac Institute, devised the Voter Education Project (VEP). It would coordinate the registration activities of the civil rights organizations, would be nonpartisan, and would cover the entire South. The VEP program cost $870,000. The main contributors were Taconic—$339,000, the Field Foundation—$225,000, and the Edgar Stern Family Fund—$219,000. Wiley A. Branton, an attorney who had represented the black children in the Little Rock school desegregation case, was put in charge. He proved to be a competent and hard-headed administrator. VEP opened shop early in 1962 and ran for two and one-half years.

NAACP, SCLC, CORE, and the Urban League entered the VEP program with a minimum of friction. SNCC, more youthful and militant, balked. Several of its leaders believed that VEP was a ploy by the Kennedy administration, probably abetted by King, to cool them off. But the lure of money was seductive. SNCC bravely chose to work in the most backward and violent areas, particularly Mississippi and southwest Georgia.

Registration went slowly in the face of formidable difficulties, but progress was made, especially in areas where there were no bars against black registration. Mississippi was entirely different: bitter resistance, intimidation, beatings, jailings, murders. The place names came to express terror: McComb, Greenwood, Hattiesburg, and Terrell County, Georgia, which might as well have been in Mississippi. The activists in the front lines demanded protection from the Department of Justice. Marshall doubled the number of attorneys handling voting rights to ten and sent them to the South. Many suits were filed. But, for the reasons noted above, the department offered little immediate protection to those at risk. Early in 1964 Branton cut off funds for the Mississippi campaign; it was futile and dangerous to try to register Negroes in that state. VEP, obviously, was not a great success. The number of black people who were allowed to register in the Deep South (there appear to be no reliable figures) as a result of its efforts seem to have been modest and they must have concentrated in the cities. The gains in small towns and rural counties were minimal at best. But, as Pat Watters and Reece Cleghorn noted, "VEP did telescope into two and one-half years a Negro registration increase which could not otherwise have been predicted in less than ten years. It moved Negro registration off dead center, where it had been for most of the previous decade, and reestablished momentum."[6]

Mississippi was special. It seemed hardly to be part of the United States. The winds of change that had swept across the South after World War II became a gentle zephyr as they slowed over the Magnolia State.

The basic industry remained cotton, as it had been before the Civil War.

There was very little new industry. Most of the businesses that moved to Mississippi were seduced from the North by the attraction of a surplus of low-skilled, low-wage, nonunion labor. There was only one city, Jackson, and its population was less than 150,000. For the rest, the state was sprinkled with a handful of towns. Most people remained in agriculture, much of it marginal.

Mississippi was extremely poor, underdeveloped like the Third World. Comparative statistics by state usually ranked Mississippi last. It had the lowest per capita income, in 1960 only 53 percent of the national average. Even this depended heavily on federal support; in 1958 Mississippi paid $385 per capita to the federal government and received $668 in return. Expenditures on public services were miniscule. The state spent less per pupil on education than any state, and its teacher salaries were the lowest. Half the adults over twenty-five had less than nine years of schooling—eleven for whites and six for blacks. Because of lack of employment three-quarters of the college graduates drifted away. In fact, out-migration, white and black, was a chronic hemorrhage. A number of county seats retained fewer people than they had had in 1900. Whole counties seemed to be disappearing. Life in Mississippi, Walter Lord found, had "a stagnant quality," a "strange emptiness," "an air of isolation." The nation's culture seemed to stop at the state's borders. The only bookshop in Jackson was run by the Baptist Church. Newstands rarely offered well-known national magazines. Oxford, "the center of learning," had no established bookstore. Twenty-seven counties were without real libraries.

Thus, Mississippi turned inwards upon itself, becoming in James W. Silver's phrase "a closed society." "Mississippians are obsessed by their sense of the past, but this does not insure the accuracy of their historical picture; they see legend rather than history." Silver continued,

> There are parallels between the 1850's and the 1950's which remind us that Mississippi has been on the defensive against inexorable change for more than a century, and that by the time of the Civil War it had developed a closed society with an orthodoxy accepted by nearly everybody in the state. The all-pervading doctrine, then and now, has been white supremacy, whether achieved through slavery or segregation, rationalized by a professed belief in state rights and bolstered by religious fundamentalism. In such a society a never-ceasing propaganda of the "true faith" must go on relentlessly, with a constantly reiterated demand for loyalty to the united front, requiring that non-conformists and dissenters from the code be silenced, or, in a crisis, driven from the community. Violence and the threat of violence have confirmed and enforced the image of unanimity.

Ole Miss was the state's crown jewel. Founded in 1848, the University of Mississippi had a beautiful campus in Oxford. The state's young people from the best white families came there, a few to get an education, more to have a good time. Despite a dreadful academic history, Chancellor J. D. Williams, who arrived in 1946, had significantly upgraded its quality. By the early sixties it was possible to get a decent education—"if you want to." But few did. First things came first at Ole Miss. It provided a high-spirited social life around an imposing fraternity and sorority house system. Ole Miss had an outstanding football team. In a four-year span it produced a Miss Dixie, a Maid of Cotton, four Miss Mississippis, and two Miss Americas.

Kosciusko, in Attala County, was in the center of the state. Outside town Moses Meridith scratched a living from an 84-acre cotton and corn farm. He had ten children, the seventh J.H. (given intitials to prevent whites from calling him by his first name). Meredith was one of 34 Negroes out of 5,179 in the county of voting age who was allowed to cast a ballot. He had never gotten beyond the fourth grade and was determined that his children be educated. J.H. went to the local segregated schools. He was quite bright, determined and courageous, and inner-directed, which gave him tunnel vision. As he grew older he became quietly enraged about segregation. When he graduated from high school he yearned to go to college. But he knew that Negro colleges were of low quality and that he could not get into Ole Miss. In 1951 he volunteered for the Air Force because it was the most integrated branch of the service. He filled out his name to James Howard Meredith.

In his nine years in the Air Force Meredith took many courses toward a college degree and concluded that he would finish up at Ole Miss. In the summer of 1960, after a three-year hitch in Japan, he left the service and returned to Mississippi with his wife and child. That fall he enrolled in all-black Jackson State College. He had some catching up to do. While he knew about the Armed Forces Integration Program, he had never heard of CORE and was barely aware of the NAACP. He followed the presidential election and heard Nixon speak in Jackson. Since important candidates never campaigned in Mississippi, he assumed that the Republicans were in trouble.

Meredith was now even more determined to get into Ole Miss. He wrote:

> Everything that I have done, I did because I had to. A very long time ago some force greater than myself placed before me my life's role, the mission that I was to accomplish. How my divine responsibility was to be carried out was left to me. Obviously I was returning to Mississippi to work at the task of fulfilling my mission.

Kennedy's election, which he partially misread, was the turning event. "I was firmly convinced that only a power struggle between the state and the federal government could make it possible for me . . . to gain admission to the University of Mississippi." He was impressed by "the civil rights platform which Kennedy had insisted upon at the Democratic convention." Since the election had been very close, the Negro vote was "the decisive factor" and he expected the administration to respond to this pressure. He was less confident about Mississippi. Ross Barnett, an extreme racist, had been elected governor in 1959. But the "loyal Democrats," led by moderate racist James P. Coleman, might take the governorship in the next election. In any case, the whites were split on race. Meredith would gamble. "Once he had decided," Marshall said later, "all sorts of things flowed from it."

On January 21, 1961, the day after Kennedy's inauguration, he wrote to the registrar at Ole Miss for an application for admission and a catalogue, which arrived a week later. He sent off to the other colleges for transcripts of his records. On January 29 he went to see Medgar Evers, the NAACP field secretary in Mississippi, who encouraged him to notify Thurgood Marshall of the NAACP Legal Defense Fund, which he did at once. Marshall wanted to know a great deal about Meredith.

On January 31, Meredith filed his application with R. B. Ellis, the registrar, for admission in the spring semester. It required a photograph and he stapled one to the form—a neat young man with a polka-dot tie, unmistakably Negro. He wrote, "I am not a white applicant. I am an American—Mississippi—Negro citizen." The application was complete except for the names of six alumni references. He explained that alumni were all white and that he did not know any. Instead he presented certificates from several Negroes regarding his moral character. On February 4, Ellis wired him that no applications received after January 25 would be considered.

That day Meredith wrote to the Justice Department. He summarized his personal history, including his application to Ole Miss, and stated, "I have not been accepted and I have not been rejected." He asked that the government insure compliance with the laws. "I feel that the federal government can do more in this area if they choose and I feel that they should choose." His letter seems to have reached Marshall. Meredith then went to see William Higgs, a recent graduate of the Harvard Law School, who had returned to Jackson to practice civil rights law. Higgs said that he had talked to Burke Marshall, who was expecting them to call. Marshall told him that the civil rights division was interested in his case and would help in any way it could.

Thurgood Marshall, convinced that Meredith was acting in good faith and was qualified for admission, assigned Constance Baker Motley as counsel to his case. On May 25, Ellis denied Meredith's application because the university would not accept a transfer of credits from Jackson State since it was not a member of the Southern Association of Colleges and Secondary Schools.

Motley on May 31 filed suit in the federal district court before Judge

Sidney C. Mize, alleging that the denial of admission was unconstitutional. She asked for a speedy trial so that Meredith could enter the summer session on June 8. Mize did not make his decision until December 12, 1961, holding that the university had not discriminated against Meredith because of his race.

He filed at once with the court of appeals in New Orleans, asking for speed by injunction so that he could enter for the term starting February 8, 1962. In a split decision on January 12, Judge John Minor Wisdom held that Mississippi maintained a system of racially segregated schools and colleges, found the record "muddy," and ordered Judge Mize to hold "a full trial on the merits." Judge Elbert Tuttle dissented; he would have granted the injunction so that Meredith could enroll for the spring semester.

On February 3, Mize denied Meredith's admission on two grounds. The first, delivered with a straight face, was that "the University is not a racially segregated institution." The second was that the registrar would be justified in rejecting Meredith because he was "a rather unstable person."

Meredith returned to New Orleans and there was another split before a somewhat different court. On June 25, 1962, Judge Wisdom in a carefully reasoned and comprehensive opinion delivered a total victory to Meredith. While the university had no published policy of excluding Negroes, "the hard fact to get around is that no person known to be a Negro has ever attended the University." Wisdom continued,

> Reading the 1350 pages of the record as a whole, we find that James H. Meredith's application for transfer to the University of Mississippi was turned down solely because he was a Negro. We see no valid, nondiscriminatory reason for the University's not accepting Meredith. Instead, we see a well-defined pattern of delays and frustrations, part of a Fabian policy of worrying the enemy into defeat while time worked for the defenders.

Judge D. A. DeVane dissented on the basis of Mize's finding that Meredith was a troublemaker. He was concerned about the impact of the decision on the citizens of Mississippi and he felt it was the duty of the court to prevent another Little Rock.

A nasty controversy now erupted in the Fifth Circuit. The state asked Mississipian Judge Ben F. Cameron to grant a stay order until the state filed an appeal with the Supreme Court. Cameron issued the order and the circuit overruled him. They went through this charade three times. The circuit court then sent the issue to the Supreme Court.

By now the Meredith case was the major public issue in Mississippi and was attracting national attention. On September 4, 1962, the board of trustees of the university announced that it had withdrawn authority from any official of the institution to enroll Meredith. "The entire power, authority, duty, responsibility and prerogative with regard to action on the application and/or admission of said James Howard Meredith should be and the same to be hereby expressly reserved exclusively unto this Board of Trustees. . . ."

On September 10, Justice Hugo L. Black vacated Judge Cameron's stay order. The alternative would "only work further delay and injury to movant." He thought there was "very little likelihood that this Court will grant certiorari." Though convinced that he had the authority to act alone, Black, recognizing the importance of the Meredith case, took no chances. "I have submitted it to each of my Brethren, and I am authorized to state that each of them agrees . . . that I have the power to act, and that under the circumstances, I should exercise that power as I have done." On September 13, Judge Mize surrendered and ordered the University of Mississippi to admit Meredith. Motley told the press, "This is the end of the road for the University." But she failed to take account of Ross Barnett.

That night the governor, whose usual manner was that of a simple, kindly country boy, delivered a fiery speech on statewide television. "We will not surrender to the evil and illegal forces of tyranny," Barnett proclaimed. "We must either submit to the unlawful dictates of the federal government or stand up like men and tell them 'NEVER.'" In a formal proclamation he dusted off the hoary southern doctrine of interposition, long since rejected by the Supreme Court, that a state could "interpose" itself to protect its citizens against the federal government. "Mississippi, as a Sovereign State, has the right under the federal Constitution to determine for itself what the federal Constitution has reserved to it." The state legislature unanimously backed the governor. As Silver had noted, Mississippi had now closed in upon itself and had snuffed out any expression of moderation. When the governor said "interposition," Burke Marshall would later recall, the Department of Justice knew that it was in for trouble.

On September 19 the state court in Jackson enjoined anyone from enrolling Meredith. Since he had lived in both counties, in his application Meredith had created an ambiguity over whether his residence was in Attala (Kosciusko) or Hinds (Jackson) County. For this he was charged with a crime. The legislature on September 20 enacted a law making anyone accused of criminal behavior ineligible for admission to a state institution of higher learning. The United States moved at once to prohibit enforcement of this law. Also on September 20 the board of trustees granted the governor full power to deal with Meredith's admission.

The Kennedy administration observed the unfolding of these events with foreboding. The last thing the President wanted was another Little Rock, with Ross Barnett playing Orval Faubus and John Kennedy in the role of Dwight Eisenhower. Robert Kennedy's worst dream was to have soldiers and

Mississippians shooting at each other. He desperately hoped to avoid sending troops into Oxford. But Barnett seemed to foreclose every other option. The Meredith case was not like a voting rights question which could be left to the courts to dispose of in leisurely fashion. Here the President faced time constraints and direct defiance of the federal courts—district, circuit, and Supreme Court. Kennedy had taken an oath to uphold the Constitution and it was his responsibility to enforce the court orders. The people in the Department of Justice and the White House waded in.

Obviously, it would not do to sit back and allow Meredith to be lynched. This was no idle concern; violence was a way of life in Mississippi. The Ku Klux Klan could strike at any moment. While the Citizens' Council, now a formidable force, denied that it engaged in violence, there were many skeptics. Major General Edwin A. Walker, the right-wing extremist, made several trips to Mississippi from his home in Dallas to encourage "patriots" to "rally to the cause of freedom" and to carry "firearms." Motley called Meredith in Kosciusko and, because he was reluctant to leave the state, as he put it, "tricked" him into going to Memphis. A black lawyer to whom he was sent said his job was "to get you the hell out of Mississippi." He stayed with his cousins in Memphis and was later moved to nearby Millington Naval Air Station.

The administration decided to use marshals first, holding the Army in reserve. On September 14 the former began to gather at Millington under Chief U.S. Marshal James Joseph Patrick McShane, a tough former New York City cop. He took over Meredith's security.

The registration period for the fall semester at Ole Miss was September 19 to 24, 1962. Robert Kennedy decided to enroll Meredith at 4:30, half an hour after the close of registration, on September 20. He notified Barnett, who, while he hardly agreed to cooperate, promised a security escort. Two border partrol cars with Meredith, McShane, St. John Barrett, a Justice Department attorney, and two marshals left Millington in the afternoon for the 78-mile drive down U.S. 51 to Oxford. The governor flew from Jackson at about the same time. The federals turned into the Mississippi highway patrol station in Batesville, where they were met by its commander, Colonel T. B. Birdsong. A convoy of about half a dozen cars moved down the highway to Oxford and, then, by a side entrance, onto the campus to the Continuation Center. A crowd of three to four thousand had gathered, held back by state troopers.

A long table had three chairs for Barnett, Ellis, and Meredith. Ellis read a statement transferring his duties as registrar to the governor. Meredith said that he wanted to enroll. Barnett read a proclamation which stated: "James H. Meredith, you are hereby refused admission as a student to the University of Mississippi." He handed it to Meredith. Barrett warned the governor that legal action would be taken against him. The federal party then left, escorted to the Tennessee line by the highway patrol.

The next day, September 21, the government asked Judge Mize to hold the chancellor, the dean, and the registrar in contempt for refusing to register Meredith. He refused to do so. An appeal was made to the Fifth Circuit and eight judges heard the matter on September 24. The university officials stated that they were ready to register Meredith. The court ordered the trustees to revoke the delegation of the registrar's duties to the governor. Barnett then issued a proclamation declaring that any action by a federal official that interfered with Mississippi law was a crime. On September 25 the court enjoined all officers of the state from interference with the enrollment of Meredith. That day the board of trustees withdrew the registration authority of the governor and offered to enroll Meredith at 4:00 p.m. the next day. To forestall Barnett, the federal lawyers got a sweeping order from the court forbidding the governor from blocking the registration by any means.

About noon that day the governor walked from the Capitol in Jackson to the trustees' offices in the state building. A border patrol Cessna with Meredith, McShane, and Doar landed at the Jackson airport at 3:25 and a patrol car took them to the state building, arriving at 4:34. Meredith walked through a large crowd which shouted, "Nigger, go home!" and was led to the tenth floor. The TV cameras whirred. The governor stood in the doorway and read a proclamation: "I, Ross R. Barnett, Governor of the State of Mississippi . . . do hereby deny you admission to the University of Mississippi." The federal party returned to the airport and left town. That night the court cited Barnett for contempt and ordered him to appear in New Orleans on September 28. The Attorney General immediately informed Barnett that Meredith would be in Oxford the next day.

The Cessna landed at 9:29 a.m. on September 26 with Meredith, McShane, and Doar. They set out in a highway patrol convoy for the university. Two blocks from the campus they were stopped by lines of state troopers, county sheriffs, and patrol cars. Before the television Lieutenant Governor Paul Johnson read a proclamation of interposition on behalf of the governor. Doar and McShane argued to no avail. Again, they returned to the airport.

On September 27 Kennedy and Barnett had a series of telephone calls in which they agreed that Meredith and company would proceed to Oxford, the marshals would draw guns, and the state would back down in the face of superior force. But by now Barnett had so aroused Mississippi that a huge mob had gathered in Oxford, many armed, that the governor could not control. At 3:50 p.m. Barnett and Johnson called Kennedy to report the dangerous situation and to ask him to call off Meredith. Kennedy refused. At 4:20 Barnett called to say that the situation had eased. A convoy of thirteen border patrol cars sped down the highway with twenty-five marshals with empty guns, a border patrol plane overhead. Birdsong now reported that the mob was out of hand. A worried Barnett called Kennedy to say that a lot of people were going to get killed. Kennedy ordered the convoy to stop at Como, fifty miles from Oxford and, once again, to turn back.

September 28 was a fateful day. The circuit court held Barnett in contempt and, unless he registered Meredith by October 2, subjected him to a fine of $10,000 a day. Kennedy, having talked to Barnett twenty times since September 15 and having sent Meredith on four occasions for registration, now concluded that there was no way to deal with the governor. He passed responsibility for the Meredith case to the President. He began to send a team of Justice Department officials to Memphis.

That afternoon he met with Cyrus Vance, Secretary of the Army, and General Maxwell Taylor, chairman of the Joint Chiefs of Staff, at the Pentagon. He had an eerie feeling in the war room when maps of Mississippi were rolled out. The military plan had three stages. In the first the President would federalize the Mississippi National Guard, which would place Mississippians face-to-face with fellow Mississipians. In case that was insufficient, two military police battalions would move into Oxford. Finally, two battle groups of the 82nd Airborne Division would be placed on reserve in Memphis. Major General Creighton Abrams would be in overall command; Brigadier General Charles Billingslea would be in charge in Oxford. The operation, in fact, would move through all three stages. By the morning of October 3 the Army had 9,827 troops in Oxford, 4,833 in Memphis, and 5,449 on alert in Columbus, Georgia.

Saturday, September 29, was the tense lull before the storm. Marshals, border patrolmen, and federal prison guards, a total of 541, poured into Millington from across the nation, the majority from the South. An advance team scouted Oxford. Three days earlier General Walker on Shreveport radio had called for 10,000 men to "rally to the cause of freedom" by standing with Governor Barnett. "I will be there," Walker declared. KKK Klaverns mobilized and convoys of cars and pickups from many points in the Deep South set out for Oxford. On Saturday, Walker flew into Jackson from Dallas.

That afternoon the President spoke to Barnett. "I'd like to get assurances from you," Kennedy said, "that the state police down there will take positive action to maintain law and order." The governor: "I don't know whether I can or not."

In Oxford the campus was deserted because the students had moved to Jackson for the Ole Miss–Kentucky football game on Saturday night. The packed stadium formed a sea of Confederate flags, and at halftime, with Mississippi ahead 7 to 0, there was a great shout: "We want Ross!" The governor went onto the field, raised a clenched fist, and declaimed, "I love Mississippi! I love her people! I love our customs!" The crowd roared deliriously. Some compared it with a Nazi rally.

Shortly after midnight the President ordered the Secretary of Defense (1) to enforce the court orders; (2) "to use such of the armed forces of the United States as he may deem necessary;" and (3) to nationalize the Mississippi National Guard.

On Sunday morning, September 30, Barnett and the Attorney General negotiated frenziedly over the former's bizarre proposal to have federal forces confront the state's headed by the governor, who, after a show of defiance, would ask his men to step aside. When Barnett wavered, Kennedy said the President would expose his scheme on national television. The governor, shaken, blurted out: "Why don't you fly him in this afternoon?" That was it.

That morning Nicholas Katzenbach had appeared on a network TV show and had then returned to the Department of Justice. Kennedy asked whether he had plans for the afternoon. Katzenbach said he was free. The Attorney General told him to go down to Oxford as head of a department team to get Meredith into Ole Miss any way he could. His final words: "If things get rough, don't worry about yourself; the President needs a moral issue." The Jetstar on which they flew needed more runway than the airport provided, and the pilot had to jettison the fuel. It was, Katzenbach said, "a very exciting landing."

The first airlift of 170 marshals arrived at the Oxford airport at 2:35; Katzenbach was there by 4:00. He took some of them with him and left the others to await Meredith. Katzenbach's convoy set off for the university and arrived at the Lyceum, the physical and administrative center, where he assumed the registration would take place. He stationed his men around this graceful antebellum building, and the cars returned to the airport to pick up the other marshals. Katzenbach also arranged to have Meredith lodged in Baxter Hall at the west edge of the campus. He then went to the phone booth in the Lyceum to inform the Attorney General to send Meredith to Oxford. He left the line open, which would run up a staggering phone bill. Meanwhile, students were returning from Jackson and were gathering around the marshals, who were reinforced in the late afternoon. As the crowd grew in size, it became increasingly ugly.

Katzenbach drove to the airport at 6:00. The plane arrived a few minutes later and Meredith, Doar, and McShane stepped out. A car with Meredith, Katzenbach, and Doar, escorted by the state highway patrol, drove to Baxter Hall. Meredith moved into the counselor's apartment and started to study. Katzenbach left twelve marshals, later increased to twenty-four, to guard him.

As darkness gathered, the crowd at the Lyceum grew, reaching between 2000 and 4000, and was joined by tough outsiders, many armed. They taunted the marshals, threw bricks, and set fires. A delegation representing the governor, headed by state Senator George M. Yarborough, looked over the dangerous situation and informed Katzenbach that they were withdrawing the highway patrol. There were hurried calls to Washington and Jackson and the decision was rescinded. But by 7:40 the patrol had disappeared. Mississippi would leave the preservation of order to the federal forces.

The mob now turned violent. Cameramen and reporters were beaten, rocks and bricks rained on the marshals, cars were smashed and set afire. Yarborough and the highway patrol reappeared, the latter to push the crowd back. It was briefly orderly and then broke into violence again. At 7:58 McShane gave the order to use tear gas and the fusillade covered the patrolmen as well as the mob. There was now a full-scale riot at Ole Miss.

At 8:00 the President announced on national television that he was sending troops into Oxford. He appealed to the citizens of Mississippi to preserve order and to uphold the law. He praised the state for its "tradition of honor and courage, won on the field of battle and on the gridiron." But he insisted that the courts had spoken and that it was his constitutional duty to enforce their decision. If the speech had any effect on the mob in Oxford, which is unlikely, it was to make it more violent.

Enraged at having been gassed, the highway patrolmen again left the campus. About 9:00 General Walker arrived. Though he did not endorse violence, his mere presence encouraged it. Paul Guihard, a reporter for Agence France-Presse, was shot in the back at point blank range and died instantly. Ray Gunter, a resident of the Oxford area, was killed by a bullet to the forehead. A number of others were wounded by gunfire, rocks, and gas. They lay bleeding on the floor of the Lyceum. At 10:00, Katzenbach, who had hoped desperately to avoid the use of troops, phoned the "Command Headquarters" in the Cabinet Room at the White House to report that the situation was out of control. Kennedy ordered the Army to take over.

The first troops, about fifty or sixty unarmed Mississippi national guardsmen under Captain Murray Falkner (nephew of the novelist, who spelt his name differently), joined the line alongside the marshals. Both took a fearsome beating. It was a maddening three hours, which enraged the President, before the first regular troops arrived at 2:04 a.m., units of the

503rd Military Police Battalion, helicoptered to Oxford from Millington. They had the dirty and dangerous task of routing out the hard-core rioters. Later that morning they were joined by the main body of the 503rd and the 716th MP Battalion, sent in by motor convoys. General Billingslea declared the area secure at 6:15 a.m. on Monday, October 1.

The toll was formidable—two dead, 160 marshals requiring medical care, 27 of them shot by gunfire, hundreds of people gassed, the Ole Miss campus a dreadful mess. Of 200 arrested, only 24 were students. At 7:55 a.m. on Monday, Katzenbach, Doar, McShane, and several marshals escorted Meredith to the Lyceum and Ellis registered him as a student at the University of Mississippi.[7]

3

Civil Rights: Year of Decision, 1963

BIRMINGHAM, Alabama, was the most segregated big city in the South. While local boosters prided themselves on their Confederate heritage, the town was founded six years after Appomattox. As a major manufacturing center, particularly of steel, it resembled the industrial cities of the Midwest rather than its southern sisters. The dominant firm was the Tennessee Coal, Iron & Railroad Company, a subsidiary of United States Steel.

About 140,000 blacks, close to 40 percent of the population, were the victims of overpowering discrimination. Few voted; there were no black policemen; almost none participated in the administration of justice. The labor market was rigidly segregated by race. Negro men were confined to hard, dirty, low-skilled work in the mills; boys to messenger jobs; and women to work as janitresses, maids, and cooks. Examples as of 1963: TCI had 1200 white-collar employees of whom 8 were black; Hayes Aircraft had a work force of 3,340 with 178 Negroes; Ingalls Iron Works employed 313 blue-collar workers with 96 black and had 169 white-collar people of whom none were Negroes; Blue Cross–Blue Shield employed 202 with 6 black.

Birmingham was a major center of federal employment, but, according to Robert Kennedy, the situation was "disgraceful." The VA and post office aside, of 2000 federal employees, only fifteen were Negroes. City parks and playgrounds, lunch counters, toilets, drinking fountains, and department store fitting rooms were rigidly segregated. Harrison Salisbury wrote in the *New York Times:*

> Every channel of communication, every medium of mutual interest, every reasoned approach, every inch of middle ground has been fragmented by the emotional dynamite of racism, reinforced by the whip, the razor, the gun, the bomb, the torch, the club, the knife, the mob, the police and many branches of the state's apparatus.

In the wake of the *Brown* decision in 1954 the attitude of white Alabama and white Birmingham had hardened. Extremists took command and the voices of moderation were stilled. In the face of a federal court order, the city closed its parks and playgrounds rather than desegregate them. It abandoned its professional baseball team rather than play integrated clubs in the International League. It refused to allow blacks to sit in the municipal auditorium, and the Metropolitan Opera and Broadway road companies stopped coming to Birmingham. Between 1957 and 1963 there were eighteen race-related bombings and over fifty cross-burning incidents.

Eugene "Bull" Connor was the local czar of race relations, the hard-nosed, short-fused, frog-voiced commissioner of public safety. "Long as I'm poleece commissioner in Birmingham, the niggers and the white folks ain't gon' segregate together in this man's town." Whenever blacks demonstrated, Connor's cops hauled them off to jail. They arrested a United States senator who entered a door marked "Colored." Connor trumpeted defiance at the Supreme Court and challenged the Attorney General to a fist fight. He warned that "blood would run in the streets" before the city integrated.

More serious, since he was much smarter and much more powerful, was George Corley Wallace, who had been elected governor of Alabama in 1962. In the 1950s Wallace had come across as a back country populist, a fairly liberal follower of Governor "Big Jim" Folsom, and—like his mentor—a moderate on race. But in 1958 Wallace had lost a hard-fought election for governor to John Patterson, who ran an all-out racist campaign calling for the banishment of the NAACP and currying the favor of the Klan. Wallace told his friends: "John Patterson out-nigguhed me. And boys, I'm not goin' to be out-nigguhed again." He ran against integration in 1962. The white crowds liked him when he called federal judge Frank Johnson "a low-down, carpetbaggin', scallawagin', race-mixin' liar." They loved him when he pledged that, if the federal government tried to integrate a school in Alabama, "I am going to place myself, your governor, in the position so that the federal court order must be directed against your governor. I shall refuse to abide by any such illegal federal court order even to the point of standing in the schoolhouse door." Wallace was elected with the largest majority in Alabama gubernatorial history. On inauguration day in January 1963 he proclaimed that he stood where Jefferson Davis had stood at his inauguration. "From the cradle of the Confederacy, this very heart of the great Anglo-Saxon Southland, I draw the line in the dust and toss the gauntlet before the feet of tyranny. And I say, Segregation now! Segregation tomorrow! Segregation forever!"

Fred Shuttlesworth, the pastor of Bethel Baptist Church and the head of the SCLC affiliate in Birmingham, the Alabama Christian Movement for Civil Rights, had been directing a low-key campaign to desegregate buses, schools, parks, and lunch counters since 1956. Connor had jailed Shuttlesworth and his followers and homes had been bombed, including the pastor's. Shuttlesworth was eager for SCLC to mount a major campaign in Birmingham.

King called the SCLC inner circle to a two-day retreat at Dorchester in eastern Georgia on January 10, 1963, for an important policy session. Among those present were King; Shuttlesworth; Andrew J. Young, the executive vice president; Wyatt T. Walker, the executive director; and Stanley T. Levison, King's trusted adviser. Through undercover agents in the Communist party and telephone wiretaps, the FBI became convinced that Levison and a black SCLC employee, Jack O'Dell, were Communist agents and passed this information to the Attorney General. Kennedy, Wofford, Marshall, and Seigenthaler urged King to fire O'Dell and to terminate his relationship with Levison. King suspended O'Dell temporarily, but refused to break off with Levison.

At Dorchester the SCLC confronted three main issues: a review of the failure of the just concluded Albany, Georgia, campaign, a drive on Birmingham, and, because the latter was certain to be tough, planning for federal intervention. The Albany venture had dragged on for much of two years and SCLC had been outsmarted by Laurie Pritchett, the police chief, who adopted a policy of nonviolence and thereby gave the federal government no reason to intervene in the town. The defeat had shaken SCLC morale and had evoked criticism from militants in the civil rights movement.

King drew two main lessons from Albany. The first was that SCLC had attacked segregation in general rather than on specific issues, such as lunch counters or buses. The black community needed a victory on a particular matter in order to sustain its support. Second, SCLC had brought its pressure on the city commission, which did not respond because very few Negroes voted. It should have boycotted downtown stores, where the economic pain would have been felt at once.

Shuttlesworth was the driving force behind the Birmingham campaign. He claimed a core of backing from his own group and the congregations of four young pastors, including A. D. King, Martin's brother. For Martin Luther King, Jr., the raw injustices and tensions in the city created a moral imperative to act, one that, despite his forebodings, he could not evade.

There were great potential rewards. If SCLC succeeded in Birmingham, the city's size, importance, and racist reputation virtually guaranteed a spillover impact on the entire South.

Shuttlesworth and Walker regarded Bull Connor as their greatest asset. The exact opposite of Pritchett, Connor was certain to use brutality and violence. This would bring out the press and the TV cameras and the nation and the world would learn about the real Birmingham. More important, violence would suck in the Kennedy administration.

At the close of the retreat Levison warned that Connor was a formidable adversary. He had smashed a much stronger labor movement with force. As Levison recalled later, this led King to say, "Everyone here should consider very carefully and decide whether he wants to be with this campaign. . . . Some of the people sitting here today will not come back alive from this campaign. And I want you to think about it." These words left a mark on Young. King, he felt, did not want to confront Birmingham. "He went to Birmingham because Fred Shuttlesworth pleaded with him to do it." Young continued, "Every time he made a commitment to something like this he was committing his life. . . ."

In November 1962 the voters of Birmingham had changed the form of their city government from a commission system to a mayor-council plan. The three commissioners, of whom Connor was one, would not serve out their terms to 1965. The election of the new mayor and councilmen was set for March 5, 1963. The race for mayor was between Connor and Albert Boutwell, a moderate on racial issues. The latter won a slender plurality. Connor and the other commissioners immediately filed suit to prevent Boutwell from taking office. Until the Alabama Supreme Court resolved the issue, Birmingham had two powerless city governments.

King had delayed the Birmingham drive until after March 5 to avoid clouding the election, fixing the opening date for March 14. But since neither Boutwell nor Connor had won a majority, a runoff was scheduled for April 2. King again postponed the campaign. Boutwell won decisively, but Connor continued as police commissioner. Many local blacks and whites, as well as the Department of Justice, urged King to hold off until the court ruled. King refused to do so.

On April 4 small groups of demonstrators, mainly students from local Miles College, sat in at downtown lunch counters. This impact was so slight that some stores did not even bother to call the police. For those that did, the police made arrests.

After a few days King realized that he was unable to recruit demonstrators in the numbers needed. One reason was that many blacks expected Boutwell to win over Connor in the Supreme Court and did not want to embarrass him. The other was Shuttlesworth, who was widely disliked in the black community because of his dictatorial style. As one Birmingham black put it, "Shuttlesworth sees himself as taking orders only from God who speaks to him and through him." King then met with local leaders to persuade them to support the campaign. He was largely successful, particularly with A. G. Gaston, the black millionaire real estate, insurance, and mortuary tycoon.

The city obtained an injunction against the SCLC prohibiting marches and protests. King and Abernathy defied the order on April 12, Good Friday, by leading a small march on City Hall, and were promptly arrested. Eight prominent white clergymen—Protestant, Catholic, and Jewish—had recently issued a statement calling the demonstration "unwise and untimely." In his five days behind bars King wrote a famous reply—"A Letter from Birmingham Jail"—in which he called on the churches to speak out for freedom and justice.

Through the remainder of April the demonstrations were modest because there simply were not enough marchers. King worried that the press would leave town for lack of news. As Walker put it, "We had run out of troops. We had scraped the bottom of the barrel of adults who could go." SCLC decided to take the unprecedented and potentially dangerous step of calling out black high school and even elementary school students. They responded enthusiastically and by the end of April Walker could organize sizable marches. Connor responded with mass arrests, stuffing the students into crowded jails. The tension in the streets mounted.

On May 3 a large group of demonstrators, mainly students, gathered in Kelly Ingram Park in the Negro business section. Police and firemen infiltrated the park with police dogs straining at their leashes and high-pressure fire hoses. The young blacks taunted them. Suddenly a Negro boy yelled, "Freedom! Get the white dogs!" Connor ordered, "Let 'em have it!" The hoses pinned demonstrators against walls and ripped off their clothes. The dogs lunged at demonstrators, and several people, including three children, were bitten. While most of the blacks were nonviolent, one group rained bricks, rocks, and broken bottles on the officers and several were injured.

The next day was worse. Tough blacks who hated the police and had never heard of nonviolence came out of the bars to join the youthful demonstrators. In the melee the blacks almost broke the police lines. Fearing that the violence was getting out of hand, SCLC called off the demonstration for the day and told its followers to go home.

The two days of violence had put Birmingham on every front page and TV screen in the nation. There were graphic scenes of streams knocking people down and of police dogs attacking. The American public was shocked and the President said that the photographs made him "sick." He was also deeply concerned because there seemed no lawful solution to the mounting crisis. The administration's top lawyers had reviewed the options and, as Marshall put it, "ended up with a nothing." The President sent Marshall down to Birmingham on May 4 to mediate.

As he later described them, the obstacles were formidable. The daily demonstrations kept growing in size, and while the police maintained some control, this was with mass arrests, hoses, and dogs. By now, Birmingham had become a world-wide press and television sensation.

King was a problem. At the outset Marshall asked him the demands. "He didn't really know." Marshall emphasized that there was no city government to deal with. Eventually King came up with some proposals for the downtown stores—opening lunch counters and providing jobs above the janitor level. The problem with King, Marshall concluded, was that he wanted "a success for himself and he wanted the success for his people." To Marshall, King was both sincere and cynical, cynical in that he hoped to use Birmingham to make himself "*the* Negro leader."

Marshall had even more trouble with the whites. All he had going for him was credibility. Earlier Vincent Townsend, the editor of the *Birmingham News,* had asked him to persuade King to defer the demonstrations until the city government question was resolved. While Marshall was certain that the overture would fail, he tried. Since the telephone line was tapped, the whites now trusted him.

But no one in the power structure would assume responsibility for dealing with racial questions. A year earlier the Chamber of Commerce had created a Senior Citizens Committee under Sidney M. Smyer, the head of a big real estate firm and a former chamber president, to deal with Shuttlesworth's campaign. Nothing seems to have happened. Marshall knew David Vann, who had been Justice Black's law clerk, represented one or several stores, and was assisting Smyer. Through Vann, Marshall tried to involve Smyer in negotiating over lunch counters and jobs. Smyer said he would meet with local blacks, but not with King.

Worse yet, Smyer said his committee was powerless, that the town was run by "the big mules": the heads of TCI, Hayes International, the two main banks, and the phone company. The stores could not desegregate their own lunch counters unless the mules approved. Marshall called the Attorney General, who called the President, who called Secretaries Dillon, McNamara, Wirtz, and Hodges, who called the mules. There seemed to be some softening. Dean Eugene V. Rostow of the Yale Law School was asked to call Roger Blough, a law school alumnus and the head of the Yale capital fund drive. The chairman of U.S. Steel, Blough had recently aroused

President Kennedy's ire in a confrontation over steel prices and may have been eager to improve his standing in the White House. He called Arthur Wiebel, the head of TCI, who began to urge netotiations with the Negroes. Since Wiebel was the biggest mule, the rest of the team began to move.

The combined pressure from Marshall, the administration, and violence in the streets brought the white power structure to heel. On May 7, Smyer called his committee, Boutwell and his city council, and Sheriff Melvin Bailey together in an emergency meeting. Marshall found Bailey to be a professional officer. The latter reported that local law enforcement was "strained to the utmost." Unless the demonstrations stopped, martial law must be declared. Frightened, the committee authorized its negotiators to strike a deal.

That evening a secret meeting was held at the home of John Drew, a wealthy black insurance man. Andrew Young spoke for SCLC and he seemed to know what he wanted: desegregation of the downtown stores; upgrading black employment in the stores; establishment of a biracial committee to discuss racial issues; and amnesty for arrested demonstrators. The white businesmen were tough bargainers. They were willing to integrate the fitting rooms and to create a biracial committee. They insisted on only token black employment in the stores and Young gave in. The question became whether each store would hire one black clerk or cashier or whether the stores as a whole would take on one or two. The whites were adamant against amnesty, stating that they had no authority to act. For the SCLC amnesty was critical because it had no funds to post bond for and defend 2600 people. Nevertheless, the negotiators were encouraged by the progress.

The next day, May 8, was a seesaw. During the morning King announced a moratorium on demonstrations that day in order not to undermine the negotiations. In Washington the President, encouraged by a report from Marshall, expressed gratification over "the progress in the efforts by white and Negro citizens to end an ugly situation in Birmingham." He commended Marshall and urged the leaders on both sides "to continue their constructive and cooperative efforts."

This optimism was short-lived. Shuttlesworth, who had been in the hospital, was released that afternoon. He became enraged when he learned that King had called off the demonstration. He denounced Marshall as the architect of perfidy. To King, he said, "You're mister big, but you're going to be mister S-H-I-T." He walked out.

Connor, who was losing control of the city, struck back. His officers informed King and Abernathy that the bonds on which they had been released from custody three weeks earlier had been raised to $2500. They refused to pay and were hauled off to jail. Marshall and the other black leaders persuaded Gaston to write out the checks for their release.

Marshall felt that an agreement was impossible unless the SCLC put up the bail for those arrested. He passed the problem to the Attorney General, who turned it over to Walter Reuther, the president of both the UAW and the AFL-CIO Industrial Union Department, and to Joseph L. Rauh, the UAW's Washington counsel. They quickly raised $160,000, $40,000 each from AFL-CIO, the Industrial Union Department, the UAW, and the Steelworkers. Governor Rockefeller also contributed an amount variously recalled as $25,000 or $100,000, and Mike Quill of the Transport Workers chipped in $50,000.

The negotiators returned to the table that evening, May 8, the next day, May 9, and into the morning of May 10, 1963. There were no demonstrations during that period. Under Marshall's watchful eye, Vann and Young worked out the Birmingham Truce Agreement. The essentials were as follows:

1. Fitting rooms desegregated within 3 days after the end of demonstrations.
2. The removal of Jim Crow signs from washrooms, restrooms, and drinking fountains within 30 days of the establishment of the new city government.
3. Lunch-counter desegregation within 60 days of the new city government taking office.
4. "A program of upgrading Negro employment" [the fact that it would involve only one sales person or cashier was not published].
5. Creation of a Committee on Racial Problems and Employment within 15 days of the cessation of demonstrations.
6. The release of those who had been arrested [nothing was announced about the bail arrangement].

That afternoon, Friday, May 10, SCLC called a press conference at the Gaston Motel, its headquarters and the place where King was staying. Shuttlesworth was allowed to read a laundered version of the truce agreement. King then hailed it as "a great victory" and the first step toward the total desegregation of Birmingham. He promptly left for Atlanta. Smyer announced tersely that his committee had made an agreement, but carefully noted that he did not speak for any city official, the courts, or the board of education. Robert Kennedy stated that the agreement was "a tremendous step forward for Birmingham, for Alabama, and for the South generally." Marshall went back to Washington.

This optimism was not shared by the opposition. Connor said the Negroes "didn't gain a thing" and urged white people to boycott the stores that agreed to desegregate. Arthur Hanes, the "old" mayor, denounced the agreement as "capitulation by certain weak-kneed white people under threats of violence by the rabble-rousing Negro, King." He added ominously, "I certainly am not bound by the concessions granted to the terrorist, King, and I have no intention of doing one thing to implement or facilitate these agreements." Boutwell made a noncommittal statement, though he let word out privately that, once in power, he would help effectuate the agreement. In any case, there was now quiet in the streets, and the state troopers sent in by the governor left town.

On Saturday night, May 11, about a thousand berobed Klansmen from Alabama and Georgia gathered in suburban Bessemer's Moose Club Park for a rally. As Michael Dorman wrote, "Two flaming crosses sent eerie shafts of light shimmering across the park. Racist speeches were shouted into the night." The Klansmen began leaving the rally about 10:15.

At 10:30 a bomb thrown from a moving car hit the front of the Reverend A. D. King's house and made much of it a shambles. Fortunately, King, his wife, and their five children were in the rear and escaped unharmed. At 11:58 another bomb exploded at the Gaston Motel in the room immediately beneath the one Martin Luther King, Jr., had occupied. He had offered the room to the Reverend Joseph Lowery, who planned to stay there but later changed his mind and went home to Nashville. The motel was severely damaged, but injuries were minor.

An angry crowd of blacks gathered outside the motel. Since this was Saturday night, many had been drinking. As the local police and firemen arrived, the blacks attacked with rocks and bricks. Soon Colonel Al Lingo, Wallace's tough enforcer, arrived with his heavily armed state troopers. Connor sent in his dogs and his special six-wheeled armored riot car. Fires started and a full-scale riot erupted, which raged for four hours. By 4:30 Sunday morning the police had sealed off the area. Hedrick Smith wrote in the *New York Times:*

> Much of the nine-block area looked as if a vicious storm had struck. Smashed and disabled police cruisers were abandoned in the streets. Seven stores and homes lay charred by fire. There was a hole in the brick wall of the A. G. Gaston Motel, caused by an explosion. Plate-glass windows were shattered in store after store in the Negro area.

Hospitals treated more than fifty people who had been wounded.

Burke Marshall, resting at his farm in West Virginia, was awakened at 2:00 a.m. with the disastrous news from Birmingham; the Attorney General shortly thereafter at his home in McLean, Virginia. The President, spending the weekend at Camp David, was briefed when he awoke. Helicopters brought all three to Washington. On late Sunday afternoon the President met at the White House with the Attorney General, Marshall, Katzenbach, McNamara, Vance, and Army Chief of Staff General Wheeler.

The immediate question was whether the aroused and tough blacks would renew the riot Sunday night. A fear was that they might kill a policeman. King had now returned to Birmingham, and Marshall phoned to ask whether he could control his people. King thought he could if there was not another bombing. His concern was the men in the bars, but on Sunday night drinking was likely to be lighter than on Saturday. King said he would go into the bars and pool halls to calm the toughs, but he did not do so until Monday night.

There were three obvious steps for the President to take immediately. The first was to send Marshall back to Birmingham to re-establish communication between the white and black communities and to preserve the truce agreement. If the latter blew up, more Negro violence seemed inevitable. The second was to establish a point for possible military action in the city. A Colonel Keller was already there, had touched base with the Department of Justice and the FBI, and had started planning. General Colby was sent in to take over that function. Finally, the President distrusted Governor Wallace and wanted to deny him use of the Alabama National Guard. Papers were prepared calling for its federalization on a moment's notice.

The main question was whether the President should send troops into Birmingham to restore order. With sharp irony, he made it clear to the Department of Defense leaders that he wanted no repetition of Oxford. They, obviously, had anticipated the criticism and had large forces available. Kennedy decided not to send soldiers to Birmingham that night but, rather, to Ft. McClellan and Maxwell Air Force Base, both within a short distance of the city. Between 600 and 700 men were quickly dispatched to these bases, and, evidently, 1000 more were to follow shortly. Elements of the 2nd Division at Ft. Benning and the 82nd Airborne at Ft. Bragg were put on alert. General Wheeler assured the President that he could have 15,000 troops available for duty in Birmingham on Monday, if needed. In fact, 17,793 men were to be assigned to what came to be known as Operation Oak Tree.

Sunday evening Kennedy met the press with a prepared statement. He expressed deep concern over the bombings and the rioting. "This government," he stated, "will do whatever must be done to preserve order, to protect the lives of its citizens, and to uphold the laws of the land." He lauded the truce agreement. "The federal government will not permit it to be sabotaged by a few extremists on either side." He then ordered three "initial steps." First, Marshall would return to Birmingham at once. Second, McNamara would send military units to bases "in the vicinity of Birmingham." Finally, the Alabama National Guard would be made promptly available for federal service. He expressed the hope that the citizens of the city would make it unnecessary for him to send in troops.

Wallace was incensed and called Kennedy a "military dictator." But the city calmed down. King went into the bars and pool rooms to quiet the blacks. Marshall met with the leaders of both races. On May 15 the Senior Citizens Committee reaffirmed its support for the truce agreement and for the first time published the names of its members, the prominent business leaders. Smyer released his version of the agreement, which showed differences with King's, but both quickly agreed to meet to resolve the disagreements. On May 23 the Alabama Supreme Court unanimously ruled that Boutwell was the lawful mayor. He expressed public support for the truce agreement for the first time. Connor and Hayes were gone. The city was quiet—temporarily.

The Birmingham crisis was decisive in making civil rights the central domestic issue of the decade. This was not because of the event itself, which was no more grave than the Freedom Rides or Ole Miss. Rather, it was Bull Connor and his fire hoses and snarling dogs playing on television. Americans generally, white and black, viewed the brutality of the struggle in the Deep South on their living-room screens for the first time and were shocked. The President wryly told the black leaders that they owed Connor a great debt.

Despite the seemingly happy ending, Birmingham hardly healed racial conflict in Alabama. To the contrary, it ripped open a wound that would bleed again at the University of Alabama, a second time in Birmingham, and, somewhat later, in Selma.

Birmingham released deep passions within the black community. Earlier the civil rights demonstrations had been controlled by leaders and organizations dedicated to nonviolence. Now blacks who had never heard of nonviolence were taking to the streets to vent their rage against Jim Crow. The black comedian Dick Gregory quipped that even his maid was sassing him. This new attitude, as the Department of Justice and the President noted with deep concern, was evident not just in Birmingham but also in other cities in the South as well as the North.

Birmingham also transformed the civil rights organizations, which gained greatly in public support and were energized. Equally important, SCLC became the most renowned and King, to the annoyance of his competitors, emerged as the foremost black spokesman. He toured the country in triumph and money poured into the SCLC treasury.

Most important, Birmingham convinced John and Robert Kennedy as well as Burke Marshall that they must ask Congress for a comprehensive civil rights law, that the policy of executive action now and legislation later was no longer viable. One example, Marshall noted later in relation to Birmingham, was that the federal government had no legal means to deal with the lunch-counter issue. As the President said in general, it was necessary "to bring this problem under law." He reached this conclusion, according to his brother, despite his feeling that it would be his "political swan song."[1]

On the long night in September 1962 when the troops had finally arrived in Oxford, according to Theodore Sorensen, a wearied President Kennedy asked his brother whether there would be "any more like this one coming up soon." The Attorney General had bad news. A lawsuit now in the courts involving the University of Alabama would likely reach the critical stage in the spring of 1963. "Let's be ready," the President said grimly.

The problem in Alabama was Governor Wallace and his pledge to stand in the schoolhouse door. The Kennedys tried to deal with him. On April 25, 1963, the Attorney General, accompanied by Marshall and Ed Reid, an old Alabama politico, talked to Wallace at the state capitol in Montgomery. The visit was superficially civil: Kennedy: "I just came by to pay my respects to you as governor." Wallace: "We're the courtesy capital of the nation." "Yawl want anything?" A secretary got the Attorney General a Coke. But Wallace, who detested Kennedy, did not mask his feelings. He had the United Daughters of the Confederacy place a wreath over the iron star on the floor of the capitol porch at the spot where Jefferson Davis had been sworn in as president of the Confederacy, and a stern Daughter, with arms folded, stood guard. "He didn't like the idea of Bobby maybe steppin' on it," an aide explained. State troopers swarmed on the lawn outside the governor's office. As Kennedy told it, when he tried to shake hands with one of the troopers, he put his club in his stomach and "belted me with the stick." At the start of the interview, Wallace ostentatiously turned on the tape recorder. "We might wanna save this conversation for posterity." Kennedy asked, "Will you follow the orders of the court?" Wallace replied, "I will never submit to an order of the federal court ordering the integration of the school system."

On May 18 the President, who was commemorating the TVA dam at Muscle Shoals, talked to Wallace on a helicopter flight to Huntsville, Alabama. Kennedy concentrated on Birmingham, particularly getting the downtown stores to hire blacks. Wallace was confident that he could maintain order in the city and denounced "outside leadership," particularly the "faker" Martin Luther King. The President did not even bother to bring up the university.

The relationship between Washington and Montgomery was tense and nasty. On the helicopter, according to Pierre Salinger's summary of the conversation, Wallace said "that the Reverend Martin Luther King and the Reverend Shuttlesworth vie with each other to see who could go to bed with the most nigger women, and white and red women too. They ride around town in big Cadillacs smoking expensive cigars." As the university crisis gathered, Michael Dorman, who was covering civil rights in the Deep South for *Newsday,* was tipped off that Wallace had "a history of mental troubles." In fact, during the war he had been a navigator on a B-29 in the Marianna Islands and, after viewing several crashes, was grounded for "flight fatigue, anxiety state." He was still rated 10 percent mentally disabled and continued to draw disability pay from the VA. Dorman did not use the story.

In 1963 the University of Alabama was the only public institution of higher education in the nation that refused to admit Negroes. In 1956, following a court order, the university had enrolled Autherine Lucy, a black woman. But a mob had followed her to class and she had been showered with eggs and rocks in a riot. She left and then accused the university of being partly responsible for the disturbance. She was expelled.

In the spring of 1963, Constance Baker Motley of the NAACP asked U.S. District Judge Seybourn Lynne to revive the Lucy rule requiring the university to admit qualified Negro applicants to the summer sessions at the Tuscaloosa and Huntsville campuses. There was a total of five black applicants, three of whom wanted to enroll for the summer term: Vivian Juanita Malone, a graduate of all-black Alabama A & M, who hoped to study at the School of Commerce at Tuscaloosa; Jimmy A. Hood, a star high school athlete and big man on campus, who wanted to major in psychology at Tuscaloosa; and Dave Mack McGlathery, a graduate of A & M, a former mathematician at a naval weapons lab, and presently a mathematician at the Marshall Space Flight Center in Huntsville, who desired graduate study in math at the Huntsvile Campus.

On May 16, Judge Lynne extended the Lucy rule and ordered the university to admit these qualified students. The board of trustees immediately announced its willingness to comply, but Wallace, an ex officio member of the board, refused with defiant statements. Lynne, a devoted alumnus of the law school, was deeply concerned about the possibility of disorder at the university. He forbade Wallace to obstruct the enrollment of the black students. Lynne then sent word privately to Wallace that, failing compliance, he would put him in the federal penitentiary for two years for contempt. If he served the time in Alabama, he could remain as governor; if he were sent to another state, he would have to give up his office. Wallace was appalled to learn that Attorney General Kennedy would decide where he went to prison. Worse still, the federal pen in Atlanta had the highest concentration of virulently anti-white Black Muslim inmates in the nation.

According to the schedule, Malone and Hood were to enroll at Tuscaloosa on Tuesday, June 10, McGlathery at Huntsville on Wednesday. The university tightened security at Tuscaloosa. Students were put on a 10:00 p.m. to

6:00 a.m. curfew. At Ole Miss pop bottles and bricks had been weapons. Here dispensing machines were changed, paper cups replacing Coke bottles. Building contractors trucked away their bricks. The city police, headed by professional lawman W. M. Marable, manned roadblocks and rode patrols. Anyone with a concealed weapon was automatically arrested (although Imperial Wizard Bobby Shelton of the Klan was busy bailing them out of jail).

On Sunday the governor rolled his forces into Tuscaloosa. Al Lingo had 825 men—state troopers, local police, and deputized game wardens and liquor agents. Lingo installed his headquarters' tractor-trailer on the campus. Taped to the door was a cartoon showing Burke Marshall telling some blacks, "The NAACP sent me down here to desegregate you trash." Wallace also called out the Alabama National Guard, and on Sunday afternoon 500 troops were in the armory in Tuscaloosa.

The federal forces, however, were much superior. General Abrams arrived early and established his headquarters at the Army Reserve Training Center in town. On Sunday a large Justice Department team headed by Katzenbach began to gather. A small hand-picked group of marshals who had served with distinction in Oxford was charged with protecting the black students. About 7000 national guardsmen from Alabama and Mississippi were in training exercises nearby, several thousand conveniently at Ft. McClellan. Brigadier General Henry V. Graham, a respected soldier with the Dixie Division, was chosen to command the federalized Alabama guard, if called out. About 2000 riot-control specialists from the 2nd Infantry Division had moved from Ft. Benning, Georgia, to McClellan and were on standby alert. Signal Corps units set up communications at a number of key Alabama locations. Marine Corps helicopters were at Dobbins Air Force Base in Georgia to ferry troops to Tuscaloosa, if needed. There was a lot of equipment in town—radio cars, airplanes, helicopters, and even a military boat for patrol on the Black Warrior River, which flowed by the campus.

On Monday morning university officials secretly informed Katzenbach of the plan for the next day. General registration would begin at 8:00 a.m. in Foster Auditorium and Malone and Hood would be enrolled at 10:00, at which time the campus and the building would be sealed off. The students in the dormitories to which the Negroes had been assigned had been notified and had accepted the news calmly. Wallace had been given an office in the lobby of the auditorium and would have a public address system. Only one of the three doors would be left unlocked and fifteen or twenty highway patrolmen would be in the lobby. Katzenbach got the keys and floor plans of the dormitories, had them searched and secured in advance, and got permission to station his men in the auditorium. That afternoon Wallace arrived, reviewed the Alabama guard run through riot-control exercises, and inspected the campus. He also received a telegram from the President urging him not to appear the next day in the interest of preserving order. Wallace responded, "My presence here guarantees peace."

At 9:55 on Tuesday morning the President issued a proclamation "commanding" the governor to desist from unlawful obstruction. It was intended more for Katzenbach than for Wallace. At 10:44 a convoy arrived at Foster Auditorium, including a brown Ford with Malone and Hood in the care of Doar. They stayed in the car.

Wallace stood at a lectern in the doorway to the auditorium. Katzenbach, accompanied by Macon Weaver, the local U.S. attorney, and Peyton Norville, the district's U.S. marshal, walked up to the door. TV cameras ground away. The hot Alabama sun beat down on them and was reflected off Katzenbach's bald pate. He started to tell the governor that he had a proclamation from the President commanding him not to resist the court order. Wallace cut him short and proceeded to read a five-page proclamation of his own. He attacked "the unwelcome, unwanted, unwarranted and force-induced intrusion upon the campus of the University of Alabama today of the might of the Central Government." I "do hereby denounce and forbid this illegal unwarranted action."

"Governor Wallace," Katzenbach said, "I take it from that statement that you are going to stand in the door and that you are not going to carry out the orders of the court, and that you are going to resist us from doing so. Is that correct?"

"I stand according to my statement," Wallace replied.

"Governor," Katzenbach insisted, "I am not interested in a show. I don't know what the purpose of this show is. I am interested in the orders of these courts being enforced. There is no choice for the federal government. I would ask you once again to responsibly step aside. If you do not, I'm going to assure you that the orders of these courts will be enforced. From the outset, Governor, all of us have known that the final chapter of this history will be the admission of these students. These students will remain on the campus. They will register today. They will go to school tomorrow."

Wallace, his shoulders thrown back, stood defiantly in the doorway. In accordance with the Justice Department plan to give the governor his moment on television, Katzenbach, Weaver, and Norville withdrew to the car. Katzenbach and several marshals escorted Malone to Mary Burke Hall and saw her properly installed in the dormitory. Doar took Hood to his dormitory.

At 11:35 the President issued Executive Order 11111 federalizing the Alabama National Guard. General Graham, who was piloting a reconnaissance plane over Ft. McClellan while observing the Dixie Division on maneuvers, was ordered immediately to Tuscaloosa to take command. He did so and learned that he had a new commander-in-chief.

The plan was now to register the students at 3:30 p.m. at the auditorium under the protection of Alabama guardsmen, not federal troops. Graham, an airborne general wearing paratrooper's jump boots, fatigues, and a billed cap, had his men secure the auditorium and the dormitories. At 3:31, a minute behind schedule, Malone and Hood were escorted from the dormitories.

Graham and four special forces guardsmen strode to the door of the auditorium. Wallace, who was inside, had his necktie straightened by an aide and walked outside. Leaning down to the little governor, the general said, "It is my sad duty to ask you to step aside." Wallace read a statement in which he thanked the people of Alabama for their "restraint." "We must have no violence. . . . God bless all the people of Alabama, white and black." The governor snapped to attention and gave the general a salute, which was returned. Wallace then stepped aside.

Malone and Hood entered the auditorium and were routinely registered, soon followed by other students. University officials and Doar took the blacks to the bursar's office to pay their fees. They went outside to face several hundred newsmen and the cameras. Hood said, "We are very happy our registration has taken place without incident. We hope to get down to our purpose—study." It was all over at 3:45.

The Huntsville registration was moved to June 13. Charles Morgan, Jr., who represented McGlathery, described the event:

> Two days later in Huntsville—in accord with an agreement that alleviated George Wallace's logistical problem—my client, Dave M. McGlathery, left work and drove to the University Center. He walked from his car into the school alone. Nearby, federal officials, including Mr. Katzenbach, waited to see that the registration went off smoothly. The governor, having been turned down in a late request that McGlathery come to Tuscaloosa two days before so that only one stand would be necessary, passed up the opportunity for an encore before a North Alabama audience.
>
> Inside the building, a Negro maid greeted Dave in subdued tones. And two janitors stopped work to watch. But there was really nothing special to see. Just another student registering at the University of Alabama.

On the afternoon of June 11 there was a good deal of amused satisfaction at the White House and the Justice Department. The students were registered and now there were no public universities that refused to admit blacks. No one had been killed or wounded; in fact, the only casualty was a slight sunburn on the top of Nick Katzenbach's head. The only force employed had been Alabama's own National Guard, not the U.S. Army. Judge Lynne had not been compelled to hold the governor in contempt and sentence him to jail. To be sure, Wallace had been allowed to play out his farcical political game, but that was much better than the alternative.

When Burke Marshall returned to Washington from Birmingham, he found the mood transformed.

> Everybody's mind was turned to the future and they thought that this pattern of Birmingham had been established, that it would recur in many other places. And it did that summer. And the President wanted to know what he should do—not to deal with Birmingham, but to deal with what was clearly an explosion in the racial problem that could not, would not go away, that he had not only to face up to himself, but somehow to bring the country to face up to and resolve. And during the week after that, that's what he decided to do, and so that's what led to the legislation.

Kennedy knew, Marshall pointed out, that a comprehensive civil rights bill would tie up Congress for the rest of the year and make other legislation impossible. He knew it would be "very tough" to get anything. He knew that he would lose southern support. "He knew how much was riding on it for him, politically and historically."

Shortly after noon, Washington time, on June 11, when Kennedy federalized the Alabama National Guard and he did not yet know what Wallace would do, he thought he should explain his action to the country and asked the networks to clear air time at 8:00 that evening. By 5:00 he knew that there would be no crisis in Tuscaloosa and could have cancelled the clearance. But he decided to use the time to address the nation on the broad topic of civil rights. Sorensen drafted a speech, but Kennedy did not like the draft. He, Sorensen, the Attorney General, and Marshall then brainstormed for ideas. The President kidded his brother, Robert later recalled, that "I'd gotten him into so much trouble." Answer: "Burke Marshall had gotten both of us in trouble." The President also kidded Marshall: "Come on now, Burke, you must have some ideas." Sorensen dashed to his office to rework the speech and did not deliver it until a few minutes before 8:00. In fact, Kennedy had to extemporize the conclusion. Later that evening he told Sorensen, "For the first time I thought I was going to have to go off the cuff.'" Nevertheless, many Kennedy watchers regard that speech, certainly the feeling and eloquence with which it was delivered, as the finest he ever made.

He opened briefly with the University of Alabama, commending the students for their constructive behavior. When American troops are sent abroad "we do not ask for whites only." Equity demanded that blacks receive the same treatment at home. "Every American ought to have the right to be treated as he would wish to be treated, as one would wish his children to be treated."

But this was not the case. A black baby had only half the chance of a white baby of completing high school, one-third of finishing college, twice the chance of becoming unemployed, a life expectancy seven years shorter, and the prospect of earning only half the income.

"This is not a sectional issue." Discrimination was a problem in every city and every state. There was "a rising tide of discontent that threatens the public safety." Nor was it a partisan issue. Americans should rise above party to resolve it. "It is better to settle these matters in the courts than in the streets."

The nation now confronted "a moral issue." It was "as old as the scriptures and as clear as the American Constitution." At heart it was "whether all Americans are to be afforded equal rights and equal opportunities, whether we are going to treat our fellow Americans as we want to be treated." It was a century since the Emancipation Proclamation had freed the slaves. "Yet their heirs, their grandsons, are not fully free." Thus, "redress is sought in the streets, in demonstrations, parades and protests which create tensions and threaten violence and threaten lives."

Congress must act to deal with this moral crisis. Next week, the President said, he would send up a bill "giving all Americans the right to be served in facilities which are open to the public." He would also ask for authority for the federal government to participate in lawsuits to end segregation in the public schools and for the right to vote.

He asked Negroes to be responsible and to uphold the law. But, in return, blacks had the right to expect that "the law will be fair; that the Constitution will be color-blind."

A few hours after the President finished speaking, Medgar Evers, the NAACP field secretary in Mississippi, arrived at his home on the north side of Jackson. He parked the car in the driveway, got out, and started toward the house. A sniper's bullet caught him just below the right shoulder blade. Keys in hand, he staggered to the doorway and collapsed. He was soon dead. The Evers murder grimly punctuated the President's message.[2]

The administration had actually introduced a civil rights bill in 1962. Robert Kennedy and Marshall thought they ought to do something and they selected voting rights, a sixth-grade literacy test for qualification, as the least controversial issue. The President considered it politically hopeless and was uncomfortable about sending up such a bill as a gesture. Senator Mansfield, the majority leader, told Marshall that it would never pass and that a Democratic President had special difficulty with cloture. Marshall agreed, as he later put it, that in 1962 "the Negro and his problems were still pretty invisible to the country as a whole."

Kennedy accepting the Democratic nomination for President at the Los Angeles Memorial Coliseum, July 15, 1960. Here he announced the New Frontier. *Jacques Lowe/Woodfin Camp & Associates*

Kennedy campaigning in 1960 in the New York City financial district. ABOVE: With his wife, Jacqueline, in a ticker tape parade up Broadway. BELOW: Interested construction workers viewing the parade. *Burton Berinsky*

Kennedy campaigning among coal miners during the West Virginia primary in 1960. *Hank Walker/Life Magazine © Time Inc.*

Kennedy defeated Nixon, with a razor-thin margin. The early returns, as this photo shows, put Nixon ahead. LEFT TO RIGHT: (*standing*) Pauline Fluet (Steve Smith's secretary); (*seated*) Eunice Shriver, Bill Walton, Pierre Salinger; (*on stairs*) chauffeur; (*standing*) Ethel and Robert Kennedy; (*seated*) John Kennedy; (*standing*) Angie Novello (Robert's secretary). *Jacques Lowe/Woodfin Camp & Associates*

Kennedy had a great appeal to young people. ABOVE: Campaigning in 1960 at the Fargo, North Dakota, airport. *Stan Wayman/Life Magazine © Time Inc.* BELOW: Memorial Stadium, University of California, Berkeley, Charter Day, March 23, 1962, where 92,000 attended. Many more were on the side of Strawberry Canyon and he was met with huge crowds at the Alameda Airport and along the route to the campus. Berkeley supplied more Peace Corps volunteers than any other university. *Bancroft Library*

The Democratic congressional leadership.
SEATED FROM LEFT: Speaker Sam Rayburn and Senate Majority Leader Mike Mansfield.
STANDING FROM LEFT: Majority Leader John McCormack, House Whip Carl Albert, Senate Whip Hubert Humphrey.
Kennedy Library

Kennedy with Theodore C. Sorensen, his special adviser, trusted assistant, speech writer, and, later, biographer.
Kennedy Library

Kennedy throwing out the first ball to open the ill-fated Washington Senators' season. At left is Lawrence O'Brien, who handled congressional relations for the White House. Secretary of Labor Arthur Goldberg (partially hidden) is behind him. Senator Mansfield is just left of the President. *Kennedy Library*

Attorney General Robert F. Kennedy.
George Tames/NYT Pictures

The Attorney General with Burke Marshall, assistant attorney general in charge of civil rights. *National Archives*

The President with Harris Wofford, White House adviser on civil rights, later head of the Peace Corps in Africa. *Kennedy Library*

On May 14, 1961, the Ku Klux Klan attacked and set afire the Greyhound bus carrying the Freedom Riders near Anniston, Alabama. Several were severely beaten. *UPI/Bettmann Newsphotos*

On September 26, 1962, Justice Department officials escorted James Meredith to the University of Mississippi for enrollment. Two blocks from the campus Lt. Governor Paul Johnson (*left*), backed by state troopers, stopped them and barred Meredith (*right*) from registering. Here John McShane, the chief U.S. Marshal and John Doar of the Civil Rights Division, argue in vain with Johnson. *Flip Schulke/Black Star*

McShane's marshals (he is at left) took heavy fire in the riot at Ole Miss. *Kennedy Library*

As part of the strategy of dealing with civil rights by executive action, Kennedy established the Committee on Equal Employment Opportunity. Its main program, Plans for Progress, was a pledge by employers to desegregate their plants. Lockheed Aircraft Corporation was the first to sign on for its Marietta, Georgia, operations. FROM LEFT: Vice President Johnson, chairman of CEEO; Courtland E. Gross, president of Lockheed; Kennedy; and Secretary of Labor Goldberg, vice chairman of CEEO. *Kennedy Library*

Police violence in May 1963 in Birmingham, Alabama, and the uproar at the University of Alabama a month later transformed civil rights policy. Photographs like this of the Birmingham violence, Kennedy said, made him "sick." *Charles Moore/Black Star*

In addition to dogs, the Birmingham police used high pressure hoses on demonstrators. *Charles Moore/Black Star*

The racial crisis in Alabama convinced Kennedy that his policy of executive action alone would no longer work. In one of the most momentous decisions of his presidency, Kennedy on June 11, 1963, addressed the nation on television, calling for passage of a comprehensive civil rights law. *Kennedy Library*

RIGHT: The immense crowd at the March on Washington on August 28, 1963, from the Lincoln Memorial. Here Martin Luther King, Jr., delivered his "I Have a Dream" speech. *Fred Ward/Black Star*

After the March, its leaders came to the White House. FROM LEFT: Whitney Young of the Urban League; M. L. King, Jr., of the Southern Christian Leadership Conference; Rabbi Joachim Prinz; Archbishop Patrick A. O'Boyle; Philip Randolph, organizer of the March; the President; the Vice President; and Walter Reuther of the UAW. *Kennedy Library*

Nevertheless, the President went along and Mansfield introduced the voting rights bill. Robert Kennedy later said that it "never had any chance of passing. . . . Nobody was interested in it." They needed 67 votes in the Senate for cloture and were unable to get 50 to shut off a filibuster.

Now in the late spring of 1963 the situation was transformed and Kennedy responded. These seem to have been the reasons for the change: First, the policy that had been in place for more than two years—executive action now and legislation later—had exhausted itself. While much had been accomplished under that policy, diminishing returns had set it. The time had come to move to the next phase.

Second, Birmingham not only had made civil rights visible to Americans generally, white and black, but had made it the nation's decisive domestic issue. It was of the utmost political significance that the growing recognition of the problem was accompanied by a marked increase in sympathy for the Negro cause in the white community, among Republicans as well as Democrats.

Third, black unrest was sweeping the nation and the nonviolent civil rights organizations were losing control. A hot summer was about to begin. It was indispensable, the President thought, to get blacks off the streets and into the courts. The federal government must give a sign to Negroes that it was sensitive to their legitimate demands and was trying to satisfy them. As Robert Kennedy put it, "There's obviously a revolution within a revolution in the Negro leadership. We could see the direction going away from Martin Luther King to some of these younger people, who had no . . . confidence in the system of government we have here . . . , that the way to deal with the problem is to start arming the young Negroes and sending them into the streets."

Fourth, if the President forfeited leadership on this issue, he would wind up in the ridiculous position of being a follower. Northern Democrats, liberal Republicans, and Adam Clayton Powell, the Harlem congressman, had bills in the hopper which were gaining political support. For Robert Kennedy this seems to have been the decisive factor, leaving the President no option but to assume leadership.

Finally, as the President stressed in his address of June 11, 1963, there was a moral imperative to civil rights. But this was not new; as he said, it went back to the scriptures and the Constitution. All Americans, black and white alike, had the right to equality of treatment under law.

The President called the first conference on legislation on May 20, 1963. The Attorney General, Marshall, and Sorensen, among others, were present. They discussed the items that should be covered, particularly public accommodations, education, and the Powell Amendment (forbidding discrimination in federally financed programs). The tone was exploratory and it was clear that this would be an omnibus bill. Thus, the question was what should be covered and what should be left out.

On June 1 the President met with the Vice Pesident, the Attorney General, Marshall, Sorensen, Secretary of Labor Wirtz, and others. Marshall, who thought that opening public accommodations to blacks would affect the greatest number, raised the question of drawing the line on size of establishment to meet the interstate commerce test of constitutionality. If small hotels and restaurants were excluded, how would one define small? Both the President and Johnson stressed the importance of education for young blacks. Kennedy explored the possibility of giving priority to school lawsuits, but Marshall thought this would not be helpful. Discrimination in employment was a conundrum. Should the bill merely embrace the executive order to legitimize CEEO or should it include a full-fledged fair employment practice commission to cover all substantial employers? The former would produce only the minor gain of providing CEEO with its own appropriation. But both Kennedy and Johnson were concerned about the politics of FEPC. The President thought Truman had erred in "throwing FEPC at the Congress." Further, he knew that the cloture vote in the Senate would require Republican support, and no one could expect Everett Dirksen of Illinois, the minority leader, to back a broad fair employment title. Johnson thought it had little or no chance. House whip Carl Albert informed him, "We can't pass an FEPC. Halleck [the minority leader] told me under no circumstances would he vote for it."

Johnson, moreover, was profoundly disturbed about the way the Kennedy administration was handling the civil rights bill and poured out his heart to Ted Sorensen in a mammoth phone call on June 3. The President, he said, will be "cut to pieces with this and I think he'll be a sacrificial lamb." He thought it possible to get a good bill, but it would have to be "thought through." The votes had to be counted before the President sent up his message. The congressional leaders of both parties must be brought into the process immediately. "Hell, if the Vice President doesn't know what's in it how do you expect the others to know what's in it?" Kennedy had "plenty of time . . . right up to September." He had heard that Mansfield's plan was to wait till the House acted and then give its bill to the Judiciary Committee with instructions to report in a month. "My hunch is, that would be the best way." Most important, Johnson urged, "I would sure get the Republicans in on this thing." Dirksen was the key and he must be stroked. Otherwise, "Everett Dirksen and Dick Russell [the southern leader in the Senate] will be sitting around . . . over a mint julep."

Lyndon Johnson's views had an impact. The submission of the bill was delayed 10 days. Robert Kennedy and Marshall talked to many members of

Congress. On June 5 the President met with the Republican leadership and on June 11 with the leaders of both parties.

On June 13, Mansfield and Dirksen reached tentative agreement on the items the bill would cover: (1) voting rights based on literacy; (2) extension of the Civil Rights Commission, probably indefinitely; (3) school desegregation; (4) the Powell amendment; and (5) statutory authority for CEEO, "the purpose being to counteract F.E.P.C. proposal." While they did not discuss a civil rights conciliation service to assist communities that wanted to desegregate, Mansfield thought Dirksen would not be "adverse." They were unable to agree on public accommodations, though Dirksen said he would try to work out language.

Marshall delivered a copy of the administration bill to Dirksen the next day. On June 17 the President met the leaders of both parties. Dirksen, "speaking for myself," said he could accept the bill, including a conciliation service, called the community relations service, excepting only public accommodations.

On June 18, Senators Mansfield and Hubert Humphrey on the Democratic side and Dirksen and Tom Kuchel on the Republican agreed to the handling of the bills. Mansfield and Dirksen would introduce the administration bill less Title II, public accommodations, and it would be S. 1750 and be referred to the Judiciary Committee. Humphrey, Kuchel, and perhaps others from both parties would introduce the entire administration bill, S. 1731, also to be sent to Judiciary. Senator Warren Magnuson of Washington, chairman of the Commerce Committee, would join with Mansfield to introduce Title II separately, S. 1732, and would take it in his committee.

Also on June 18, Mansfield wrote a memorandum to the President in which he laid out the basic political strategy for the Senate. The goal would be to win as much "as is possible" rather than aiming for "all" and risk losing everything. Supporters of the legislation must work under the dictatorship of arithmetic "by counting 67 votes on cloture for whatever bill is pushed." Any act or word that subtracts from that number is to be avoided. "As it now stands, the short side of 67 lies in the public accommodations title."

Reaching 67 demanded "complete cooperation and good faith with respect to Senator Dirksen." Further, he needed to persuade half a dozen Republicans to join him on cloture, and public accommodations was the handle. His power to persuade his colleagues would be lost "if the impression develops that the Democratic Party is trying to make political capital."

The Mansfield memorandum laid out part of the Democratic strategy on the bill that would produce the longest debate in the history of the Congress and a law that many consider the most significant ever enacted. Both parties were divided on civil rights. Northern Democrats supported a strong bill;

southern Democrats opposed any bill. Liberal Republicans, who were not numerous, agreed with the northern Democrats. Conservative Republicans wanted a bill but opposed public accommodations and fair employment practices. In other words, civil rights was not a partisan issue. Both parties had adopted strong planks in their 1960 platforms.

From the administration's view, the House was the obvious place to start because it was certain to pass a bill and it might go all the way with the two disputed issues. There were two reasons for this: the absence of the filibuster and a sizable majority of northern Democrats, liberal Republicans, and committed conservative Republicans. The Senate would be the battleground. Here the southern Democrats were stronger and possessed the outstanding leadership of Senator Russell. More important, they had the filibuster, which could be shut down only by 67 votes for cloture. Historically, southern Democrats and western senators from both parties had an understanding that the westerners would vote against cloture on civil rights in return for southern support for resource issues dear to the West. This had worked so effectively in the past that no filibuster had ever been shut off on a civil rights bill.

The crucial question for the administration, therefore, was to gain Republican support for cloture. Thus, as Mansfield stressed, Dirksen was the man, and he was an anomaly. Ostensibly, Everett Dirksen was a traditional conservative parochial Republican from Pekin, Illinois. In fact, he was a great deal more; he was, as Neil MacNeil put it, a "professional." He was smart, well-read, hard-working to the point of endangering his health, and a master of the Senate, its rules, its legislative tricks, and its personalities. He had a sense of history and introduced the French phrase that civil rights was "an idea whose time has come." Like the actor he was, Dirksen throve in the limelight and worked hard at making himself distinguishable and rememberable. He affected baggy clothes and mussy hair. Richard Strout wrote that the top of his head resembled "the kelp of the Sargasso Sea," and *Newsweek* observed that his expression had "the melancholy mien of a homeless bassett hound." But his most notable feature was his oratory. Early on, MacNeil wrote, it was "a mongrel mix of grand opera and hog-calling." But over time he refined it with pomposity, bizarre language, deliberate mispronunciations, mixed metaphors, "a golden thesaurus on his larynx," and circumlocutions. Sample: "I shall invoke upon him every condign imprecation." He hated to use a written speech that required him "to flounder around the piece of paper trying to find where the hell you were." Rather, "I love the diversions, the detours. Without notes you may digress." His wife said, "He just loves to talk. I don't pay any attention because I've heard it all before."

Although a true-blue Republican, Dirksen was not narrowly partisan. He and Kennedy got along famously. The President gave him his Illinois appointments even when Mayor Daley of Chicago protested and barely opposed his re-election in 1962. Kennedy expected legislative help in return. But Dirksen's best Democratic relationship was with Mansfield. They were exact opposites and fitted together like Yin and Yang. Mansfield was modest, retiring, orderly, laconic, and straightforward. He shrank from the limelight. If Dirksen wanted center stage in the great debate to play out his metamorphosis from politician to statesman, Godspeed!

During June the Justice Department drafted the civil rights bill. Norbert A. Schlei and Harold H. Greene were the lawyers responsible for this task. At the same time the President, the Attorney General, Marshall, and Wirtz met under forced draft, usually at the White House, with separate groups of leaders of business, labor, the clergy, lawyers, women, and educators to persuade them to support the prospective bill and, where appropriate, to open jobs and union programs to blacks immediately. Kennedy and Johnson also met with Eisenhower to try to enlist his support. The former President was not much help. He still thought the major challenge was to change minds, not laws, with the possible exception of voting rights. Kennedy deliberately avoided meeting with the civil rights leaders until after the bill went to Congress because he did not want the appearance of their having dictated its provisions.

The President sent the civil rights bill up on June 19, 1963. S. 1731 and H.R. 7152, the complete measure, was called the Civil Rights Act of 1963 and asserted constitutionality in the commerce clause and the Fourteenth and Fifteenth amendments. It had the following major provisions:

Title I, Voting Rights, would establish completion of the sixth grade of school as *prima facie* proof of literacy for the right to vote in federal elections.

Title II, Public Accommodations, would provide to all persons without regard to race, color, religion, or national origin access to hotels, motels, lodging houses, places of entertainment, retail establishments, restaurants, and related facilities.

Title III, Desegregation of Public Education, would require the commissioner of education to undertake a variety of programs to desegregate the public schools and would empower the Attorney General to bring suit against a school board for failure to achieve desegregation.

Title IV, Community Relations Service, would create such a service to assist communities in resolving disputes over discriminatory practices.

Title V, Commission on Civil Rights, would extend that agency's life for four years.

Title VI, Nondiscrimination in Federally Assisted Programs, would permit the federal government to refuse to fund any program which provided for discrimination based upon race, creed, color, or national origin.

Title VII, Commission on Equal Employment Opportunity, would codify the executive order creating CEEO.

The civil rights leaders, particularly King, basking in the victory of Birmingham, were exhilarated by Kennedy's commitment to legislation. With NAACP leading the way, these organizations and some seventy other labor and church groups formed the Leadership Conference on Civil Rights to lobby for the bill. They wanted it strengthened, particularly by including all public accommodations, by giving the Attorney General plenary power to file suits, and by establishing a genuine FEPC. Walter Reuther of the United Automobile Workers (UAW) offered office space in Washington and Arnold Aronson, a civil rights veteran, took over the operation. Joseph L. Rauh, the UAW counsel, became the chief lobbyist, working with Andrew Biemiller of AFL-CIO, Jack Conway of the Industrial Union Department, Walter Fauntroy of SCLC, and Clarence Mitchell of NAACP. Further, on June 21 the March on Washington was announced publicly in order to draw attention to black unemployment and to lobby the Congress for the civil rights bill.

The President met with the black leadership at the White House on June 22, three days after the bill went up. He gave them a tough appraisal of the prospects in the Senate, particularly on cloture. As for the March on Washington, "We want success in Washington, not just a big show at the Capitol." He was concerned that the march would create "an atmosphere of intimidation" that would give some senators an excuse to vote against cloture. Johnson echoed Kennedy's views. Philip Randolph, who had initiated the march, James Farmer, and King insisted that it must go forward, but the caution Kennedy and Johnson stressed had an effect. Thereafter, as Roy Wilkins of the NAACP and Reuther urged, they focused on a demonstration at the Lincoln Memorial, not at the Capitol.

At the end of the meeting Kennedy asked King to stay and took him for a walk in the Rose Garden. He was blunt. Naming Jack O'Dell, a staff member of the Southern Christian Leadership Conference, and Stanley Levison, King's trusted adviser, he said, "They're Communists. You've got to get rid of them." Both the civil rights movement and the civil rights bill were at stake. "If they shoot *you* down, they'll shoot us down too." King agreed on O'Dell but demurred about Levison. He did fire the former and, to his surprise, Levison agreed with Kennedy's assessment. Their relationship was temporarily suspended, though King could not bring himself to sever it completely.

The administration bill was referred to the House Judiciary Committee as H.R. 7152. The chairman, Emanuel Celler, the liberal Brooklyn Democrat who was probably the chamber's leading spokesman for civil rights, sent it to

Subcommittee No. 5, which he also headed. He had stacked that subcommittee with six other liberal Democrats, including Peter Rodino of New Jersey. There were only four Republicans, all conservatives on economic issues. But the senior Republican, William McCulloch of Ohio, was a moderate on civil rights. He had recently introduced a civil rights bill for twenty-four Republicans that was quite strong, though it had no public accommodations title. John Lindsay, the liberal New York Republican, then brought in a public accommodations bill for the same number of Republicans.

When the hearings opened before the subcommittee on June 26, Robert Kennedy was the lead witness. It was not an auspicious start. As Charles and Barbara Whalen put it, he turned "churlish and combative" under questioning and admitted that he had not even read Lindsay's bill. The latter was outraged and aired the cloakroom rumor that the administration had made a deal with the South to get a weak bill by scuttling public accommodations. Celler had to intervene to prevent verbal bloodshed.

The next day Bobby Baker, the secretary to the Senate Democrats, made two polls of the leanings of senators and they were shockers. The President's bill with public accommodations would be defeated 49 to 47. The Mansfield-Dirksen bill without Title II would pass 68 to 22. The westerners and the Republicans were the swingers. Baker wrote, "It is virtually impossible to secure 51 Senators who will vote for the President's bill." The only possibility of winning on cloture, he pointed out, was to "find a formula in the Public Accommodations area that would be acceptable to Senators Aiken [Vermont], Hickenlooper [Iowa], and Dirksen." Aiken, it may be noted, was the inventor of Mrs. Murphy, the doughty widow who rented rooms to support her family. Neither Aiken nor Dirksen wanted the long arm of the federal government to compel Mrs. Murphy to accept customers she did not want even if her boardinghouse was in interstate commerce.

Clearly, the administration needed Republican support and the Justice Department again turned to Burke Marshall. Congress had gone into recess over the Fourth of July and McCulloch was back home in his law office in Piqua, Ohio. While there were almost no blacks in his district, McCulloch early on had practiced in Jacksonville, Florida. He had concluded that Jim Crow was both cruel and unconstitutional. He had played an important role in the passage of the 1957 and 1960 civil rights laws. He was decent and respected. Marshall went off to Piqua to see him.

McCulloch could not resist keeping Marshall waiting. The latter, who was repairing his house, used the time to shop for nails in a hardware store. When he finally got in, Marshall and McCulloch hit it off at once. McCulloch's bill and H.R. 7152 were quite similar and Marshall persuaded him to accept the administration bill and to work to get other Republicans to go along. But the canny Republican exacted his price. He insisted on a veto over any amendments the administration might offer in the Senate to prevent the gutting of the bill and a commitment by the President to give the Republicans equal credit for passage of the measure. He got his way.

According to Katzenbach, McCulloch's demand that the administration hold to the House version in the Senate instead of giving pieces of it away, as had occurred in 1957 and 1960, radically changed the strategy. "We began to think more seriously of cloture than we ever had before." Katzenbach and Marshall talked to between forty and forty-five uncommitted senators to learn what they would approve. "The only possible way of getting the bill through was to get a bill that on our soundings seemed acceptable in the Senate."

Between July 10 and August 2 the subcommittee heard from 100 witnesses in 22 days of hearings. A chastened Attorney General performed more effectively. Roy Wilkins for the Leadership Conference on Civil Rights and Representative James Roosevelt of California urged an FEPC title.

In mid-August, McCulloch secretly worked out a bill with the Justice Department that both could accept and it received Celler's approval. But the President asked Celler to stall in order not to incite the southern Democrats on the Ways and Means Committee, headed by chairman Wilbur Mills of Arkansas, to bottle up the important tax bill. On August 28 members of Congress watched the great throng gather in the March on Washington. At its conclusion the President invited the civil rights leaders to the White House, where he and Johnson told them that they needed sixty Republican votes to carry the House. The implication was that there would be a price to pay: no public accommodations and no FEPC.

On September 10 Ways and Means approved the tax bill. Celler was now free to move and he had devised a whacky strategy: to bring a very strong bill out of the subcommittee which could be traded down in the full committee. The trouble with this plan was that it violated both the McCulloch-Marshall agreement and the McCulloch-Justice bill which he had approved.

On September 15 a bomb went off at the Sunday school of the Sixteenth Street Baptist Church in Birmingham and four little girls were killed. In the national outrage, the liberals on the subcommittee determined to strengthen the bill. A whole series of amendments sailed through, several of exceptional importance. Title II would cover many additional forms of business, including private schools, law firms, and medical associations. Only Mrs. Murphy was excluded—rooming houses with five or fewer rooms. In a new Title III the Attorney General would have power to initiate or intervene in civil suits charging discrimination against state or local officials. Celler's fellow opera-lover, Rodino (they sneaked up to New York together to attend performances at the Met), introduced the amendment that eliminated the CEEO Title VII and substituted for it a broad FEPC bill that would apply to all firms with twenty-five or more employees. McCulloch walked out of the September 25 meeting dazed, seething, and betrayed. He told the press that the House would not pass the new bill.

Two hours after his statement the House adopted the tax bill. Celler now moved forward with his strategy. On October 1 he railroaded what the Whalens called "a vastly strengthened" bill through the subcommittee. McCulloch called it "a pail of garbage." Katzenbach was beside himself over Celler. "We had everything under control, and then he collapsed in the face of the liberals. Spine was not Manny's strong point." The President was outraged. "Can Clarence Mitchell [of NAACP] and the Leadership group deliver . . . 60 Republicans on the House floor? McCulloch can deliver 60 Republicans." The 14 Republicans on the Judiciary Committee met with Charlie Halleck, a self-described "gut fighter" as far as Democrats went, and vowed not to take the blame for the emasculation of H.R. 7152.

The full committee met on October 8. Now Celler's task was much tougher. The committee consisted of 17 liberals, 9 southern conservatives, 9 moderate-to-conservative northerners, and 1 maverick. That day Halleck and Katzenbach met with Speaker John McCormack. Halleck said the GOP would not do Celler's dirty work for him. They agreed that the bill must be weakened. Halleck offered to propose half the amendments but insisted that liberal Democrats offer the other half.

Robert Kennedy called in Celler and denounced him in purple language for violating the agreement with McCulloch. He also directed Celler to arrange a meeting with McCulloch, Katzenbach, and Marshall. The Ohioan agreed to make another try, but only if the administration made the Democrats support the trimmed bill. Kennedy put out the word that no one in the administration could take credit for the bill if it passed. All the kudos would go to McCulloch and Celler, in that order. The Republicans, to protect themselves, asked Kennedy to appear before the Judiciary Committee in executive session to state his views of Celler's bill. He agreed to do so.

The Attorney General's statement of October 15 had been cleared in advance by the President and his political aides, O'Brien and O'Donnell. He went over the subcommittee bill title by title, asking for the elimination of all changes from the version McCulloch, Celler, and the Justice Department had earlier agreed on. He returned the next day to answer all questions. It was a bravura performance.

On October 22 the Judiciary Committee blew up in Celler's face. Arch Moore, the West Virginia Republican, in frustration moved to report out the subcommittee bill in order to kill it. He had the votes. Liberal Democrats would vote for it out of conviction; southern Democrats would vote for it to destroy it; and Republicans would vote for it in retaliation for the Democratic double-cross. Celler was saved by the noon bell and adjourned the committee.

The President was appalled by the mess and realized that he must step in. He met with Johnson, McCormack, Albert, Halleck, minority whip Les Arends, Celler, and McCulloch in the Cabinet Room on the evening of October 23. The Attorney General, Katzenbach, and Marshall waited nearby, if needed. Kennedy made a strong plea for a bill everyone could agree on. McCulloch said, "All the king's horses and all the king's men could put humpty-dumpty together again." The President jumped at the suggestion and led a quick tour through the bill that showed that the differences were not great. The next day Kennedy and Halleck worked on rounding up votes to defeat the Moore motion. The President made little progress with the Democrats, but Halleck delivered the Republicans. The crusty minority leader himself came out for civil rights despite the fact that "the colored vote in my district didn't amount to a bottle of cold pee." He was annoyed because his black driver could not enter the restaurants when they drove down to Warm Springs. "Once in a while a guy does something because it's right." McCulloch hammered out a compromise with Lindsay to hold the liberal Republicans. He then reached agreement with Katzenbach and they wound up with a revised bill.

Kennedy, who needed ten votes from his own party, met with the Democrats on the Judiciary Committee to twist arms. He got nine and one possible. Halleck delivered another vote.

The tide had turned. While seventeen votes were needed to defeat the Moore motion, there were more. On November 20 a chastened Celler put the compromise bill through the Judiciary Committee 20 to 14. It differed from the original bill and had the following significant features:

Title I. Voting Rights, limited to federal elections.

Title II. Public Accommodations, retail stores and personal services not covered.

Title III. Public Facilities, the Attorney General would intervene only after an individual had filed a complaint alleging discrimination and had demonstrated that he was unable to pursue the suit on his own.

Title IV. Public Education, no change.

Title V. Community Relations, eliminated. The Civil Rights Commission became Title V. It was made permanent and was authorized to investigate voting fraud.

Title VI. Federally Assisted Programs, limited to federal grant, contract, and loan programs.

Title VII. Equal Employment, EEOC retained, but powers reduced and could compel action only through a federal court.

The President had hoped to have the civil rights bill passed during 1963 so that the issue would not cloud the 1964 presidential campaign. That was now impossible. It must still clear the Rules Committee whose chairman was eighty-year-old, arch-conservative, arch-segregationist Howard Smith of Virginia. As the Whalens put it, Smith "wrapped his iron grip around H.R. 7152 two days before John Kennedy was killed and . . . planned to do everything in his power to keep it captive."[3]

On a December afternoon in 1962, Bayard Rustin, as was his custom, dropped by the headquarters of the Brotherhood of Sleeping Car Porters in Harlem to visit with his dear friend, A. Philip Randolph. Though Randolph was seventy-three and Rustin fifty-two, they were much alike: tall, slender, handsome, dignified, graceful former athletes, scarred veterans of the civil rights wars, flirters with Marxism. Each had another major commitment—Randolph to the labor movement, Rustin to the peace movement.

They talked with satisfaction about the gains made by street demonstrations in the South. But, Randolph said, this was fragmentary. There was the need to pull everything together in a national demonstration that combined the demands for civil rights with those for jobs. Did Rustin have any ideas? He suggested a March on Washington. Phil Randolph had been the key figure in the aborted first March on Washington in 1941. As the price of calling it off, the blacks got FDR to issue the executive order establishing the first FEPC. Randolph rather wished the march had taken place. He told Rustin that the idea was splendid and asked him for a detailed plan. Rustin, who conferred with others, took several months to respond. He then proposed a demonstration of 100,000 in May 1963 for economic demands. Randolph said there must also be "a demand for freedom." "Fine," Rustin responded, "we'll call it a march for jobs and freedom." They needed the support of all the civil rights organizations.

Randolph sounded out King first, who was enthusiastic. Wilkins and Young were cool at the outset, but came around later. The idea gained strength from Birmingham, the introduction of the civil rights bill, and the meeting with Kennedy on June 22. The black leadership met in New York on June 25 and agreed on the date for the march—Wednesday, August 28. Wilkins asked who would be in charge. Randolph said it would be Rustin, of course. Wilkins, evidently concerned that Rustin's radical background would be a liability, demurred. Young proposed that Randolph become director and then name Rustin to take charge. Everyone accepted the compromise.

In late June, Rustin rented a tenement from the Friendship Baptist Church at 130th Street and Seventh Avenue for his headquarters. The facilities were minimal. Rustin quickly recruited a staff, mainly young people from CORE and SNCC. He had only eight weeks and a budget of $120,000 supplied by labor unions, particularly the UAW, churches, individuals, and the civil rights organizations. Among the last, the NAACP made the largest contribution and SCLC gave nothing, much to the annoyance of Roy Wilkins.

The organizers anxiously sought the support of the labor movement. George Meany was now solidly behind civil rights. He was pushing the building trades to clean up their constitutions and apprenticeship programs and firmly backed the civil rights bill with a strong FEPC. But a street demonstration was just not his style. Despite the urging of Reuther and some of the smaller unions, Meany opposed the march and there were hardly any votes of endorsement in the AFL-CIO Executive Council.

As the plans developed, almost everyone got scared. The organizers at first feared that they could not deliver a large crowd. Later, when widespread church support came in, their concern was that too many white people would march (they wanted a ratio of three blacks to one white). The District of Columbia establishment was scared of a riot. The police planned for mob control and insisted on using their large K-9 Corps. Robert Kennedy, with Bull Connor fresh in memory, ordered them to keep their dogs in their kennels. The department stores, expecting looting, moved stocks to warehouses and locked up. The Justice Department was extremely concerned about the preservation of order and the effect that a breakdown would have on the civil rights bill. After the President on July 17 endorsed a peaceful demonstration at the Lincoln Memorial, the Attorney General worried that a small turnout would make his brother look bad.

Robert Kennedy had another concern, that the march was "very, very badly organized," an assessment with which Marshall agreed. This, of course, was a serious criticism of Rustin's group and it seems to have been half true. The Harlem organizers concentrated on persuading people to come to Washington and they were spectacularly successful in doing so. They seem to have paid little attention to how they would be handled once they were in town. This is what troubled Kennedy and Marshall.

In July the Attorney General put John Douglas in charge of the march. The son of Senator Douglas, he had just joined the department and he was assisted by Alan Raywid, also an attorney at Justice. They worked for five weeks on planning. Raywid's first assignment was to draw up a list of things that might go wrong. He came up with more than seventy possibilities and they then worked up contingency plans for each. In accordance with administration policy, Douglas insisted that the demonstration be confined to the area between the Washington Monument and the Lincoln Memorial; the Capitol was out of bounds. An exception was made for the "Big Ten" leaders to lobby Congress. Douglas worked out police plans to handle trouble. Justice Department spotters would be posted on government buildings and would monitor the flow of auto traffic from Baltimore. While the police chief cruised, Douglas would station himself at police headquarters. He insisted that the march committee rent a first-class and expensive sound system with forty-six huge speakers which could be controlled by his people in an emergency. Executive orders were drafted so that, in case of a riot, the President could order out 4000 troops stationed at nearby Ft. Myer and 15,000 paratroopers from the 82nd Airborne at Ft. Bragg, North Carolina. There were plans for food, water, toilets, and emergency medical care (50 doctors and 100 nurses at 15 first-aid stations, plus a volunteer group of New York doctors at the Willard Hotel). The Washington Senators cancelled the night games scheduled for D.C. Stadium on August 27 and 28 (since they were in tenth place, no one seemed to care).

There were two last-minute problems involving the program. Anna Arnold Hedgeman, who represented the National Council of Churches on Rustin's committee, protested that no woman was scheduled to speak at the Lincoln Memorial. She was largely ignored and even the wives of the leaders were not invited to the White House after the march, enraging Coretta King. Two minor compromises were made. While the Big Ten, all men (Wilkins, Young, King, Floyd McKissick of CORE because Farmer was in jail in Louisiana, John Lewis of SNCC, Randolph, Reuther, and Protestant, Catholic, and Jewish leaders), would lead the main march with locked arms down Constitution Avenue to the Memorial, five black women were allowed to do the same with a much smaller crowd down Independence Avenue south of the Mall. In addition, Daisy Bates of Central High School in Little Rock was permitted to say a few words at the Memorial.

John Lewis, who was very young and inexperienced, had a speech written at SNCC headquarters in Atlanta that was loaded with inflammatory rhetoric. On the evening of August 27 a copy was delivered to Archbishop Patrick O'Boyle of Washington, who would give the invocation. Lewis was slated to say, "We will march through the South, through the heart of Dixie, the way Sherman did. We shall pursue our own scorched earth policy and burn Jim Crow to the ground." If there was any Catholic prelate in America who exceeded Father Hesburgh in his commitment to civil rights, it was Archbishop O'Boyle. He refused to stand on the same platform with a man who seemed to condone violence. A battle raged for a day over whether "Sherman" and "scorched earth" would be edited out of the Lewis speech. The resolution took place in the storage room at the Memorial behind the statue of Lincoln with Reuther looking up and saying, "Abe, I need your blessing and your help." The offensive words were removed just before the archbishop delivered the invocation.

These annoyances were quickly forgotten in the magical quality of August 28. The weather was perfect—though, typical for a Washington summer day, it turned hot and humid during the afternoon. Despite the size and complexity of the operation, virtually everything moved smoothly, and *mirabile dictu,* on time. There was, of course, no riot, no violence, no disturbance of any sort. The immense throng between the Monument and the Memorial, along with millions of people around the world who viewed the spectacle on television, were enthralled, entertained, moved, and uplifted. As Arthur Schlesinger wrote, "Certainly that afternoon in August had a purity that no one present could ever forget."

At 8:00 a.m. chartered trains began to roll into Union Station from New York, Philadelphia, Pittsburgh, Cincinnati, Detroit, Gary, Chicago, Miami, and Jacksonville. An immense number of chartered buses poured into the city. Many people came by air and many, many more by car. By 9:30 about 90,000 people were spread across the lawn at the Monument. They were entertained by Joan Baez, Odetta, Josh White, the Freedom Singers, and Peter, Paul, and Mary. Near noon the Big Ten led the march down Constitution Avenue to the Memorial, with entertainment stars Harry Bellafonte, Charlton Heston, Burt Lancaster, Marlon Brando, and Sidney Poitier close behind, followed by the huge crowd.

At noon the police estimated the turnout at 200,000. Thomas Gentile, the historian of the march, thought, considering the ebb and flow, that as many as 400,000 might have heard the afternoon program. They were about three-quarters black and one-quarter white. An on-the-spot survey indicated that the Negroes were significantly middle-class, white-collar, high-income, and educated and that many were students. The press tent at the Monument gave out 1,655 special press passes, not counting the 1200 regular passes of Washington correspondents. Television beamed the event around the world. CBS and BBC provided complete coverage; NBC and ABC showed the event selectively.

The main program took place on the steps of the Lincoln Memorial during the afternoon. The two grand ladies of Negro song, Marian Anderson and Mahalia Jackson, performed. It was a special moment for Anderson since she had been denied the right to sing in Constitution Hall by the Daughters of the American Revolution in 1939 because of her race, causing Mrs. Roosevelt to resign from the DAR. She sang "He's Got the Whole World in His Hands." Mahalia Jackson sang the spiritual, "I've Been 'Buked and I've Been Scorned." Each of the Big Ten was allotted seven minutes and several delivered powerful addresses. The capstone, of course, was the "I Have a Dream Speech" by Martin Luther King, which took nineteen minutes.

King's rich, resonant, and powerful voice rang out over the enormous throng. The Negroes of America, he said, had come to Washington "to cash a check." With the Declaration of Independence and the Constitution the nation had issued a promissory note to all of its citizens. But the check given to blacks had been returned for "insufficient funds." They now demanded that the check be honored with "the riches of freedom and the security of justice." Not tomorrow, "*now* is the time."

He asked black people to conduct themselves with "dignity and discipline" and not drink from "the cup of bitterness and hatred." Negroes, he urged, must trust "our white brothers." "We cannot walk alone."

King had begun by reading the speech he had written. But, as he spoke and the crowd responded powerfully, he got caught up. He had earlier developed the dream peroration and had used it at a mass meeting in Birmingham in April and at a huge civil rights rally in Detroit in June. "I just felt that I wanted to use it here. I don't know why. I hadn't thought about it before the speech." He pushed his text aside and went on extemporaneously.

> I say to you today, my friends, so even though we face the difficulties of today and tomorrow, I still have a dream. It is a dream deeply rooted in the American dream. I have a dream that one day this nation will rise up and live out the true meaning of its creed—we hold these truths to be self-evident, that all men are created equal.

His dream was that descendents of slaves and slaveowners would sit down together in brotherhood in Georgia, that Mississippi would become an oasis of freedom and justice, that little children of both races would join hands in Alabama. He dreamed of "a beautiful symphony of brotherhood."

> When we allow freedom to ring, when we let it ring from every village and hamlet, from every state and every city, we will be able to speed up that day when all of God's children—black men and white men, Jews and Gentiles, Protestants and Catholics—will be able to join hands and sing in the words of the old Negro spiritual, "Free at last, free at last; thank God Almighty, we are free at last."

When he stepped back, dripping with perspiration, King received an enormous ovation.

As the crowd slowly drifted out, Phil Randolph stood alone at the end of the platform. This had been, he would say, "the most beautiful and glorious" day of his life. Bayard Rustin walked over and put his arm around the tired old man's shoulders. Rustin said, "Mr. Randolph, it looks like your dream has come true." Rustin could see the tears streaming from his eyes.[4]

4

Keynesian Turn: The Tax Cut

ALTHOUGH ALL American Presidents have confronted economic problems, none had been trained in economics. John F. Kennedy was no exception.

During the presidential campaign Nixon, who was not averse to playing the kettle calling the pot black, called Kennedy an "economic ignoramus." The latter seems to have enjoyed acting in that role, at least in private with sophisticated economists. He recalled his grade in elementary economics at Harvard as a gentleman's C; in fact, he received a B. When Heller's name came up for chairman of the Council of Economic Advisers, Kennedy said that he was the only economist he had ever met who was not from Harvard and that he had not known that there were any. With his knowledgeable economic advisers, he pleaded inability to recall the difference between monetary and fiscal policy. They suggested that he associate *m*onetary with the name of the chairman of the Federal Reserve, William McChesney *M*artin, Jr., who controlled monetary policy. When Martin's term was about to expire and Kennedy was considering not reappointing him, he wondered how he would remember the difference when Martin was gone. Heller had the answer: another member of the Fed was named *M*itchell.

Nixon was half right. Kennedy was hardly an "ignoramus" in the subject, but there were large gaps in his knowledge. The course at Harvard, given by a leftist instructor, left him with no particular understanding of either Adam Smith or Karl Marx. When he was in the Senate during the fifties, Kennedy came to deal with two economists of note. This was because he represented a state in a region that was running downhill and was eager to get federal aid for New England. He consulted with Seymour E. Harris of Harvard, a devout Keynesian, who had studied the decline of the region. Harris, whose literary output was prodigious, supplied Kennedy with copies of his books, but it is unlikely that he read them.

Senator Paul H. Douglas of Illinois was the other. He had been a distinguished member of the University of Chicago economics department, was considered by many to be the nation's premier labor economist, had been president of the American Economic Association, and, when the Democrats took control of the Senate in 1954, became chairman of the Joint Economic Committee. Douglas, concerned about the decline of the coal industry in southern Illinois, wrote a depressed areas bill and Kennedy became floor manager in 1956. Later Kennedy was a member of the Joint Economic Committee but was so busy running for President that he seldom attended hearings. In fact, Paul Samuelson was amazed to learn later that Kennedy was on the committee because he was never present when the economist testified. When Douglas and Kennedy stumped Illinois together in 1960 (the senator was up for re-election), Douglas recalled, "he would ask very profound questions on monetary and banking policy."

In early August 1960 Kennedy asked Archibald Cox of the Harvard Law School to bring some good economists down to Hyannis Port. He took Paul A. Samuelson of MIT, Harris, John Kenneth Galbraith of Harvard, and Richard A. Lester of Princeton sailing for the day on his boat, *Marlin.* They talked mainly about economic growth, which much interested Kennedy because he was concerned about unemployment.

In fact, he made growth a major issue during the presidential campaign. But he never discussed the question in detail, relying instead on the pat slogan of getting the country moving again and the goal of a 5 percent growth rate. Samuelson concluded that neither he nor his political advisers had thought the problem through.

After Kennedy became President, he resumed his education in economics. Given his keen intelligence, his extraordinary command of speed reading, and his high motivation, he proved an unusually adept pupil. Dealing with economic issues on a daily basis, he constantly faced policy options. They, of course, included conservative positions, policies based on traditional neo-classical theory, the body of doctrine that Galbraith called "the conventional wisdom." Samuelson and Joseph A. Pechman of the Brookings Institution thought that he had been influenced in that direction by his father, characterized by Pechman as "a very conservative man in economics." But Kennedy was now more interested in "the new economics," the ideas Keynes had set forth in *The General Theory* in 1936 that had become by 1961 widely accepted in Europe and had deeply penetrated American thinking, particularly in the universities.

Samuelson, who was consulted often, wrote the pre-inaugural task force report and was the author of a number of memoranda to Kennedy. Harris spent a day with the President at the America's Cup races off Newport on September 22, 1962, talking economics. Galbraith, until he went off to India as ambassador (Kennedy called the mission his "period of penance"), was often in the White House and thereafter aimed intercontinental missives at the President with economic advice from New Delhi.

But for his education Kennedy relied mainly on the three distinguished professors of economics he appointed to the Council of Economic Advisers—Walter W. Heller of Minnesota, James Tobin of Yale, and Kermit Gordon of Williams—Keynesians all. Until the Bay of Pigs in April 1961, the President was not especially busy and met frequently with the full Council. Thereafter he mainly saw Heller and was on the receiving end of some 300 of his famous memoranda. While none was ever acknowledged, Heller would hear him make a point that could have come from no other source. Mrs. Evelyn Lincoln, the President's secretary, repeatedly told him, "Mr. Heller, President Kennedy reads every memorandum you ever sent him from cover to cover."

While Kennedy got along splendidly with Heller, his relationship with Tobin was special. He had been criticized for appointing too many people from Harvard and was delighted to learn that the economist was from Yale. But when he met Tobin the face looked suspiciously familiar. Tobin, who had never left the Harvard Yard during his undergraduate and graduate years, pointed out that he had been in the class of '39 and had been surrounded by Kennedys, Joe in '38 and Jack in '40. He and Joe had served on the student council together. Kennedy said, "Here I thought I was not appointing a Harvard man but a Yale man and it turns out there's a Harvard man under the skin after all."

Tobin was immediately caught up by Kennedy's "obvious intelligence" and persuasive charm. His summary word: "impressive." Kennedy admired Tobin's intellect and his stature as an economist and loved to twit him. Kennedy had read an article in the London *Economist* before Tobin had seen it and, to the latter's surprise, called him "a growth man." D. H. Robertson of Cambridge had presented his students with this examination question: "Consider the growth theories of the following economists. Which one of them makes the least nonsense and why?" The list included James Tobin.

Tobin agreed to serve on the Council for a year, but he actually stayed until August 1, 1962. During the fall of 1962 he was in Washington and went to the White House to see the President alone. Tobin found the occasion "memorable" and wrote a memorandum about it. Kennedy was "extremely cordial, informal, and friendly" and called him "Jim." As usual, he joshed Tobin, this time about "the leisure and high income of the academic life to which I had returned." The President really wanted to talk economic theory, including his favorite esoteric subject, the gold drain, and he went on for more than half an hour on a busy day. He stated his views and wanted to know whether Tobin agreed. He was showing "how well he had learned his lessons," and, since he was on top of the subject, was "obviously having a good time." It was rather like a Ph.D. oral examination and Tobin had to pass him. Eventually Kenny O'Donnell had to break it up so that the President might move to his delayed next appointment. Heller agreed with Tobin's assessment of Kennedy's progress. He was "remarkably free of preconceived doctrines," and was "deeply committed to more rapid growth as an instrument of the common good and eager to put the power of modern economics to work."

Despite these tributes from his teachers, there was a quirk in Kennedy's economics that neither they nor anyone else has explained: he was profoundly concerned, some thought "obsessed" a better word, by the balance of payments and the gold outflow. He was, that is, a latter-day mercantilist. According to Sorensen, even the Treasury, which was responsible for worrying about these matters, "resisted his prodding." The Council sensibly pointed out that "the totals owed this nation by others far exceeded the claims upon our reserves," but this advice fell on deaf ears. If someone brought up devaluation of the dollar, the President warned that "he did not want that weapon of last resort even mentioned outside his office—or used." He said that "devaluation would call into doubt the good faith and stability of this nation and the competence of the President." Sorensen continued:

> "I know everyone else thinks I worry about this too much," he said to me one day as we pored over what seemed to me the millionth report on the subject. "But if there's ever a run on the bank, and I have to devalue the dollar or bring home our troops, as the British did, I'm the one who will take the heat. Besides it's a club that DeGaulle and all the others hang over my head. Any time there's a crisis or a quarrel, they can cash in all their dollars and where are we?"

Schlesinger's assessment was identical. Kennedy "used to tell his advisers that the two things that scared him most were nuclear war and the payments deficit. . . . He has acquired somewhere, perhaps from his father, the belief that a nation was only as strong as the value of its currency."

Secretary of the Treasury C. Douglas Dillon found Kennedy captivated by this problem. He insisted on getting the latest figures and, if they were late, would phone for them. Gold intrigued him "as a mental exercise as well as feeling it was very important." His concern, Dillon said, was "almost phobia." "He used to joke that he had a gold telephone that would ring on a dime—not true, but he was very interested."[1]

During the period between his election and his inauguration Kennedy made two important economic decisions—naming a task force to study the problems and making the key appointments for his administration. Paul A. Samuelson headed the task force, whose other members were Heller, Tobin, Pechman, a prominent Washington attorney Henry Fowler, Gerhard Colm of the National Planning Association, and Otto Eckstein of Harvard.

Samuelson's report, Prospects and Policies for the 1961 American Economy, was released to the press on January 6, 1961. The basic problem, he argued, was that the recession that began in mid-1960 was superimposed on an economy that was already "sluggish." Thus joblessness, presently over 6 percent, might rise even more in the next few months.

Depending upon the unemployment rate during 1961, Samuelson proposed a set of "optimistic" and "pessimistic" policies. The former assumed a level near 6 percent, the latter a "rise toward and perhaps beyond the critical 7½ percent level that marks the peak of the postwar era."

On the optimistic assumption, Samuelson primarily urged reliance on fiscal policy—both public expenditures on new programs and an acceleration of spending on old ones. The former were those proposed by the Democratic platform and the President-elect. But he distinguished between outlays which justified themselves and those intended to stimulate employment. An example was defense: Expenditures should be determined "on their own merits. They are not to be the football of economic stabilization." On this assumption, Samuelson urged higher budgets for defense, foreign aid, education, urban renewal, Medicare and other health programs, unemployment insurance, public works, highways, depressed areas, natural resource development, and housing.

Unemployment compensation deserved special attention. In theory the system was the nation's first line of defense against an economic downturn. But it had proved inadequate during the 1957–58 recession and was certain to do so again this time. Emergency appropriations were needed to allow all states to pay for at least thirty-nine weeks of joblessness because many of the reserves were depleted and they had still not repaid their 1958 federal loans. In the long run, Samuelson urged federal standards to cover the employees of all firms regardless of size, a benefit of "at least" one-half of prior earnings, and a minimum of twenty-six weeks of coverage supplemented by an additional thirteen weeks during periods of high national unemployment.

But suppose joblessness during 1961 should rise to 7½ percent. On that pessimistic assumption, "it will be the duty of public policy to take a more active, expansionary role." This should be, Samuelson concluded, a temporary reduction in tax rates across the board, perhaps 3 or 4 percent, effective immediately in order to stimulate the system and create jobs.

There are three major centers of economic policy in the government—the Federal Reserve Board, the Council of Economic Advisers, and the Treasury. A President has very limited authority over the Fed—to appoint members as their terms expire and occasionally to "lean" on the governors, to which they may or may not respond. He is free, however, to name the members of the Council and the Secretary of the Treasury as his own subordinates. Taking a leaf from FDR's book, Kennedy staffed these agencies with opposites, giving the Council to the liberal Keynesians and putting Treasury in the hands of a moderately conservative Republican banker. They were certain to clash.

The Council of Economic Advisers was headed by Professor Walter W. Heller. Heller was chairman of the economics department of the University of Minnesota. He was 45, tall and slender, exceptionally articulate with both pen and tongue. His father was a civil engineer and mathematician. Walter shone at Oberlin and did his graduate work at Wisconsin, which had a notably liberal economics department. He specialized in finance and taxation and was an early Keynesian. During the war (bad eyes kept him out of the service) he worked in the Treasury and helped install the new income tax withholding system based on his study of the Canadian model. He went to Minnesota in 1946. He was a brilliant teacher, exploiting a gift, rare among economists, to reduce abstruse and often vague economic concepts to simple English. This talent was carried over into his extremely brief written memoranda, treasured by a very busy President. He was economic adviser to the leading Minnesota Democrats—Senator Humphrey and Governor Orville Freeman.

On October 1, 1960, Kennedy had come to Minneapolis for a campaign speech and wanted to meet Heller. Humphrey brought him to the hotel, where they spent 10 minutes together while Kennedy changed his shirt.

Kennedy asked a series of rapid-fire questions which Heller answered: (1) Can we achieve a 5 percent growth rate by government action? Answer: Probably, by stimulating demand. (2) Is accelerated depreciation an effective way to increase investment? Answer: Yes, but it should be packaged with support for consumption. (3) Why has the German economy grown so fast in the face of high interest rates? Answer: Docile labor with low wages, allowing for high profits and savings. (4) Can a tax cut be a significant stimulus? Answer: Yes, if properly timed. The chemistry worked; Kennedy and Heller combined flawlessly.

Heller was asked to come to Washington on December 16 and was "carefully sneaked" into the back door of Kennedy's Georgetown house. While waiting in the dining room, he learned that Douglas Dillon was in another room, presumably to be offered the Treasury. Kennedy burst into the dining room and immediately invited Heller to become chairman of the Council. He said Heller topped the lists and, based on the ten minutes they had talked in Minneapolis, he liked Heller's "approach to economics." "I think Dillon will accept," he said. "I need you as a counterweight to him. He will have conservative leanings, and I know that you are a liberal." As they walked out, Heller raised the question of a tax cut. Kennedy said it might be a possibility, but he could not square it with the sacrifice theme he was stressing.

For Heller, who had three children, the eldest of whom had just gone off to college, the decision required sacrifices. But his family agreed: "If you say, 'no,' we'll never forgive ourselves." After all, for a policy-oriented economist, this was the best job in America. He went to Palm Beach on December 19 to see Kennedy. Heller asked for the right to nominate the other members of the Council and the assurance that there would be no competing "White House economist," such as Eisenhower had used. Kennedy agreed to both at once. Thus, Walter Heller became chairman of the Council of Economic Advisers.

He proposed James Tobin of Yale and Kermit Gordon of Williams College as the other members of the Council, to which Kennedy agreed. Tobin, 42, was Sterling professor of economics. Born in Champaign, Illinois, his father was publicity director for University of Illinois athletics and his mother was a social worker. Tobin must have been about the brightest of the best. He graduated *summa cum laude* from Harvard, had a Ph.D. from that institution, was a Junior Fellow there, became Sterling Professor at Yale at thirty-two, and would later win the Nobel Prize in economics. Tobin was a dedicated teacher and formidable scholar. He was considering a year at the Center for Advanced Study in the Behavioral Sciences in Palo Alto when Kennedy phoned. Tobin demurred, using what he considered the clinching argument: "Mr. President, I am what you might call an ivory tower economist." Kennedy topped him: "That's the best kind! Professor, I am what you might call an ivory tower President." Tobin said, "That's the best kind!"

Tobin had spent the war in the Navy, four years on the U.S.S. *Kearny*. Herman Wouk had been in his training class, and the character "Tobit" in *The Caine Mutiny* is said to have been based on Tobin. He was a man who cared deeply. At Yale he was regarded as an old Roman because of his sense of civic duty and virtue, but there was an Irish twinkle in his eye. He was a child of the Great Depression, and when he enrolled at Harvard in 1935, he studied economics because he hoped to help the world avoid another such disaster. Keynes published *The General Theory* in 1936 and Tobin was hooked immediately. He would later try to fill some of the potholes that Keynes had left behind. Tobin was particularly concerned about the welfare of the poor and blacks. Aside from being a Keynesian, Tobin was an across-the-board liberal economist. He urged policies to redress the balance between the public and private sectors, to establish full employment, to utilize the power of government to assist those in need, and to redistribute income.

Kennedy's attitude toward these Keynesians at the outset was highly ambivalent. On the one hand, he wanted a two-track system, with the Council advocating positions that he, for political reasons, was unable to support publicly. He told Heller to use his position "as a pulpit for public education." He expected the Council to be ahead of adopted policy, to test out ideas, to inform the public. As Heller put it, he wanted them to be "a little beyond the frontier." The issue came up when Senator Douglas invited Heller to testify before the Joint Economic Committee on March 6, 1961. Arthur F. Burns, who had been chairman of the early Eisenhower Council, had refused to testify without a transcript except in executive session, which angered Douglas. His successor, Raymond J. Saulnier, adopted the same policy. Kennedy, by contrast, instructed his Council to testify in open session.

The Council took this mission very seriously and Tobin wrote a statement for Heller that Kennedy cleared. Aside from its public impact, Heller considered it "one of the milestones in the education of the President on economic matters."

The testimony addressed three basic questions. The first was the status of the Council under the Employment Act and its relationship to the President and the Congress. Heller reviewed the history, starting with the conflict in the late forties between Chairman Edwin G. Nourse (no testimony) and Vice-Chairman Leon V. Keyserling (open policy advocacy) on testimony before Congress. The members of the present Council, Heller stated, limited only by their confidential advisory relationship to the President, "are glad to discuss, to the best of their knowledge and ability as professional economists, the economic situation and problems of the country, and the possible alternative means of achieving the goals of the Employment Act and other commonly held economic objectives."

The second was the festering conflict between those who regarded unemployment as a problem primarily responsive to aggregative remedies, that is, fiscal and monetary policies, and those who viewed it as a set of structural rigidities—for example, declining industries and regions—which demanded specially tailored solutions. In the Kennedy administration, generally speaking, the Council was the seat of aggregative policy, the Labor Department the seat of structural programs. But, as Heller stressed, this difference could easily be exaggerated. The Council wanted all the remedies possible to improve the skills and mobility of the labor force. Similarly, the department was willing to increase aggregate demand.

Tobin, who took this issue most seriously, was not concerned about Labor. Rather, he directed his small fire at the President and his heavy artillery at the Federal Reserve Board. Of the structural analysis, he emphasized, "We wanted to nail it." "My personal opinion," Tobin said, "is that all nonprofessional economists, all laymen in economic matters, are instinctively attracted by the idea of structural unemployment." Kennedy, who never stopped worrying about joblessness, had this instinct. Concerned about abandoned textile workers in Massachusetts, for example, he would ask, "How are we ever going to get these people back to work?" The Fed was more cynical. When unemployment rose in the late fifties, the Board argued that this was wholly due to structural causes and that the country must learn to accept a higher level of joblessness. Tobin found this exercise in deflection "shocking." Robert M. Solow, a brilliant young economist from MIT, who was on the Council's staff, compared structural unemployment in 1953 and 1957 with 1961, and showed that, in fact, it had not grown worse.

The Council called the final issue "the performance gap," that is, the difference in gross national product between the system operating at an actual almost 7 percent unemployment rate and the product at hypothetical "full" employment of 4 percent. The Council reckoned that the difference in 1961 was $50 billion and 5 million jobs. The relative growth of government revenues in recent years, Heller pointed out, "brings the budgets into balance substantially below full employment at current levels of Federal expenditure and tax rates."

One solution to this problem was to cut taxes, which the Council advocated behind closed doors at the White House but not publicly before the Joint Economic Committee. This was the other side of Kennedy's ambivalence toward the Keynesians, to avoid any major challenge to the balanced budget by urging a reduction in taxes.

Chastened by his narrow electoral victory, confronting a conservative Congress, and stressing the sacrifice theme with business and labor, the President had concluded that a tax cut in 1961 was politically out of the question. When the Council told him that this was the way to get the country moving, he responded that they might be right, "but we would be kicked in the balls by the opposition." While the Keynesians continued to push, they respected his political judgment. "Suppose," he asked Samuelson, "that I ask for something, a bold program, and don't get it? I'm turned down." Samuelson said, "Then you've fought the good fight." "Yes," Kennedy replied, "but if that jeopardizes other programs, that's pure vanity." Samuelson decided that "maybe he really had something." Gordon pointed out that "when things go well, the President gets the credit. When things go badly, the President gets the blame." In retrospect he thought Kennedy was "more right than we were."

At the outset the lawyers in the Kennedy entourage—Ted Sorensen, Myer

Feldman, and Richard Goodwin—regarded the Keynesians as trouble makers who needed to be closely watched. All three sat in on the meetings of the Samuelson task force and were horror-struck when the members seriously considered a tax reduction. They make it clear, Samuelson said, "that I should not ask for a tax cut because it would be embarrassing and it would not take place." He left it out. Sorensen, with his tart tongue, scalded the Keynesians: Why talk about *un*employment? If you stressed employment, 94 percent was an A – while 97 percent was an A. Why should the President expend political capital raising the grade so little? "The last thing in the world that the President wants to have," Sorensen said, "is the reputation of a reckless spender." "The most expendable thing in this Administration is your reputations as professional economists." According to Tobin, Sorensen thought that Keynesian fiscal policy was "a bit of traditional liberal dogma with which his man may been saddled." Sorensen would come around, but it would take time.

The Treasury was so venerable, so bureaucratized, and so powerful that, as Joseph Kraft wrote, even Pennsylvania Avenue was forced to make a "symbolic detour" to get around it. It was, as well, closer to the White House than any other department. Though Kennedy knew an enormous number of people, he knew no one he thought qualified to head this formidable agency. As Presidents are inclined when faced with this dilemma, he turned to that shadowy power center that Richard H. Rovere called "the American Establishment," which, while extending its fingers elsewhere, kept its hand firmly on Manhattan, particularly in the investment banking houses and corporate law firms at the bottom of the island. While the Establishment was mainly Republican, it had little hesitancy about "serving" Democratic Presidents.

The urbane banker Robert A. Lovett was the epitome of the Establishment. Kennedy talked to him and was so bewitched that he offered Lovett his choice of State, Defense, or Treasury, particularly the last. Lovett, who suffered from bleeding ulcers, did not think he should take on this demanding portfolio. During the war he had been Assistant Secretary of War for Air and had brought in a team of management specialists from the Harvard Business School. The star, the banker thought, was Robert S. McNamara, just named president of the Ford Motor Company. Kennedy asked Sargent Shriver to look into McNamara. His record was so impressive that Kennedy offered him either Treasury or Defense. McNamara immediately turned Treasury down because he had no banking experience. After deliberation, he accepted Defense.

The finger now pointed at C. Douglas Dillon. According to Rovere, the Establishment, like the Mafia, denied its own existence, but everyone agreed that "Douglas Dillon is true blue." His father had founded the prominent Wall Street firm of Dillon, Reed & Co. During the depression Papa had bought Chateau Haut-Brion outside Bordeaux whose red Graves was one of the "immortal" wines of the Médoc. His son had graduated from Groton with high honors and *magna cum laude* from Harvard, where he studied American history and literature. He had been a member of the New York Stock Exchange, had been an international banker, and had risen to be chairman of the family firm. Dillon had served with the air arm of the Seventh Fleet in the southwest Pacific during the war and had received the Air Medal and the Legion of Merit. He had an apartment on Fifth Avenue and homes in Washington; Hobe Sound, Florida; Dark Harbor, Maine; Far Hills, New Jersey; and Versailles, France. He was an Episcopalian.

Dillon was a lifelong Republican. As chairman of the New Jersey Republican State Committee, he had actively supported Eisenhower for President in 1952. He served throughout the Eisenhower years as Ambassador to France, Under Secretary of State for Economic Affairs, and Under Secretary of State. He had contributed substantially to Nixon's 1960 campaign. It was said that, if Nixon had won, he would have offered Dillon either State or Treasury. In fact, when Kennedy proposed the latter, Dillon talked to both Nixon and Eisenhower. Nixon gave him no encouragement and Eisenhower told him not to accept unless he had a commitment in writing of a free hand. Kennedy said that the President did not make "treaties" with his appointees.

With this background, Dillon was, obviously, a conservative, which worried liberal Democrats. But he was open-minded and flexible, as Kennedy discovered when he questioned Dillon about economic growth. Later Dillon bridled when some called him a "Keynesian," but he accepted ideas from the Master supplied by Seymour Harris. The economist found that Dillon "really has an open, inquiring mind." Harris came to call him "the Alexander Hamilton of the Twentieth Century." Dillon and his wife would establish close personal relationships with both John and Robert Kennedy and their families.

Dillon proved to be an outstanding administrator. He was very well-informed and was in full command of the Treasury. He had excellent relations with Congress and was an impresssive witness. He had, as well, a keen grasp of the executive agencies. He made outstanding appointments, so that Treasury became as well-staffed as Justice, if not better.

Henry H. Fowler was Under Secretary of the Treasury. A courtly Virginian, Yale law graduate, Democrat, prominent Washington lawyer, and member of the Samuelson task force, Fowler was especially effective on the Hill. Robert V. Roosa was Under Secretary for Monetary Affairs with responsibility for the balance of payments and the gold drain. Holding a

Michigan Ph.D. in economics, he had taught at Michigan, Harvard, and MIT, and had been with the New York Federal Reserve Bank for fourteen years. Roosa was a master of international monetary matters and his defense of the dollar was a virtuoso performance. Stanley S. Surrey was Assistant Secretary for Tax Policy with responsibility for tax reform. He was the Harvard Law School's specialist on taxation and was co-author of a leading casebook. Perhaps the nation's prime authority on tax inequities, he had worked on this problem with the House Ways and Means Committee and headed the Kennedy task force on taxation. Senator Byrd, chairman of the Finance Committee, would not approve his nomination until Dillon promised that he, rather than Surrey, would take the chief responsibility for tax policy. Mortimer M. Caplin became Commissioner of Internal Revenue. He had been a brilliant student at the University of Virginia as well as the NCAA middleweight boxing champion. He had graduated at the head of his class from the Virginia Law School and during the war had served in the Navy, decorated as a beachmaster in the Normandy invasion. He taught taxation at the Virginia Law School and was among the nation's top authorities in the field. Robert and Edward Kennedy had been his students.

Dillon told the President that George Humphrey, who had been Eisenhower's secretary, thought he knew all the economics he needed and had gotten rid of the department's economists. Dillon felt that he needed economic advice. Kennedy suggested Seymour Harris. Roosa, with his banker's preference for secrecy, was upset because Harris had the peculiar habit of writing letters to the newspapers. Nevertheless, Dillon made him his economic consultant. Dillon listened to the Harvard economist and religiously attended the seminars of academic experts that Harris arranged for him. The economists's Keynesian views, though never preached, had an impact on Dillon, and Harris provided a bridge between the Treasury and the Council.

Interestingly, the Treasury, including Dillon, sought to educate the President in the politics of economic policy. On November 13, 1961, Fowler sent Kennedy a memorandum, which the Secretary would see, stressing the importance of sustaining recovery, but he covered it with a letter in longhand that Dillon would not see which dealt with politics, namely, the election of 1964, when Kennedy would be running for a second term. "Kennedy prosperity" was a rare asset that must be preserved. The duration of recent business cycles suggested that the next recession would come in 1963 or 1964. During the summer and fall of 1964 the voters, Fowler urged, must be bathed in the "glow of a well-confirmed recovery." Dillon returned to this theme in a memorandum to the President on July 12, 1963, when the big Kennedy tax bill was before the House Ways and Means Committee.

> I am certain that the use of the cyclical argument, i.e., that a recession would ordinarily be due next spring because of the passage of time, will tend to induce a negative reaction except in those Congressional circles whose firm support can already be counted on. Although I believe that the cyclical argument has validity, and I know that it is heavily relied upon by economists, the plain fact of the matter is that it is not widely accepted in conservative circles. In addition, use of the cyclical argument pointing toward 1964 indicates a major political interest on your part in the bill, which stresses the very reason for which many Republicans secretly oppose any bill at all. The same may well be true of some of the conservative Southerners whose votes we will need. Therefore, I would strongly urge that the cyclical argument be avoided in connection with the need for a tax bill this year.[2]

In his first State of Union message on January 30, 1961, President Kennedy said, "The present state of our economy is disturbing." He pointed to seven months of recession, three and one-half years of "slack," seven years of inadequate growth, nine years of falling farm income. "This Administration," he declared, "does not intend to stand helplessly by."

But, as he had with civil rights, Kennedy adopted a policy of executive action now and legislation later. On January 5, two weeks before he took office, he had told Heller that the Council's first assignment was to develop an administrative anti-recession program and to have it ready by February 2. The members worked furiously, often till 4 or 5 a.m. They conducted what Gordon described as "an undiscriminating dragnet," looking for anything "possible and defensible." Since Galbraith's departure for New Delhi was delayed, Kennedy kept him around the White House doing odd jobs, including riding herd on the Council. He brought them a "laundry list" of spending possibilities and barked at them: "You fellows don't have any imagination. Get some money spent fast!"

The President delivered his message to Congress on the Program to Restore Momentum to the American Economy on February 2. It covered three areas: executive action to accelerate federal spending, easier money, and a legislative program. The last, covering manpower training, the minimum wage, Medicare, and aid to education, will be discussed in later chapters. Kennedy, of course, did not ask Congress for tax reduction, but he did propose a tax credit for business as an incentive to expand investment in plant and equipment.

The executive acceleration program consisted of the following elements: Post office construction was speeded up, particularly in areas of high unemployment. Federal-aid highway funds were released ahead of schedule, and the President urged governors to spend federal funds more rapidly for roads, schools, hospitals, and waste treatment facilities. IRS returned refunds to taxpayers more quickly. The VA paid veterans their life insurance dividends earlier. Price support payments to farmers for crop storage were accelerated. The Secretary of Agriculture expanded free food distribution to needy families, and a pilot food stamp program was launched. The college housing and urban renewal programs were pushed forward. Procurement agencies were directed to make purchases in areas of substantial unemployment. The Council had hoped to speed up the payment of Social Security benefits, but that required legislation. This program generated no political problems.

Lower interest rates, however, raised sharp controversy. The liberals strongly favored easier money. On April 6, 1961, Heller urged the President to take such action. But the Council's strong position paled alongside Galbraith's views. He considered the Fed a roadblock to recovery and held the governors, particularly the chairman, in contempt. If Kennedy "set Bill Martin on his ass," Galbraith said, "we'd get out of the recession." After reviewing an interest rate policy the Fed and Treasury had agreed upon, he wrote in his diary, "It used one line to suggest lowering rates and twenty to tell of the dangers." He asked Roosa why "the banks and the Federal Reserve could never bring themselves to express any warmth and compassion for the unemployed." This attitude, he pointed out, had "made every central bank in the world an immediate arm of the government—I personally think a good thing, too. Bob agreed." Galbraith was outraged when the Fed, obsessed with the dangers of inflation, allowed interest rates to rise. His strong feelings on this issue, doubtless, had something to do with Kennedy's pleasure in sending him off to India. "Otherwise you and Heller would have me expounding a far too radical position." Galbraith protested that he was not a radical, "only a realist." Harris and Samuelson, like Galbraith, wanted Kennedy to get rid of Martin. Tobin later reminded Samuelson that he had said that Martin's departure was worth "x" billion dollars in gold. Neither could remember what "x" was.

Martin's term as chairman of the Fed expired in 1963, his membership in 1970. He had no intention of going quietly in order to gratify the liberal economists. He began lobbying early for reappointment as chairman. He invited Heller and Tobin to lunch at his office on May 29, 1961. He said he had once been a registered Democrat, had been appointed by Truman, and had voted for Stevenson in 1952. During the presidential campaign and since, he had heard, Tobin wrote, "of some sentiments in the White House that he should resign." He made it clear that "he did not intend to do so." On January 16, 1963, Dillon wrote the President strongly urging Martin's reappointment as chairman. "He can be a tower of strength in support of our tax program and in the balance of payments." While Kennedy disliked bankers and agreed with Heller that Martin should go, he found this politically impossible because it would alienate the financial community. Kennedy reappointed Martin.

The weakness in the liberal position on interest rates was that it concentrated on the domestic side, particularly on the impact of tight money on unemployment. The other phase of the problem was international, the balance of payments. During most of the postwar era the balance had favored the U.S., but in 1957 it turned negative, the dollar came under pressure, and gold flowed out. This remained an acute problem in 1961. Even Samuelson had written in his task force report: "The days are gone when America could shape her domestic stabilization policies taking no thought for their international repercussions." Since interest rates were higher in Europe, capital moved abroad. If the Fed reduced rates here, this would increase the flow. Both Kennedy and Dillon considered this an extremely important issue and Roosa's job was to choke off the capital movement. The President did not want to take the risks that went with lowering interest rates. "The $3 billion payments deficit 'tail,'" liberal Congressman Henry Reuss of Wisconsin observed, "has been allowed to wag the $500 billion 'dog.'"

Thus, the February 2, 1961, easy money policies were limited primarily to the government itself. There was one exception: in what came to be called "the twist," the Fed and the Treasury did nothing about short-term interest rates (less than one year) and supported long-term rates. The federal policies were the following: Agriculture released funds for operating loans and for assistance to low-income farmers. The FHA reduced the maximum rate on insured home loans from 5¾ to 5¼ percent. The Federal National Mortgage Association stepped up secondary market purchases of insured mortgages. The Federal Home Loan Bank Board liberalized the terms under which savings and loan associations could issue mortgages. The Small Business Administration reduced the rates charged to small business and to state and local government borrowers from 5½ to 4 percent. The Export-Import Bank eased credit guarantees, insurance, and financing for exporters.

The Council estimated that this package of policies raised gross national product by $3 billion. The 1960–61 recession proved to be brief and shallow. As measured by GNP in 1961 prices, the top of the preceding recovery came in the second quarter of 1960—$514.2 billion. The product then dropped steadily for three quarters to a low of $503.9 in the first quarter of 1961 as the program was taking effect. The unemployment rate followed a similar pattern with the usual lag. It peaked at 7 percent in May 1961 and then began to fall slowly but steadily.

Keynes had taught that the volume of employment was determined by the level of aggregate demand. If one were a Keynesian, therefore, a government could reduce unemployment either by increasing public expenditures or by reducing taxes. Galbraith, who favored spending, applauded the 1961 program, though he would have favored a different mix. The Council and Samuelson, who sought a tax cut, came up empty-handed.

At the turn of the year, however, the Council savored a small victory—the *Economic Report of the President* submitted to Congress in January 1962. Along with the Council's accompanying report, it was, on the whole, a ringing affirmation of its views. Kennedy had gone over it, Gordon said, with "great care," and even chose the color for the cover—green.[3]

The 1962 *Economic Report* announced two significant new policies: a 4 percent goal for unemployment and wage-price guideposts to counter inflation. The former was simple and, except for some muttering by the Labor Department, was noncontroversial. The latter was extraordinarily complex, extremely mystifying, and intensely controversial. Both the President and the Council endorsed the goal for unemployment; in his part of the *Report* Kennedy was silent on the guideposts, allowing the Council to take responsibility for them.

The President's statement about joblessness read:

> We cannot afford to settle for any prescribed level of unemployment. But for working purposes we view a 4 percent unemployment rate as a temporary target. It can be achieved in 1963, if appropriate fiscal, monetary and other policies are used. The achievable rate can be lowered still further by effective policies to help the labor force acquire the skills and mobility appropriate to a changing economy.

The Council, of course, generated the 4 percent goal. All the members agreed that the Employment Act of 1946 allowed, if it did not require, numerical goals, and there was little internal discussion over the number four. Tobin was the prime mover and he later described the reasoning as follows:

> The motivation was this, that we encountered in Washington—at least I did, and it was a new and shocking experience for me—the theory, very widespread, that we had shifted to a new stage of the economy in which it was necessary to operate at higher levels of unemployment, considerably higher than we'd operated with in the past, and there wasn't anything abnormal or disturbing about unemployment rates between 5 and 6 percent, and this was because of structural changes in the economy and so on. One evidence that was given for this was the fact that at the peak of the previous boom, unemployment had been 5 percent, so that showed that even in "good times" you don't get unemployment very far down. We were concerned, I think, to make the point that the peak of the previous boom hadn't been high enough and that what people were calling prosperity wasn't really prosperity as judged by the potential of the economy, either in production or employment.

The number four emerged almost automatically. Over the years Leon Keyserling, the Committee for Economic Development, and even Arthur

Burns had used it. One could not go below four because prices moved up at that rate at the peak of the 1955–57 boom and because, as Tobin put it, the Council did not want the figure to be dismissed as "impractical utopianism." Only the Labor Department rejected this reasoning; both Goldberg and Wirtz considered the 4 percent goal an affront to the labor movement. No figure above 4 percent made sense. It would certainly generate strong labor opposition and, more important, would excuse doing nothing to deal with joblessness. Heller had no difficulty persuading the President to adopt the 4 percent goal.

The wage-price guideposts raised far more difficult questions. Economics, a decidedly incomplete policy science, offered no answer to the problem of producing full employment and price stability simultaneously. In the fifties Sumner H. Slichter, the Harvard economist, had made a cottage industry of this theme in speeches to bankers and businessmen. Walt W. Rostow, the MIT economic historian, would later say that this was the chapter Keynes had not written. In wartime the government had fixed wages and prices with controls, notably during World War II and, to a lesser extent, during the Korean War. But the U.S. was at peace in the early sixties. If Kennedy had proposed controls, he would have butted his head against labor over wages, industry over prices, and everyone who supported collective bargaining. The authors of the guideposts insisted that they were not controls.

There was another nagging question: Why fight inflation when there was none? Prices had risen fairly rapidly between the end of 1955 and early 1958, but that modest inflation had been wrung out of the system by the 1957–58 recession. Since then, prices had advanced at a rate of slightly over 1 percent annually, essentially stable. Why tilt at windmills?

The impetus for the guideposts seems to have come from the President himself out of his "obsession" with the balance of payments. Walt Rostow supplied the linkage between that problem and wage-price policy. Among Kennedy's advisers, only Galbraith and Rostow, the latter only marginally involved with economic policy, gave primacy to inflation. Galbraith had long written about the gravity of the problem and wanted "public machinery for restraining wages and price increases." But he was diverted to India. Rostow stayed home and was both more explicit and more dogged.

During the fifties, when he studied the stages of economic growth, Rostow had concluded that "the optimum setting for growth and stability is a regime of constant money wages and falling prices." He had also decided that the wage bargains in steel and autos were decisive in attaining these goals. He proposed that unions and employers enter into "treaties" under which labor would give up increases in money wages in return for the employer's promise to convert productivity gains into price reductions.

On November 17, 1960, in a memorandum to Kennedy on military and foreign policy, Rostow added the following:

> I cannot emphasize too strongly that the capacity of the new Administration to do what it wants to do at home and abroad will depend on promptly breaking the institutional basis for creeping inflation, notably in the key steel and automobile industries; and on driving hard to earn more foreign exchange and to increase domestic productivity over a wide front. Without determined action in these areas it will be extremely difficult to bring the domestic economy back to full employment without inflation; and this means that we shall not have the federal revenues to give extra needed thrust in military and foreign policy. If we do not evoke American effort and sacrifice for communal goals at home, we run the danger of being forced by the balance of payments position and inflation into substituting rhetoric for action abroad.

Encouraged by the President, on February 1, 1961, Rostow met with Arthur Goldberg, who threw cold water on his proposal. While in favor of a "responsible" wage-price policy, Goldberg thought labor, industry, and the public did not understand "the seriousness of the world situation" and the complexity of the problem. For these reasons Goldberg opposed an immediate wage-price agreement. Rather, he suggested that the subject be put on the agenda of the labor-management advisory committee he was putting together.

Undaunted, Rostow urged Kennedy in April "to go for a wage freeze and price cut" when the auto contracts expired on August 31, 1961. Walter Reuther, learning of this, objected strenuously to the President. Kennedy told Rostow to try to bring him around, and Rostow and Reuther met on June 20. It was a sparring match. Reuther asked whether, if he were "damn fool enough to agree" to an auto wage freeze, Kennedy would guarantee that David McDonald would get no more for the steelworkers? Rostow granted the reasonableness of the question. Reuther went on: If both he and McDonald were damn fools, would the administration assure them that the steel industry would not raise prices? He doubted "White House professors" could handle that crowd, though he did not worry about auto prices. Rostow summarized: Auto and steel wages geared to the average increase in productivity and no steel price increase. He asked if he could tell the President that Reuther "agreed to this position." He did not agree, "but I could tell the President that Reuther would consider it." Kennedy was amused by the account and told Rostow that he was "a pretty good negotiator for a professor."

In fact, the auto settlements reached in the fall of 1961 were modest: 6 cents per hour or 2½ percent, whichever was higher, annually for the three years of the contracts plus a roughly equivalent amount in fringe benefits. The UAW estimated the cost around 12 cents. The companies pretty much held the prices of the 1962 models at 1961 levels. If one considered only the wage increase, this was well within Rostow's formula. But it is highly unlikely that Rostow, or even Goldberg and Kennedy, both of whom talked to Reuther, produced this result. The employers bargained tough because auto sales were slow and there was a good deal of unemployment in the industry.

Much more important was the Council of Economic Advisers' publication of the wage-price guideposts in the 1962 *Economic Report.* While the Council recognized that inflation was not presently a problem, the members knew that signs of upward price movements would tilt the Fed, the Treasury, and the conservatives in Congress against an expansionary fiscal policy and the reduction of unemployment.

In 1961, Kermit Gordon, like Rostow, anticipated that renewed wage bargaining in the steel industry could lead to a price increase. As will be pointed out shortly, Gordon drafted letters to the steel companies asking them to keep a lid on prices after the next round of wage increases.

The wage-price guideposts had the consistency of an overripe cantaloupe. After discussing a number of caveats and qualifiers (several others were omitted), the Council stated: "The general guide for noninflationary wage behavior is that the rate of increase in wage rates (including fringe benefits) in each industry be equal to the trend rate of over-all productivity increase." The treatment of prices was vague. They might rise, remain steady, or fall, depending upon productivity in the industry. Since the long-term gain in man-hour output was roughly 3.2 percent, this suggested that wages could rise by that amount each year without pushing up prices. But the guideposts were not based upon a law or executive order; this was wage-price "control" by exhortation, or, as it came to be known, "jawboning." The President did not like the term for its biblical origin, "the jawbone of the ass."

Heller was well aware of these weaknesses. As Lloyd Ulman has written, "Walter insisted that the normative criteria which Bob Solow had crafted so meticulously be called guideposts rather than guidelines on the grounds that the latter, however silken, might be mistaken for ties that bind, whereas there is a presumption of open space between one post and the next."

If the jawboners felt unsure of themselves, the jawbonees reacted with either wonderment or outrage. Goldberg failed at the AFL-CIO convention in Miami Beach in December 1961, when he lectured the delegates on the need to show wage restraint and on the limits of the right to strike. George Meany exploded. Reuther blasted the guideposts and warned the government to keep hands off UAW negotiations. Poor Tobin tried to persuade union economists to hold down workers' gains in order to prevent inflation and straighten out the balance of payments. Industry was ambivalent. While many employers welcomed the administration's help in pressuring the unions, they would brook no government intervention in determining prices.

Nor did the President get anywhere with his Advisory Committee on Labor-Management Policy. At Goldberg's urging, he had created this tripartite body consisting of seven top leaders from both labor and management along with five distinguished public authorities plus Goldberg and Luther H. Hodges, the Secretary of Commerce, who rotated as chairman. The executive order explicitly asked the committee to advise the President on "sound wage and price policy." He tried to turn this around to get the committee to endorse administration policy. It did not work. George W. Taylor of the University of Pennsylvania, the dean of American arbitrators and the author of the Little Steel wage stabilization policy during World War II, was chairman of the subcommittee on sound wage and price policy. Taylor thought the guideposts served no useful purpose and might, as a target, encourage higher wage increases. And Taylor asked, Why worry about inflation when there was none? His subcommittee never reported.

Later Heller wrote that the guideposts, while too complex for the public to understand and "a poor instrument of consensus, . . . have been a good instrument of education." The government, he argued, had to do something about inflation, and the guideposts, "in spite of their imperfections and the dents and scars they bear," were better than nothing. No one could accuse Heller of looking at the dark side.[4]

John Kennedy was baffled and frustrated in his relations with that immensely powerful American institution—business. He was the target of hostile attacks by industrial spokesmen and the business press as well as scatalogical abuse in golf club locker rooms.

Kennedy thought this unfair. His father, after all, had been an extremely successful businessman and he, like his brothers and sisters, was a beneficiary of the business system. Intellectually he had no trace of Marxism—in fact, he was locked in conflict with the Soviet Union—or even of native populism. He had offered State, Defense, or Treasury to Lovett, had named an investment banker to Treasury, the president of Ford to Defense, Luther Hodges, a former textile executive, to Commerce, and J. Edward Day of Prudential Insurance to the post of Postmaster General. Despite misgivings, he would reappoint Martin in 1963 as chairman of the Fed. He had come out for a balanced budget. He fussed endlessly over the balance of payments. He rebuffed Heller's proposal to lower taxes in order to increase employment, but he continued Eisenhower's policy of reducing business taxes to increase investment. This was in face of the fact that many large corporations were swimming in cash. Nevertheless, IRS relaxed the rules on the depreciation of machinery and equipment, and Kennedy's 1961 tax proposal, enacted in 1962, gave industry a credit against taxes for investments. Together they added $2.5 billion annually to corporate cash flow. Liberal Democrats wondered whether their man was Nixon in disguise. "Of all the myths in current political and economic literature," Bernard D. Nossiter wrote, "one of the most imaginative and furthest removed from reality portrayed President Kennedy as anti-business."

The Kennedy versus business issue first came to a head in 1961 in a spat between Luther Hodges, the courtly and elderly North Carolinian who liked to say that he was the administration's "only tie with the nineteenth century," and the Business Advisory Council. Hodges told the National Press Club that his department, while "primarily concerned" with business, would not be "its tool and automatic spokesman." This immediately locked him into conflict with the BAC.

The Council, formed in 1933, was, Hobart Rowen wrote, "America's most powerful club." It consisted of a small number of the chief executives of the largest corporations. BAC met quarterly in Washington to consult with high government officials and twice annually for "work-and-play" sessions at posh resorts with swank golf clubs. Newsmen were excluded and the Council operated virtually in secret. Eisenhower, who was addicted to golf and to hobnobbing with big businessmen, had enjoyed BAC meetings and had recruited three members of his cabinet from its ranks.

But now the Council was in trouble. Ralph J. Cordiner, chairman of General Electric, was the BAC chairman. GE, along with lesser electrical firms, had just pleaded *nolo contendere* to price-fixing and bid-rigging in the most notorious such case in the history of the Sherman Act. While Cordiner insisted that he knew nothing of these illegal activities and allowed his subordinates who went to jail to twist in the wind, he smelled, to mix the metaphor like a long expired fish. Many thought he was a liar and the others wondered how a competent chief executive could not have known. Hodges insisted that Cordiner must go. In addition, he thought no private organization should have a special channel to the government, that secrecy was improper when federal officials were present, and that the BAC should broaden its membership to include smaller firms.

In late February 1961 Cordiner resigned and was succeeded by Roger M. Blough, the chairman of U.S. Steel. Hodges then demanded that BAC extend its membership and agree that the government should name the members. The cozy fraternity voted him down. Blough recognized that Hodges could not be ignored and reached a settlement with the secretary which went part way. But the BAC spring meeting in May at the Homestead in Virginia under the new rules was a disaster. Hodges was the only high government official who showed up and the meetings, now open to the press, were dismal. BAC cut its ties to the government.

Thomas J. Watson, Jr., the chairman of IBM, a friend of the President, and a member of BAC, arranged for Blough to see Kennedy on July 5. The President hoped to restore relations with the Council. But Blough, when he arrived, placed documents on the desk announcing BAC's divorce from the government. He left Kennedy with nothing to say. Hodges then decided that the Commerce Department would keep the name Business Advisory Council, and the other organization changed its name to Business Council.

While this seemed a tempest in a teapot, Kennedy, again denounced for being "anti-business," soon regretted his acceptance of the result. Robert Kennedy, Watson, and Heller urged him to make peace with BC and, wanting business support, he did. Small teams of Council members were assigned as liaison groups to key government agencies, including the White House. Blough himself went to the Council of Economic Advisers where he struck up a first-name relationship with Heller. Hodges was left out to dry. Blough must have derived a good deal of satisfaction from having made the President back down; he may have concluded that Kennedy was spineless.

Blough soon had another and far more important confrontation with the President arising out of collective bargaining in the steel industry. Dismayed by the seemingly endless 1959 strike, Eisenhower's Secretary of Labor, James P. Mitchell, had assembled an expert team headed by E. Robert Livernash of the Harvard Business School to make a study. *Collective Bargaining in the Basic Steel Industry* was published in January 1961, just as the new administration took office.

It was not an edifying account. During the thirties most of the industry bitterly resisted unionization and compliance with the Wagner Act, culminating in the great Little Steel strike of 1937. Even following recognition many of the companies and the union were unable to reach a first agreement, and the National War Labor Board had to impose one in 1942. In the 10 negotiations between 1946 and 1959 five resulted in strikes: 1946 (26 days), 1949 (45), 1952 (59), 1956 (36), and 1959 (116). No other important domestic industry or foreign steel industry came close in the propensity to strike. Steel stoppages were big because the industry was huge: half a million basic steel workers walked out. And since these companies owned support facilities organized by the Steelworkers, many more men in the iron mines, on the ore carriers, and in steel fabricating mills joined the strike. The captive coal mines and the coal-carrying railroads were affected immediately. "Thus," Livernash wrote, "the strikes are in some degree multi-industry in effect and nationwide in extent."

Steel was the classic oligopoly with administered prices and U.S. Steel was the price leader. The industry's pricing had been the subject of the Kefauver Committee investigation of the late fifties. It had exploited the wage agreements with the Steelworkers as occasions for simultaneous increases in prices, thereby creating the public impression that the union had pushed up the price of steel. In 1946 wages rose 18½ cents an hour and prices increased $5 a ton. In 1947 the numbers were 15 cents and $5; and in 1948, 13 cents and $10. And so on for 1949, 1950, 1953, 1954, and 1955. The 1956 agreement was for three years and the increases in the first year were almost 20 cents and $8.50. The 1959 agreement (because of the strike not signed till January 4, 1960) was again for three years and provided about a 34-cent increase in wage costs, but none of it in the first year. The industry, therefore, had no wage excuse for raising prices in 1960. Since Nixon had mediated the settlement, rumor had it that he insisted on no increase in prices until after the presidential election.

Thus, the target date in steel became June 30, 1962, when the 1959 agreement expired. The Kennedy administration was determined that there must be no rerun of the 1959 disaster. Arthur Goldberg had battered his way through the negotiations of the fifties as the union's principal spokesman and wanted to break the pattern of rotten relations and strikes. Further, he visualized himself as an activist Secretary of Labor who leaped into labor disputes to mediate settlements—the New York City tugboats, the Metropolitan Opera, the flight engineers on the airlines. Why not steel, as well, about which he knew far more than he did about the opera house?

Under the 1959 agreement steel wages and fringe benefits were scheduled to rise on October 1, 1961. Rumor had it that the industry would also increase prices. On August 2 Heller wrote Kennedy that prices generally were steady, but "steel is the major threat to price stability over the next several months. Steel bulks so large in the manufacturing sector of the economy that it can upset the applecart all by itself." Senators Paul Douglas of Illinois and Albert Gore of Tennessee, primed by the administration, urged the President to prevent an increase. On September 7 he addressed identical letters that Kermit Gordon had drafted to the chairmen of the twelve basic steel companies, urging them to absorb the prospective added labor costs. He pointed out that during the postwar era steel prices had risen much faster than wages and that the industry had enjoyed handsome profits. He warned that a rise in steel prices would spread through much of the economy, would significantly increase the cost of military procurement, would endanger the balance of payments, and would hamper economic recovery.

> If the industry were now to forego a price increase, it would enter collective bargaining negotiations next spring with a record of three and a half years of price stability. It would clearly then be the turn of the labor representatives to limit wage

> demands to a level consistent with continued price stability. The moral position of the steel industry next spring—and its claim to the support of public opinion—will be strengthened by the exercise of price restraint now.

On September 14 Kennedy wrote McDonald asking him to act in the public interest. "This implies a labor settlement within the limits of advances in productivity and price stability."

The message was clear. "The plain meaning of Kennedy's letter to the steel companies," Rowen wrote, "was that if they played ball he—the President—would turn the heat on labor. Of McDonald, he demanded statesmanship, and gave him a measure—the guidepost productivity limits—of how far he could go." The steel executives, on the whole, gave no assurances on prices. However, Joseph L. Block, the chairman of Inland, wrote that, while profits were too low, if the public interest demanded a price freeze, it should include a wage freeze until June 30, 1963. Blough disagreed with the President's analysis of profits, insisted that they lagged, and gave no clear response to the President's implied offer.

Goldberg met McDonald and R. Conrad Cooper, U.S. Steel's vice president for industrial relations, in Miami in the first part of December during the AFL-CIO convention. The Secretary of Labor stressed early negotiations leading to a quick noninflationary settlement without a strike. With McDonald privately he urged modest economic demands and warned that, if a strike occurred, the Steelworkers could expect no help from the government. On January 23, 1962, Blough, McDonald, and Goldberg met secretly with Kennedy at the White House. The President took the same line, pushing for early negotiations to gain an agreement that would not justify an increase in prices. Blough said that he was willing to begin bargaining early but would not negotiate over prices, which, of course, no one had asked him to do.

The bargaining opened on February 14 in Pittsburgh. The union's initial proposal, reflecting heavy unemployment in the mills along with government pressure, was quite modest—17 cents per hour, none in wages, all for job security. The talks recessed on March 2. Goldberg then met with Blough, who said that 17 cents was not acceptable because it was inflationary. Goldberg told McDonald that he would have to reduce the demand. The negotiations resumed on March 14 and led to agreement later that month and the signing of a new contract on April 6, 1962, almost three months before the expiration of the old one. It was, Rowen wrote, "an amazingly cheap settlement, so much so that McDonald refused to put an official value on the package." It came to 10 cents an hour, none in wages, everything to improve pensions and to mitigate unemployment. It was, obviously, within the wage guidepost.

The President, Goldberg, and Heller were jubilant. Kennedy congratulated the negotiators for their "high industrial statesmanship" and said the agreement was "obviously noninflationary and should provide a solid base for continued price stability." The business community and the press generally were enthusiastic, stressing that there would be no increase in steel prices. Kennedy, glowing, thought that steel was now safely behind him.

But on Tuesday afternoon, April 10, Blough called the executive committee of the U.S. Steel board into session in New York. The operating executives had proposed a 3½ percent price increase. The committee approved; the public relations department prepared a statement; and Blough phoned Kenny O'Donnell to make an appointment with the President. He got one at 5:45 that afternoon and boarded a corporate plane for Washington. Kennedy was surprised to learn that Blough was coming and could not imagine what he had in mind.

"Perhaps the easiest way I can explain why I am here," Blough said, "is to give you this and let you read it." He handed Kennedy the mimeographed statement which at that moment was being released to the media. The first paragraph, which the President read carefully, was as follows:

> For the first time in nearly four years, United States Steel today announced an increase in the general level of steel prices. This "catch-up" adjustment, effective at 12:01 a.m. tomorrow, will raise the price of the company's steel products by an average of about 3.5 percent—or three-tenths of a cent per pound.

The remainder of the release, which Kennedy raced through, was a muddy justification for the action.

The President was furious. He protested, "I think you have made a terrible mistake." He then asked Evelyn Lincoln, his secretary, to summon Goldberg "immediately!" The Secretary broke the record for cabinet officers sprinting to the White House. Never at a loss for words, he launched into an argument against a price increase as he came through the door. "Wait a minute, Arthur," the President said. "Read the statement. They've raised the price. It's already done."

Goldberg was stunned and enraged. He asked Blough why he had come. He answered that it was a matter of courtesy. Goldberg said it was hardly courteous to present the President with an accomplished fact. He then lectured Blough, calling the action a disservice to the nation, to U.S. Steel, and to the future of collective bargaining, as well as a deliberate double-cross of the President. Blough was somber as he left.

Sorensen, McGeorge Bundy, and acting press secretary Andrew Hatcher, who had been waiting for a scheduled meeting, came into the Oval Office and were soon joined by Robert Kennedy, Heller, Gordon, and, somewhat later, Tobin. The President was seething. One aide said he had never seen him so angry. Another said there should have been a speedometer on his rocking chair. Kennedy then made the famous denunciation which leaked to the *New York Times:* "My father always told me that all businessmen were sons of bitches, but I never believed it until now." (He later said he meant steel executives, not "all businessmen.") Kennedy and Goldberg agreed: "This is war." The President felt that he had been betrayed, that the office of the presidency had been denigrated, and that the national interest had been flouted. This demanded a confrontation: the U.S. government versus U.S. Steel.

Given the pattern of administered pricing in the steel industry, no one at the White House had much hope that the other companies would refrain from raising prices. Nevertheless, Kennedy launched a massive government-wide 72-hour mobilization against U.S. Steel.

The President had a press conference scheduled for Wednesday, and Sorensen, Goldberg, and Heller prepared the statement. Experts from BLS and the Council worked through most of the night gathering the numbers for a "White Paper." The Attorney General ordered the antitrust division to investigate the price increase. After Bethlehem raised prices, the Pentagon awarded a $5 million armor plate contract to Lukens, which had not. Senator Kefauver, "shocked" by U.S Steel's action, said that his subcommittee would make an investigation. McNamara stated that the hike would raise defense costs "in excess of $1 billion a year." A Bethlehem executive had told the press on April 9, "There shouldn't be a price rise." FBI agents summarily interrogated the reporters who had covered the story. Hodges said that he was "shocked and disappointed" and charged that Blough's act was "anti-business." Solicitor General Archibald Cox, lecturing at the University of Arizona, was put to work drafting legislation.

Kennedy's statement on Wednesday was, a U.S. Steel executive granted, "a barnburner." In bitter fury he denounced the "irresponsible defiance of the public interest." In the inaugural address "I asked each American to consider what he would do for his country, and I asked the steel companies. In the last 24 hours, we have had their answer."

Anyone in the administration who knew an executive of a steel company was instructed to phone him and ask that his firm not follow U.S. Steel's lead. The phones hummed. But the early returns were discouraging. Bethlehem, Republic, Jones & Laughlin, Youngstown, and several smaller firms fell into line by Wednesday night. Five of the lesser companies, together 14 percent of the industry's capacity had not yet acted—Inland, Armco, Kaiser, Colorado Fuel & Iron, and McLouth.

Joseph Block of Inland held the key. The Chicago firm was probably the most efficient and profitable in the industry and would not have considered a price increase on its own. Block was friendly to the administration, close to Goldberg, and hardly an admirer of Blough. Vacationing in Japan, on Friday he issued this statement from Kyoto: "We did not feel that it was in the national interest to raise prices at this time. We felt this very strongly." Kaiser and Armco quickly followed Inland.

Kennedy had watched Blough on television the day before. If another major producer did not raise prices, he said, it would be "very difficult" for U.S. Steel. "I don't know how long we could maintain our position."

Kennedy sensed an opening and made his move, asking Clark Clifford, the noted Washington lawyer, to represent him. He met with Robert Tyson, vice president of the steel corporation, on one of its airplanes at Washington's National Airport that night. Clifford thought the conversation desultory, but Blough sent word the next morning that the meetings should continue.

Meantime, there were other and encouraging signs for the administration. In the struggle for public approval the President was winning. Few people, including businessmen, welcomed an increase in steel prices. Many of those who thought that the steel corporation had the right to fix its own prices condemned Blough for the way in which he did it. Blough did not help himself by stressing the technically correct point that U.S. Steel had never promised not to raise prices. The public recognized that the corporation had led everyone to believe that it would not.

More important, number two Bethlehem was shaky. While its intentions were a mystery, there was an obvious inconsistency between the preliminary statement and the price action. Moreover, Bethlehem, an eastern and far western marketer, was eager to enter the Midwest. Here Inland and Armco were formidable competitors, now with lower prices.

On Friday, Clifford and Goldberg flew to New York on a military plane where they met Blough, Tyson, and other U.S. Steel executives in a large suite at the Carlyle Hotel. Two phones rang almost simultaneously, one for Goldberg, the other for Blough. It was the same message: Bethlehem had caved in. Blough appeared "shaken." Goldberg poured it on—the White House would release the "White Paper," Dillon would attack the "greed" of the steel companies, and so on. At 5:00 p.m. Blough surrendered. At 5:28 the tickers reported: "The United States Steel Corporation today announced that it had rescinded the 3½ percent price increase made on Wednesday, April 11." The companies that had followed Big Steel up now followed it down.

Kennedy did not gloat over his victory. Aside from damage control, there were good reasons not to do so. As Grant McConnell pointed out, this was more a triumph for the presidency than for this particular President.

Further, everyone knew that U.S. Steel would have gotten higher prices if it had gone about it quietly and selectively, as it did the next year. Kennedy met with Blough on April 17 and assured him that he held no grudges. That evening he informed Schlesinger, "I told him that his men could keep their horses for the spring plowing." At the next press conference he said, "Nothing is to be gained from further public recriminations." The White Paper remained in its file. The antitrust division and Senator Kefauver folded their tents. But this was the President's public face. Privately he was deeply affected by the confrontation, as he was by another business event which took place the next month.

Since December 1961 stock prices had been slipping. Mutual funds unloaded shares and the price of a seat on the New York Stock Exchange dropped markedly. The pace of decline accelerated sharply in the week of May 21, 1962: the Dow-Jones industrial average dropped a whopping 38.83 points. May 28, "Black Monday," was a disaster, exceeded only by the great crash of October 28, 1929. The average plummetted 34.95 points. Tuesday was wild. Prices fell sharply in the morning and recovered dramatically in the afternoon. Volume was the second highest in history. Wednesday the exchange was closed for Memorial Day. Thursday was another very busy day which ended on the plus side. Over this week the Dow-Jones declined from 611.78 at Friday's close to 563.24 at midday on Tuesday and then back up to 613.36 at the final bell on Thursday. These gyrations generated selling waves in London, Paris, Amsterdam, and other world markets and many feared another Great Depression. Prices on the New York exchange sagged ominously throughout June.

Why did the market go into this tailspin? Many businessmen had a ready answer: Kennedy. His anti-business attitude, they said, particularly the confrontation with Blough, had caused the collapse. Thousands of letters poured into the White House making this accusation about "the Kennedy market." Businessmen wore buttons which read, "S.O.B. Club." Bumper stickers stated: "Help Kennedy Stamp Out Free Enterprise" and "I Miss Ike—Hell, I Even Miss Harry."

Dillon was not much help. He said that stocks fell because they were too high. The New York Stock Exchange study was wholly descriptive, avoiding any discussion of causes. The special study group making a massive analysis of the markets and the regulatory system added a chapter on the 1962 market break. "Neither this study nor that of the New York Stock Exchange was able to isolate and identify the 'causes' of the market events of May 28, 29, and 31."

Keynes could have saved them the trouble. He had argued that stock traders based their decisions on ignorance and mass psychology. How could one apply rational analysis to irrational behavior? Heller and Galbraith agreed. The former pointed out that "the market is fallible as an economic indicator, especially in giving out false downturn signals." It was a better forecaster of upturns. But he added, "I'd feel better if the market were going up." The question whether "stock movements foretell economic movements," Galbraith wrote, "is a footless one." But he sent Kennedy his book on the 1929 crash which, "by a kind of foresight that can only inspire confidence, has just become available in a new and inexpensive edition."

The confrontation with Blough and the stock market crash transformed Kennedy's outlook on economic policy. While henceforth he was correct in his relations with business, it was clear to him that it was a no-win situation. If he offered favors, businessmen gave him no credit and demanded more; if he was tough, they called him anti-business. "It's hard as hell," he said, "to be friendly with people who keep trying to cut your legs off." According to Heller, Kennedy had never believed in the balanced budget; he had struck this pose as a political gesture to appease conservative businessmen, bankers, and politicians. He no longer needed to do so and was now free to take a new tack.[5]

On a Sunday in May 1962 the President took André Malraux, the noted French writer, to Glen Ora, his place in the Virginia countryside, for lunch. They talked about a topic that Kennedy had been mulling over as a result of his recent business encounters, the persistence of mythology in contemporary affairs. Malraux said that in Europe in the nineteenth century the ostensible issue was monarchy vs. the republic, while the real conflict was capitalism vs. the proletariat. Now the ostensible question was capitalism vs. the proletariat. "What," he asked, "is the real issue now?" Kennedy said that it was the management of an industrial society, a problem not of ideology but of administration. J. K. Galbraith had almost finished the draft of a book on this theme, which he deposited in a bank vault when Kennedy packed him off to India, and which later became *The New Industrial State*.

The conflict between myth and reality, sharpened by the conversation with Malraux, was now much on Kennedy's mind. Adopting the format of the popular TV show "20 Questions," he asked Heller to prepare a list of twenty economic myths, showing why each was fallacious. The Council soon wrote the memorandum and sent copies to the President and Sorensen. They concluded that it was the basis for a major speech and Kennedy decided that it should be the topic of his commencement address at Yale on Sunday, June 11, 1962. In New Haven, as Heller put it, "he would have a rather literate audience in financial and economic matters."

Kennedy took this speech very seriously, indeed, and it went through several drafts. Schlesinger and Galbraith, who was visiting Washington for a few days, produced one; Sorensen, assisted by Heller, wrote another; and, finally, McGeorge Bundy and Schlesinger did a third. Kennedy was still not satisfied and worked on the speech intensively at the White House and on the plane to New Haven. Heller thought Kennedy personally wrote a quarter to a third of it. The result was a sharply honed address.

Yale was bestowing an honorary degree upon the President. "I have the best of both worlds," he said, "a Harvard education and a Yale degree." But, he noted, there seemed to be "a certain natural pugnacity developed in this city among Yale men." Perhaps it would help to become an Eli because "as I think about my troubles, I find that a lot of them have come from other Yale men. Among businessmen, I have had a minor disagreement with Roger Blough, of the law school class of 1931." He offered "to smoke the clay pipe of peace with all my brother Elis, and I hope they may be friends not only with me but even with each other."

The world had changed, the President observed. It was no longer ruled by "sweeping issues." Contemporary policy questions were subtle, complex, and obstinate.

> As every past generation has had to disenthrall itself from an inheritance of truisms and stereotypes, so in our own time we must move on from the reassuring repetition of stale phrases to a new, difficult, but essential confrontation with reality.
>
> For the great enemy of the truth is very often not the lie—deliberate, contrived, and dishonest—but the myth—persistent, persuasive, and unrealistic. Too often we hold fast to the clichés of our forebears. We subject all facts to a prefabricated set of interpretations. We enjoy the comfort of opinion without the discomfort of thought.
>
> Mythology distracts us everywhere—in government as in business, in politics as in econonics, in foreign affairs as in domestic affairs. But today I want to particularly consider the myth and reality in our national economy.

Kennedy then addressed three myths, the first that "government is big and bad—and steadily getting bigger and worse." While it was true that the number of actual dollars spent by the federal government had risen for a generation, this was basically a function of the growth of the nation. When one compared the rate of growth of the federal sector since the war relative to other sectors, it had expanded less than industry, commerce, agriculture, higher education, "and very much less than the noise about big government."

The second myth related to fiscal policy. Here the government was corseted by the administrative budget, which no business or European government would tolerate. It omitted the trust funds (Social Security), neglected changes in assets and inventories, did not distinguish between expenditures and loans, and, most significant, failed to treat operating expenditures and long-term investments separately. "This budget, in relation to the great problems of Federal fiscal policy which are basic to our economy in 1962, is not simply irrelevant; it can be actively misleading."

Another fiscal myth was that the federal debt was growing at "a dangerously rapid rate." In fact, since the war it had increased only 8 percent, compared with 305 percent for private debt and 378 percent for state and local governments. Moreover, debts—private or public—"are neither good nor bad, in and of themselves."

The final myth Kennedy addressed was confidence. It was absurd to blame "unfavorable turns of the speculative wheel" on lack of confidence in the government. "Corporate plans are not based on a political confidence in party leaders but on an economic confidence in the Nation's ability to invest and produce and consume."

The reaction to the Yale speech was mixed. Kennedy reported to Bundy and Heller, as the latter wrote, that it "had received a rather chilly reception from the Commencement audience . . . , and he muttered something about their well-heeled status and conservative nature." There may be a simpler explanation. Carlyle was right when he called economics the dismal science and speeches on economic topics were and remain notoriously soporific. For the Keynesians, however, it was a triumph and a call to arms. They had finally convinced the President of the soundness of their analysis. Heller wrote that the address was "a truly remarkable document—undoubtedly, the first complete speech on economic policy—and modern economic policy at that—that a President had ever made."[6]

At a press conference on June 7 1962, the President stated:

> Our tax structure as presently weighted exerts too heavy a drain on a prospering economy, compared, for example, to the net drain in competing Common Market nations. If the United States were now working at full employment and full capacity, this would produce a budget surplus at present tax rates of about $8 billion this year. It indicates what a heavy tax structure we have, and it also indicates the effects that this heavy tax structure has on an economy moving out of a recession period. . . .
>
> A comprehensive tax reform bill . . . will be offered for action by the next Congress, making effective as of January 1 of next year an across-the-board reduction in personal and corporate income tax rates which will not be wholly offset by other reforms [Heller persuaded Sorensen to insert the following words]—in other words, a net tax reduction.

What did this cryptic statement mean? Two conclusions seemed clear. The first paragraph demonstrated that Kennedy now accepted the performance gap analysis. In the second he promised to introduce a bill in the next

Congress calling for a general cut in personal and corporate income taxes on January 1, 1963, that would not be revenue neutral, but would be, in Heller's words, "a net tax reduction."

But the statement raised more questions than it answered. Would the Council or the Treasury, who had sharp differences, be primarily responsible for working out the tax reduction? How big would the cut be? Who would be the main beneficiaries—the rich or the poor? Since the Treasury was working on tax reform, would the bill provide only for reduction or be a reduction-reform package? When would the bill be introduced? How could Congress be expected to approve such a proposal when both Wilbur Mills, chairman of the House Ways and Means Committee, and Harry Byrd, chairman of the Senate Finance Committee, opposed tax reduction, the latter violently?

The President's statement and the Yale speech pitted the Council and the Treasury against each other in a tenacious contest. While it would be fought under Marquis of Queensberry rules, the antagonists were in earnest. One would have thought that the Treasury, by far the more powerful, would have won handily. In fact, it did not. A major reason was that the Council knew exactly what it wanted and the Treasury did not.

The performance gap was the core of the Council's analysis—the difference between actual GNP and the hypothetical product at "full" employment, that is, 4 percent unemployment. In early 1961 the Troika (the Council, the Treasury, and the Bureau of the Budget) had worked out a sophisticated econometric model which produced these numbers. The difference was $40 billion in GNP. Further, if the country had had a 4 percent jobless rate in 1961, there would have been a $10 billion budget surplus instead of a deficit.

Performance gap budgeting yielded another important dividend: it was now simple to build in growth. A static model was useful only for the year it described. But the economic system was dynamic. Aside from cyclical swings, its basic state was long-run growth, a moving target. The labor force expanded continuously and would do so dramatically in the sixties as the baby boomers entered the labor market; advancing technology and other factors constantly pushed up productivity, thereby displacing labor. Insofar as the unemployment rate was concerned, therefore, the system had to run in order to stand still.

The problem for fiscal policy, the Council reasoned, was to close the performance gap and keep it shut as growth advanced. The income tax system imposed a restraint. As Heller put it, the "automatic stabilizing effect is a mixed blessing." It was good in cushioning recessions, but it could curtail a recovery before the unemployment goal was reached "by cutting into the growth of private income, which is bad." This would impose "a fiscal drag" on the economy as federal revenues rose. The remedy, "the fiscal dividend," was either to reduce taxes or to increase federal expenditures.

In Washington, whether at the Council, in the Treasury, at the White House, or on Capitol Hill, this was an academic choice. Americans, like everyone else, hated to pay taxes and the political appeal of tax reduction was irresistible. Insofar as this was a problem, therefore, it was confined to the American embassy in New Delhi.

In *The Affluent Society,* published in 1958, Galbraith had argued that a modern economy must maintain a social balance between its public and private sectors. In the U.S., "public poverty competed with . . . ever-increasing opulence in privately produced goods." The inevitable tendency was for "public services to lag behind private production." The answer, Galbraith argued, was to increase public investment.

Thus, for Galbraith, a reduction in taxes on individuals and corporations would aggravate social imbalance because the savings would be spent overwhelmingly in private markets. He put it this way:

> I am not sure what the advantage is in having a few more dollars to spend if the air is too dirty to breathe, the water too polluted to drink, the commuters are losing out on the struggle to get in and out of the cities, the streets are filthy, and the schools so bad that the young, perhaps wisely, stay away, and hoodlums roll citizens for some of the dollars they saved in taxes.

In letters from India, Galbraith "weighed in heavily" for social spending rather than a tax cut.

According to Schlesinger, Kennedy agreed with Galbraith and, "political conditions permitting, would have preferred the policy which would enable him to meet the nation's public needs." The conditional clause was decisive: political conditions did not permit. Nevertheless, the President insisted on a Council analysis of "the Galbraithian alternative" and Heller did his best.

He made some large concessions. "True, our cities need renewal, our colleges and universities have no place for the flood of students about to inundate them, our mass transportation system is in a sad state, our mental health facilities a disgrace, our parks and playgrounds inadequate, housing for many groups unsatisfactory." Further, a $9 billion increase in expenditures would have the same effect as a $10 billion tax cut.

But, Heller argued, it was impossible to spend $9 billion quickly and, if the government pushed speed, this "would lead to waste, bottlenecks, profiteering, and scandal." Further, to try to boost expenditures, if not impossible, would be politically damaging. Moreover, the world financial community would find a tax-induced deficit more acceptable and "far less likely to set off new gold outflows." Finally, Heller offered hope for the Galbraithian alternative. The tax cut would stimulate growth and "an atmosphere of prosperity and flushness in which government programs can vie much more successfully for their fair share of a bigger pie."

In the summer of 1962 Dillon put pressure on Kennedy to avoid an immediate tax cut and to wait a year. On June 7 he sent a memorandum by Seymour Harris to the White House that reviewed papers prepared by CEA, Samuelson, and Abramovitz, which urged "quick action." While Harris agreed that the economic impact would be desirable, he was worried about the politics. Dillon called Kennedy's attention to the fact that Harris "thoroughly agrees that there should be tax reductions next year." He played on Kennedy's most responsive chord on July 12. The European central bankers had just met in Basel and were nervous because of "lack of confidence in the future of the dollar and of the willingness of the Administration to take the fiscal and monetary measures needed to protect it." On July 27 the secretary informed the President that the New York bankers and the Treasury's advisory committees of bankers and investment bankers unanimously opposed a tax cut now, with the exception of one investment banker.

Thus, during the summer and fall of 1962 the President dragged his feet on taxes. He delayed in order to look at next month's economic indicators. Or, as Dillon urged, he should wait until the troubled investment incentive tax bill cleared the Senate. But it would not pass until October.

Kennedy set aside July 13 for the tax problem. In a briefing memorandum on the preceding day, Sorensen had painfully laid out the administration's ambivalence. By now Sorensen had come around philosophically to accept the Council's reasoning for a tax cut, but "I do not want us to over-react." He would favor a bill, but only if it met a laundry list of conditions. Among them: an economic downturn, assurance that tax reductions would go into consumption rather than savings, a budget deficit less than Eisenhower's $12.4 billion in fiscal 1959, the likelihood that Congress and business would accept it.

On the morning of July 13 the President met with the advocates of tax reduction—Heller, Gordon, Tobin, Gardner Ackley (who was replacing Tobin), Goldberg, Hodges and his under secretary, Edward Gudeman, Bell of the Bureau of the Budget, and Samuelson and Solow. The last two wrote the music. In early June, gloomy over the economic outlook, they had informed Kennedy, "Only an early tax cut appears to be capable of giving the economy the stimulus it needs in time to be really effective." Now, five weeks later, they considered the situation "even more unfavorable than we had expected." Remedy: "A tax cut this year can do much to avert the developing recession." But only one medication at a time: "Divorce all cuts from *any* reform."

On July 17, Schlesinger added a political argument for immediate tax reduction. "The great historic reason for the popular appeal of the Democratic party over the last generation has been that the American people regard it as the party which can be relied on to take action against recession." Even if the proposal lost, this would create "a strong issue" in the 1962 congressional elections.

On July 13 the President had lunch with a group of business leaders who favored a tax cut but disagreed over its timing. In the afternoon he conferred with Dillon and Mills, both of whom argued strongly against any immediate action. Dillon urged delay because the Treasury reform package was not ready. Mills informed Kennedy that the prospects in Congress were bleak and "I did not think I could support it." He said that the administration must educate the public. Without Mills, Kennedy realized, an immediate tax cut was out of the question. Mills, John F. Manley wrote, was not just "at the center" of the relationship between the executive branch and the Ways and Means Committee, "he is the center." "Wilbur Mills," Kennedy said, "knows that he was chairman of Ways and Means before I got here and that he'll still be chairman after I've gone—and he knows I know it."

On August 8, Kennedy had a long telephone conversation with Mills. The President was worried because the economists were forecasting a recession either in late 1962 or the first half of 1963. If it came early, the Democrats could be whipped in the November elections. If he waited to ask for a tax cut, "by the time you get a rule out of Judge Smith snow would be on the ground" and "Harry Byrd would screw us even if you put it through the House."

Mills replied that the economy was "sluggish," that a recession was coming, but he could not predict when it would occur or how severe it would be. However, neither the country nor the Congress was concerned. Thus, it was politically impossible to cut taxes now. The situation would change "if we're starting downhill, and everybody could see we're starting downhill." "Politically we can't *let* it go down." The present situation did not justify an enlargement of the deficit. But "when I become convinced that I've got a choice to make between tax reduction and increased deficit and a recession, I'm going to take the tax reduction line." Further, he had checked with Mortimer Caplin and learned that Internal Revenue could change the tax tables quickly to withhold less from taxpayers, facilitating a quick temporary tax cut.

Mills was telling the President to hold off, and given his power as chairman of Ways and Means, was extremely persuasive. As Kennedy summed up, "then we sort of put it to bed until we're ready to come out and do it." He would not send up a tax bill in 1962.

Heller, flanked by Gordon and Ackley, faced a daunting challenge when he testified before the Joint Economic Committee on August 7. He painted a somber picture of the outlook, explained it with the performance gap, and then turned lamely to the remedy. While the need for tax reduction was manifest and the President had promised one in the next year, he said, "no decision has been made on the size, composition, and timing of a recommended tax reduction."

Dillon did not go before the committee until August 15, after the President's address on taxation, which will be noted shortly. He made a strong pitch for "a comprehensive program of tax reform," neatly folding into it "a general reduction of both individual and corporate rates, effective January 1, 1963." So much for Keynesian tax cutting. Senator Jacob K. Javits, the liberal New York Republican, pressed Dillon to no avail. "We have not got a program. . . . We are not ready and have no program that is in form to be submitted, and I do not expect there will be one until Congress comes back in January."

By early August the media were full of discussion and opinion about a tax cut, though there was not much news. Prominent senators came out on both sides of the issue. The AFL-CIO and U.S. Chamber of Commerce urged a reduction, though they would give it to different income groups. The President was pressed to respond and did so on August 13 in a television address written by Sorensen and Heller. Like the latter's testimony before the Joint Economic Committee, it was anti-climactic.

Kennedy repeated the performance gap–fiscal drag analysis. The need for a tax cut, he said, was clear and he would ask for one. But not now. A bill would be presented "for action next year." Nossiter called August 13 "The Day That Taxes Weren't Cut" and a triumph for Dillon.

From India, Galbraith tried to cheer Kennedy up. After all, he had proved himself "the most Keynesian head of state in history." He should hang a portrait of the Master "in your bathroom or some other suitably secluded place." Federal taxes were not a burden. "The only people who do really feel strongly about Federal taxes are the Republican rich who are in the high surtax rates and those who would be were it not for the loopholes. The first will fight for reductions. The second will fight for their loopholes."

The administration was now stuck and Heller intervened to get it moving. There was another consideration: the Council needed the Treasury. As the author of the Council's administrative history would later write, "The Council's areas of interest overlap greatly with those of the Treasury. In many ways, the whole organization of economic policy-making puts the Treasury and the Council in a balancing position with respect to each other. The Treasury has the largest operating responsibilities in the area of economic policy; the Council has none." The Council could do the big thinking; only the Treasury was competent to perform the drudgery of working out the details of an overhaul of the tax structure. Stanley Surrey had the troops. Harvey E. Brazer, on leave from the economics department of the University of Michigan as director of the Office of Tax Analysis, headed a group of twenty-five economists and statisticians, and Donald C. Lubick, the tax legislative counsel, had fifteen tax lawyers at his disposal. They had been tied up by the investment incentive bill until October, but became free thereafter.

Heller persuaded Kennedy to establish on August 21 the Cabinet Committee on Economic Growth with himself as chairman, along with Secretaries Dillon, Wirtz, and Hodges, and Budget Director Bell. As the work progressed, Secretary Anthony J. Celebrezze of HEW and Jerome B. Wiesner, the Director of the Office of Science and Technology, joined the committee. The President asked them to submit "a general report which will enable growth policy to receive proper consideration in the 1963 budget and legislative program." While the report addressed civilian technology, the educational system, and manpower policy as means of stimulating growth, tax reduction was the centerpiece.

At the first meeting Brazer was asked to prepare a paper for discussion on taxes and growth. He presented it on October 26 at a meeting attended, among others, by high Treasury officials—Dillon, Fowler, and Dewey Daane, Deputy Under Secretary for Monetary Affairs. With reasoning similar to the Council's, Brazer concluded that a net reduction of $10 billion would be consistent with both economic growth and price stability. He felt that his paper and the ensuing discussion had "an impact on Dillon's thinking, but the change of mind was gradual, not sudden."

The report to the President on December 1, 1962, was unanimous, signed by Dillon, Gudeman, Wirtz, Celebrezze, Bell, Wiesner, and Heller. The tax recommendations were as follows:

1. A $7 to $12 billion net reduction in taxes, as promptly as possible.
2. A major reduction in all individual income tax bracket rates, with the greatest share of revenue reduction concentrated in the lowest tax brackets, a reduction in top individual income tax brackets, and some reduction in corporate income tax rates.
3. Revision of the income tax base—e.g., through modification of depletion allowances and other provisions governing taxation of natural resource industries—to permit more efficient allocation of resources and rate reductions larger than would otherwise be possible.

These recommendations, as the vague language of number three demonstrates, barely grazed tax reform—for good reason. Even as late as December 1, 1962, the Treasury still had not worked out the reform package. The Council, according to Tobin, knew what Surrey was working on but had no idea whether Dillon approved. Amazingly, the President did not see the proposals until the decisive meeting on December 26, 1962.

Since Dillon had promised reform almost two years earlier and Surrey had a large team of experts at his command, their failure to come up with a plan seemed implausible and some concluded that it was a stall. To the contrary, Dillon was committed to reforms and his people had simply not worked them out. In part this was because they presented formidable drafting problems. In addition, the top Treasury people recognized that reforms would stir the beneficiaries of loopholes and raise potentially insurmountable political barriers. They insisted on gaining advance approval from Wilbur Mills.

Fowler was the Treasury's point man in dealing with Mills, spending three and one-half hours with him in Arkansas on November 14. The chairman was now cautiously optimistic about the economy and, therefore, still opposed "quick or temporary tax reduction." But, Fowler wrote, he was convinced that "the present tax structure is a drag . . . to the economic growth and higher levels of employment of which the economy is capable." Thus, he favored "an orderly, permanent, meaningful revision downward of the tax rate structure, individual and corporate." But this must be accompanied by "a maximum effort to hold down non-defense expenditures." Mills was prepared to make such a tax bill "the first order of business" before the Ways and Means Committee. He envisioned a two-step tax rate revision taking effect on January 1 of both 1964 and 1965. He was thinking of a reduction of individual rates from the present 20 to 91 percent range to 15 or 16 to 65 percent.

While "Mr. Taxes," as Mills was often called, fancied himself as a great tax reformer and, guided by Surrey, had published a 2,382-page compendium of reform possibilities, he was, Rowen wrote, the man who "wants to keep everybody happy, and get legislation passed. . . . He is a politician who makes a shrewd assessment of what is possible." It was hard to imagine any legislative proposal less possible than tax reform.

The Internal Revenue Code was riddled with loopholes (Senator Douglas called those in oil "truckholes"). "In taxes," Philip M. Stern wrote, "'equity' has been defined as 'the privilege of paying as little as someone else.' Thus, once a loophole has been granted one group, it becomes 'inequitable' to deny it to other groups. Result: Loopholes are not only virtually imperishable; they enlarge with age." The formidable political problem was to confront and defeat the powerful and rich lobbies that stood guard over their loopholes. Wilbur Mills did not find that kind of guerilla warfare appealing.

For this reason many thought that linking reduction to reform was to invite disaster. This was the view at the Council, in the labor movement, and among liberal businessmen. On November 29, 1962, Philip L. Graham, the publisher of the *Washington Post,* wrote Dillon, "There will *not* be a tax cut early this session unless you reshape your policies." He told the secretary to stop insisting on "those petty, unneeded, unachievable 'reforms.'" But Dillon hung on, hoping that reduction would lubricate the way for reform.

Kennedy made the basic decisions at his family's residence in Palm Beach on the day after Christmas. The President brought along Sorensen, Larry O'Brien, and Myer Feldman of the White House staff; Dillon, Fowler, and Surrey were present for the Treasury; Heller represented the Council; and Gordon, who had replaced Bell, spoke for the Budget Bureau. The meeting was a victory for Dillon, who had now come to accept a tax cut. Kennedy approved his bundle of reforms, and, perhaps more important, Dillon succeeded in pressing tax reduction into the framework of the budget. Personal income tax rates would fall to the 14 to 65 percent range. The corporate tax for incomes exceeding $25,000 would drop from 52 to 47 percent. These changes would reduce revenues by $13.5 billion, $11 billion from individuals and $2.5 billion from corporations. The reforms would yield a $3.5 billion offset in revenues, but they would not come in till 1964. The $13.5 billion loss in 1963 in the administrative budget exceeded Eisenhower's $12.4 billion and was politically unacceptable. Thus, the timing of tax reduction was shoved back in two ways. Instead of starting on January 1, 1963, it would take effect on July 1, 1963. Instead of taking place in one year, it would be staged over three. "Heller," Rowen wrote, "was stunned and disappointed, but he did the only thing he could short of resigning; he defended the program with all possible vigor."

In January 1963 the tax program was the star of the State of the Union message, the budget message, and the *Economic Report.* The President fleshed out the details in a special message to Congress on January 24. The budget figures changed slightly. The overall loss of income was now estimated at $13.6 billion, $11 billion from individuals and $2.6 billion from corporations. Reforms would produce a revenue gain of $3.4 billion. Thus, the net reduction came to $10.2 billion. The drop in individual rates would be from 20 to 91 percent in 1962 to 18.5 to 84.5 in 1963, to 15.5 to 71.5 in 1964, and to 14 to 65 in 1965. Withholding rates on wages and salaries would fall from 18 to 15.5 percent on July 1, 1963, and to 13.5 percent on July 1, 1964. The corporate rate would go to 22 percent in 1963 on the first $25,000 of income. The corporate tax above that amount would be 52 percent in 1963, 50 in 1964, and 47 in 1965.

The reforms constituted the most systematic effort to restore equity since the inception of the income tax in 1916. They were numerous and many were arcane, hardly deserving of enumeration here. They fell into three broad categories: changes to relieve hardship and encourage economic growth; those to broaden the tax base and extend equity; and revisions of the treatment of capital gains. They warmed the hearts of those who abhorred loopholes. Senator Douglas embraced them enthusiastically. Even some who thought that reduction and reform should be separated, like Pechman, respected them. "The Kennedy reform proposals," he wrote, "represent a serious attempt to reverse the erosion of the income tax base—a process that has been taking place quietly but relentlessly for many years at the expense of equity and a substantial cost in revenue." These public virtues, of course, guaranteed that lobbyists who protected loopholes would be out in full force, led by the biggest of them all, the oil industry. While the Treasury did not dare to confront the depletion allowance directly (Senator Robert S. Kerr of Oklahoma, himself an oil magnate and the dominant figure on the Finance Committee, blocked the way), it did make several proposals that would have indirectly reduced it from 27½ to about 22 percent.

The publication of the reform program was, as Heller put it, "a bombshell." Kennedy was "just amazed when he saw the barrage of criticism." Heller could not resist telling him that the Council had favored a tax cut unencumbered by reform. The tax package was referred to the House Committee on Ways and Means and the Senate Finance Committee.[7]

While Kennedy urged swift action, the Ways and Means Committee moved with glacial speed. There were three reasons. First, the custom was that the President submitted recommendations, not a draft bill, so that the committee's staff wrote the measure. Because of the reforms this was no light task. The bill came to 304 pages and took five months to draft. Second, the President's main justification for quick action was to head off a recession. But, despite the gloomy prognostications of Samuelson and Solow, the economy improved steadily. GNP, which had bottomed out in the first quarter of 1961, moved into sustained growth during the remainder of that year and throughout 1962 and 1963. Unemployment, which had peaked in May 1961 at 7 percent, fell for a year and then stabilized around 5.7 percent in the latter part of 1962 and throughout 1963. Finally, Wilbur Mills's style did not allow for speed. "Given the complexity of tax law . . . ," Manley wrote, "he makes sure that the Committee is painstakingly thorough in the mark-up stage of the legislative process, that it studies the alternatives before reaching conclusions, and that it proceeds cautiously. . . ." He had been taught by the Master, Mr. Sam Rayburn, "that our whole system was to settle disputes within the committees." He did not move until he was as certain as an extremely well-informed man could be that the bill would clear both Ways and Means and the House.

The hearings on H.R. 8363, the Revenue Act of 1963, ran for 26 days between February 6 and March 27, 1963, and there were 267 witnesses. While Hodges, Wirtz, and Gordon testified briefly, Dillon carried the main load. His prepared statement ran to 571 mimeographed pages and he was questioned on virtually every feature of the bill. "The lobbyists," Senator Douglas wrote, "are as thick as flies." Practically all the interest groups—big, medium-sized, and small—marched to the witness stand, usually to denounce one or two of the reforms and, rarely, to urge rejection of the whole package. The Treasury countered on April 25 when a group of businessmen, pushed by Fowler, launched the Business Committee for Tax Reduction in 1963. It claimed 2800 members and included such luminaries as Henry Ford II, Stuart T. Saunders of the Pennsylvania Railroad, Blough, Frederick R. Kappel of AT&T, Frazar D. Wilde of Connecticut General Life, David Rockefeller of Chase Manhattan, Frederic C. Donner of GM, and Mark W. Cresap, Jr., of Westinghouse. These businessmen favored the cut and opposed the reforms. Some businessmen played games. Martha Griffiths, the liberal Democratic congresswoman from Michigan, informed the President that they were going to use their tax savings to whip the Democrats in the 1964 elections. Kennedy said, "I expect that." The Treasury also helped establish the Citizens Committee for Tax Reduction and Revision, which supported the reforms as well. But it had fewer than 100 members from labor, academic life, and small business.

The reforms were in big trouble from the outset. Though Mills protested that he supported them, he knew that he could not get a firm majority of either his committee or the House without massive surgery. The knives came out on May 20, when Ways and Means started prolonged executive sessions. Heller informed the President on June 7 that the reform proposals were "in rags and tatters." In capital gains "they're opening more loopholes than they're closing." Real estate speculators were getting a "tap on the wrist instead of a kick in the shins." And so on. In early August Mills informed Dillon that all the important reforms must be removed, and the secretary quietly accepted the inevitable. This affected the tax reduction rates because the reforms had been expected to produce an increase in revenue to offset some of the loss due to cuts. On August 12, Dillon proposed higher rates which fell slightly more heavily on low-income taxpayers, which the committee accepted.

The Republican strategy was mystifyingly ambivalent: to both approve and denounce the tax cut. "[We] favor . . . a reduction in both individual and corporate tax rates. However, we believe that a tax cut of more than $11 billion, with no hope of a balanced budget for the foreseeable future, is both morally and fiscally wrong." On two preliminary votes the Republican lines held reasonably well, but on the final vote, ordering reporting of the bill and recommending passage, Howard H. Baker of Tennessee and Victor A. Knox of Michigan joined the Democrats. While Mills preferred unanimity, he was satisfied with a vote of 17 to 8; he got all the southern Democratic votes. He was, however, concerned about the Republican charge of budget busting and persuaded Kennedy to write a letter on August 19 promising "an even tighter rein on Federal expenditures," which appeared as an appendix to the report.

The House passed H.R. 8363 on September 25, 1963, by a vote of 271 to 155, pretty much along party lines. The southern Democrats followed Wilbur Mills like sheep and, despite the fear, made no effort to make tax reduction a hostage to the civil rights bill. The tax rates on individuals would fall from 20 to 91 percent to 16 to 77 on January 1, 1964, and to 14 to 70 percent on January 1, 1965. The withholding rate would drop from the existing 18 percent to 15 in 1964 and 14 in 1965. The corporate rate on January 1, 1964, on the first $25,000 of income would decline from 30 to 22 percent and the surtax over that amount would go from 22 to 28 percent. On January 1, 1965, the surtax would fall back to 26 percent. Thus, the rate on large corporations would be cut from 52 percent to 50 in 1964 and 48 in 1965.

The Senate Finance Committee opened its hearings on October 15, 1963, and Fowler expected them to conclude on November 27. But on November 22, President Kennedy was assassinated.[8]

5

Wrestling with Structural Unemployment

DURING THE 1950s American industry introduced automatic machine controls (machines that instructed other machines) in order to increase efficiency, cut costs, and displace labor. This came to be called "automation." While some tried to preserve this fairly precise engineering definition, the word quickly became a loose synonym for structural change, worker displacement, and sometimes much more. "There are few important economic issues in our nation," William Glazier wrote, "which are not in some way attributed to automation." The comic Bob Newhart joked that there was a new smart machine that fired the other machines. Coal miners, textile workers, steel workers, longshoremen, and meat packers who lost their jobs blamed "automation." By this they meant what labor market economists were now calling structural unemployment.

This was the most complex type of joblessness, and had many causes, among them: technological change which displaced labor without providing alternative employment; exhaustion of a natural resource or a reduction in its demand; long-term decline of an industry; erosion of the need for certain skills; and employment consequences of discrimination against particular classes of workers—blacks, women, teenagers, the elderly. These results stemmed from the "structure" of the economy and the work force, hence the name. Structural unemployment tended to be especially prolonged and intractable. While the Keynesians insisted that most of the structurally unemployed would find work as the consequence of effective fiscal and monetary policies, they conceded that a hard core would remain jobless.

Robert Aaron Gordon of the University of California, Berkeley, linked structural unemployment to the concept of mobility:

The notion of structural unemployment clearly implies that two essential conditions prevail in one or more sectors of the national labor market.

First, *there must be some degree of labor immobility along one or more dimensions of the labor force.* Thus, even when there is no deficiency of aggregate demand, there will be particular sectors of the labor force from which workers cannot easily and quickly move to other sectors in search of jobs. The reasons for such immobility may be many—lack of education or training, attachment to a community or region, lack of information as to where jobs are available, restrictions on entry into an occupation, restrictive hiring practices including discrimination on the basis of race, sex or religion, and so on.

Second, *in some or all of these sectors with impaired mobility, unemployment significantly exceeds available vacancies even when there is no deficiency of aggregate demand.* Supply exceeds demand at prevailing wage rates, in some sectors of the labor market, and market forces are not strong enough to eliminate these imbalances where they exist. Hence unemployment rates are higher in these sectors than in the economy as a whole, and such differentially high unemployment rates tend to persist for relatively long periods. The persistence of these differentially high unemployment rates even at aggregative full employment means that the imbalance between demand and supply in some sectors of the labor market is not being removed by either (1) adjustments in wages, (2) reductions of supply through labor mobility outward from these sectors, or (3) an increase in the demand for the types of labor that have been in excess supply (for example, through revision of employers' hiring standards and practices or the movement of industry into a depressed area).

Concerned unions and the employers with whom they bargained came up with many ways, some quite imaginative, of containing, though not overcoming, structural unemployment. Adopting an old trade-union precept, some shortened hours in order to spread the work. Local 3 of the International Brotherhood of Electrical Workers, for example, established the 25-hour work week in New York City. The basic steel industry introduced "sabbatical leaves"—13-week vacations every five years for senior employees. Kaiser Steel and the Steelworkers, prodded by three distinguished neutral experts on their Long Range Committee, systematized the worker's share of automation gains. Historically labor had accounted for 32.5 percent of the cost of production at the Fontana, California, mill. Thus, workers received this amount of the cost savings.

Two of the collectively bargained schemes were notable—those of West Coast longshoremen and Armour meatpackers. Longshoremen did not use the term "automation." Rather, they talked about "the machine" (in a collective sense), or, in more formal moments, "mechanization."

Prior to World War II longshoremen manhandled cargo essentially as they had in the days of sailing ships. Mechanization started during the war and came on dramatically in the postwar era. Hawaiian raw sugar had been sacked and loaded by hand in the islands and unloaded the same way at the Crockett refinery on San Francisco Bay. Now it was handled in bulk. Unloading 10,000 tons of sacks had taken 6,650 manhours; in bulk it consumed only 1000. The number of men needed at the refinery warehouse plummeted from 80 to eight. Bulk handling was extended to grain and ore with similar gains in efficiency. A gang of six to eight men had brought rolls of newsprint, each weighing almost a ton, out of a ship's hold one at a time. A mechanical grab now lifted eight to the dock simultaneously under the eye of a single longshoreman. Lumber, earlier stacked by the board, was now unitized into large strapped loads. Ship turnaround time dropped from two weeks to four to five days. Scrap steel, formerly stowed piece by piece by a gang of eight, was now cut into small chunks that could be bulldozed onto a belt that dumped into the ship's hold. Most important, containerization appeared in the late fifties. Huge steel boxes were loaded and sealed at the factory; they were driven to the dock as truck bodies; a gantry crane lifted the containers onto the ship. A typical vessel held 296 containers below and 140 on deck—6500 tons of cargo. Loading and discharging that tonnage had taken over 11,000 manhours. It now consumed 850. Turnaround time dropped from 5½ days to 40 hours.

By the late fifties "the machine" dominated the bargaining between the Pacific Maritime Association and the International Longshoremen's and Warehousemen's Union (ILWU). The employers insisted on introducing the new technology and demanded the elimination of restrictive work rules (slingload limits, multiple handling, hiring unnecessary men, and so on). The union was badly split. A large minority wanted to resist mechanization by preserving the work rules. A small majority, led by Harry Bridges, the ILWU's president, argued that the rules were a temporary defense, already severely eroded and fated for final collapse. Further, Bridges urged, thus far the employers had gotten all the gains from increased productivity. Why not bargain away the work rules in exchange for "a share of the machine," including job security? The more imaginative employers, led by J. Paul St. Sure, the president of PMA, saw an opening for a deal.

On October 18, 1960, PMA and the ILWU negotiated as a supplement to their collective bargaining agreement the five-and-one-half-year Mechanization and the Modernization [M&M] Agreement. The employers won the virtually unlimited right to introduce new machinery and work methods along with substantial modifications of the work rules. The longshoremen got the M&M Fund, created by shipowner contributions for five and one-half years of $5 million annually. This fund was reserved exclusively for men who were fully registered longshoremen at the time the agreement was signed. Sixty percent was considered their "share of the machine," their slice of rising productivity. The remaining 40 percent was assumed to be the price they received for selling out their property rights in the work rules.

The agreement provided registered longshoremen with a flat guarantee against layoff. Their median age in 1958 was forty-nine. The plan offered early retirement at age sixty-two with a monthly pension of $220. At sixty-five, when Social Security kicked in, the pension fell back to $115. A man who preferred to work till age sixty-five received a lump sum of $7,920 ($220 a month for 36 months) upon his retirement. If work available dropped below 35 hours weekly, the longshoreman was guaranteed a weekly wage of $111.65.

During the fifties tonnage flowing through the West Coast ports was relatively stable and both parties to the M&M Agreement expected slow growth to continue during the following decade. If that had occurred, the number of fully registered longshoremen would probably have declined significantly in response to the M&M incentives, thereby providing jobs for the partially registered. The Vietnam War, however, enormously increased the volume of tonnage and the demand for longshoremen. Thus, the number of fully registered men declined slowly as many, particularly when the cost of living shot up, opted to work till age sixty-five and receive the $7,920 payout. They also continued to work because mechanization reduced the amount of back-breaking labor they had to perform. Thus, opportunities for partially registered men narrowed. The shipowners were the big winners. They enjoyed a phenomenal increase in productivity from a slowly shrinking work force. And, since there was plenty of work, they did not have to pay the wage guarantee for time not worked.

Meatpacking, like longshoring, was an old industry that underwent radical change in the postwar period. Historically it had been dominated by the Big Four—Swift, Armour, Wilson, and Cudahy—which had large plants in major cities, particularly Chicago with its great stockyards. They shipped by rail and marketed mainly through specialized meat markets. The popularity of processed meats, delivery by truck, and selling through supermarkets transformed the industry. Small packers operating in a livestock-producing area with low wages took a growing share of the market. The Big Four eroded into nine "national" packers and faced intense competition from the smaller independents. Chicago lost its position and was replaced by Omaha. The former Big Four shut down their obsolescent plants and laid off their workers. Mechanization swept the industry: movement of carcasses on overhead rails, pneumatic and mechanical knives, conveyors, curing by injection, stuffing and linking sausages automatically. One estimate was that between 1956 and 1960 productivity rose 20 percent and employment fell 16 percent.

Armour, number two in the industry, began serious retrenchment in 1956, closing five old plants in the next two years. In 1959 it shut down six more. In a three-year period the number of production and maintenance employees dropped from 25,000 to 15,000. The unions that represented these people—the United Packinghouse Workers and the Amalgamated Meat Cutters—were deeply disturbed and labor displacement–job security became the key issue in the 1959 negotiations.

At the urging of its imaginative labor attorney, Frederick R. Livingston, Armour proposed an Automation Fund Agreement which both unions bought. It was incorporated into the collective bargaining agreements that took effect on September 1, 1959. Armour would contribute a penny per hundredweight of meat shipped to establish a $500,000 fund. It would be administered by a nine-member Automation Fund Committee, consisting of four from Armour, two from each of the unions, and an impartial chairman. The parties were fortunate in persuading Clark Kerr, the president of the University of California, to assume this responsibility. He was a noted labor economist, had dealt with the meatpacking industry for the War Labor Board, and was gifted at handling complex problems. He would also in 1961 become a public member of the President's Advisory Committee on Labor-Management Policy, leading to cross-fertilization between the Armour experience and national policy. Professor Robben W. Fleming of the University of Illinois Law School was the committee's executive director, and in 1961, when he became provost of the University of Wisconsin, Dean George P. Shultz of the University of Chicago School of Business was named co-chairman. The committee was authorized to use the fund for "studying the problems resulting from the modernization program and for making recommendations for their solution." In 1961 this power was broadened to allow the committee to promote employment opportunities within Armour, to train qualified employees, and to pay the moving expenses of those who transferred to a new location.

The committee confronted a formidable set of problems. It was in business because Armour had closed nine plants with heavy layoffs. But the cutbacks were not over. The company soon closed five others with 5000 workers. By 1965 Armour had shut 21 packinghouses, displacing 14,000 employees. This would have been serious even if the economy produced full employment, which it certainly did not. In some towns Armour dropped its unemployed into a local labor market with virtually no available jobs. Moreover, the meatpacking work force did not fit the growing demand for employees who were educated, skilled, flexible, and nonmanual. Ninety percent of packinghouse work was heavy, dirty, low-skilled, manual labor for which there was virtually no demand from other industries. Collective bargaining, as the longshore experience demonstrated, could not create new jobs. Rather, it could rearrange jobs in order to protect senior employees. The plight of the former Armour workers also exposed the shortcomings of the nation's manpower programs—an ineffective employment service, an inadequate unemployment compensation system, virtually nonexistent retraining, and no moving allowances.

The committee contracted with labor market experts at the universities of Illinois, Wisconsin, and Chicago to make a series of studies. The first was of three plants that had closed in East St. Louis, Columbus, and Fargo, and the results were grim. A majority of the former employees, 53.1 percent, were over 45 and had many years of service with Armour; 18.8 percent were women; and blacks at East St. Louis made up 57.6 percent, at Columbus 8.6 percent, and at Fargo zero. The plant closings created "extreme hardships." A year after the shutdowns the unemployment rates among former employees were 56 percent in East St. Louis and between 25 and 30 percent in the other towns. Those with the least education (many, especially blacks, were functionally illiterate) had the most trouble finding work. As numerous plant closing studies had shown, only a tiny number, here 4 percent, left town to look for jobs. For those who found work, average wages dropped from $2.20 per hour at Armour to $1.86.

Discrimination was a severe problem. Almost half those over 45 were out of work a year after the layoffs compared with one-third of the younger workers. For women, 52 percent were unable to find jobs as contrasted with 39 percent for men. Blacks suffered most. In East St. Louis 61 percent were still out of work a year after the shutdown. The problem became acute when the large Fort Worth plant closed, idling a large number of black and Latino workers with especially heavy handicaps. The committee worked out a broad retraining program, utilizing the city's fine vocational education system. But there was an insuperable hurdle: all public and private facilities were legally segregated by race. It was virtually impossible to teach an adult Negro woman typing in Fort Worth.

As the committee learned of these problems, it moved on from fact-finding to action, starting with interplant transfers within the Armour system, which was opening a few small new packinghouses. Transfers were arranged, including a $325 moving allowance, to shift employees from Oklahoma City to Kansas City. But as they were about to take place, the latter instituted layoffs. When the isolated plant in Birmingham closed, those who lost their jobs were offered employment at distant locations. Only three accepted and two soon returned. When Sioux City shut down, the workers were offered tranfers to 12 plants, five in the Midwest, plus two new operations at West Point, Nebraska, only 50 miles away, and at Sterling, Illinois, 200 miles distant. A year after the shutdown, 234 people from Sioux City had found jobs in these other Armour plants. This success was followed by a similar program in 1964–65 when the layoffs in Kansas City threw almost 2000 out of work. A significant number were absorbed by new operations in Emporia, Kansas; Worthington, Minnesota; and Kansas City itself.

The union agreements provided two cushions against displacement—severance pay and pensions. Those unable or unwilling to transfer could opt for one of these alternatives. Severance pay was determined by the employee's wage rate and length of service, with a weighting in favor of senior workers. The average payouts in five layoffs ranged from $1,080 to $2,840. The recipients used the money primarily to liquidate debts. The pension plan allowed normal retirement at age 65 with a benefit of $2.50 monthly for each year of service. Early retirement with a reduced pension was available at sixty for men and fifty-five for women. In 1961 the parties encouraged early retirement by offering retirement at fifty-five with twenty years of service with a benefit of $3.75 monthly for each year of service; this dropped to $2.50 at sixty-five when Social Security became available. Many displaced Armour workers selected one of these options.

The committee also tried to locate Armour workers with other employers in four very different labor markets—Oklahoma City, Fort Worth, Sioux City, and Kansas City. This was very difficult in the face of unemployment and the limitations of the Armour workers. The committee relied heavily on the U.S. Employment Service to maintain the lists, give aptitude tests, counsel the applicants, and canvass local employers. In Oklahoma City and Fort Worth the local services were ineffective and the committee, with its limited resources, had to step in. In Sioux City the service was a dynamo that got the community involved in assisting the Armour displacees and got them 422 jobs. Kansas City followed this example. Those who took these jobs received much lower wages than Armour had paid.

Outside placement required retraining in these four cities. The main obstacle was that workers with the biggest handicaps needed retraining most and were those least likely to accept or absorb it. The problem was insurmountable in the face of functional illiteracy. At the outset the committee did what it could with local facilities—the employment service and vocational education programs, both of which varied in quality from town to town. With the passage of the Manpower Development and Training Act in 1962, the committee placed its people in MDTA programs as they slowly became available. Tuition was a problem because the trainees were broke and the committee was little better off. The latter paid the first $60 plus one-half the balance up to $150. There was no money for subsistence. Despite these difficulties, training programs were established in Oklahoma City, Fort Worth, and Sioux City, and those who completed them were very successful in finding work, particularly in the skills for which they had been trained.

The Armour Automation Committee faced structural unemployment problems that mirrored in microcosm those the nation was confronting. As noted, the Armour experience exposed the inadequacies of U.S. manpower programs. Thus, many in Congress and in the Labor Department were urging the development of manpower policies to deal with structural unemployment.[1]

Paul Douglas spent the Senate's Lincoln's Birthday recess in 1954 in "Egypt," the southernmost and poorest region in Illinois. Its thick bituminous seams had provided jobs for 80,000 to 100,000 coal miners. But natural gas and oil were displacing coal and the nation was in recession. Mines were shutting down. The obsolete freight car works in Mt. Vernon, which had supplied 1400 jobs, closed its doors. Marginal hardscrabble farmers who worked the clay soil were being squeezed out. Egypt, Douglas realized, had become a depressed area.

When he returned to Washington, the senator learned from other members of Congress that the same process was under way elsewhere: in the cutover timber country in northern Michigan and Wisconsin, in the iron mines near Duluth, in the anthracite and bituminous districts in Pennsylvania, in the textile towns of Massachusetts, New Hampshire, and Maine. He traveled to many of these areas and added a trip to view "the abject poverty" in the Indian villages of Arizona.

Douglas became convinced that federal action was needed to deal with this national problem. He worked the issue hard in his successful campaign for re-election in 1954. He then instructed Frank McCulloch, his administrative assistant, to form a task force to frame a bill. He was joined by Solomon Barkin, the research director of the Textile Workers Union, who was concerned about the decline of the organized sector of the textile industry; William L. Batt, Jr., who had worked on depressed areas both as a special assistant to the secretary of labor and as Pennsylvania's secretary of labor and industry; and Prentiss Brown, formerly a senator from Michigan and now the head of Detroit Edison.

Conservative economists argued that, as product markets played themselves out, stranded workers would move to labor markets where jobs were available. Douglas was certain that most of the unemployed would not move. In southern Illinois they stayed home waiting for a lucky spin of the wheel or took jobs in St. Louis or Evansville, often with punishing commutes. Douglas pointed to social capital, that is, a heavy investment in their home towns in services which would have to be duplicated at great expense elsewhere if they moved. "They had their homes on paved and lighted streets, with connections for sewers, water, and electricity." Their communities had stores, churches, schools, medical care, police, firemen, and lodges. Instead of abandoning this infrastructure by moving to jobs, Douglas reasoned, the workers should stay home and have the jobs come to them.

After study his task force told him that private firms would not hire men directly from the mine or off the farm; they would have to be trained. While reluctant to add features that might lose votes, Douglas agreed. If unemployed workers were to enter training programs, they must have an income, certainly an amount no less than their state's unemployment insurance benefit.

A former student, now in the Labor Department, told Douglas that the baby boomers would be flooding the labor market in the sixties and would overwhelm his program without a threshold cutoff. She suggested that a local labor market have at least a 6 percent jobless rate in order to become eligible for assistance.

On July 28, 1955, just before the 84th Congress adjourned, Douglas introduced the first of his "depressed areas" bills. Among the six co-sponsors were Senators Kennedy and Humphrey, both from states with troubled communities. Over the next six years the bill would go through many changes and a complicated legislative history, the latter deserving here no more than a brief summary.

On July 25, 1956, the Senate passed the Douglas bill 60 to 30, with only 3 Democrats in opposition and 17 Republicans breaking ranks in support. But, urged on by the Eisenhower administration, the House Rules Committee refused to allow the bill to reach the floor. Eisenhower's landslide re-election in 1956 seemed to mark the demise of the Douglas bill.

The sharp 1957–58 recession, however, brought it back to life. On May 13, 1958, the Senate again passed the bill, 46 to 36. This time Howard Smith, chairman of the Rules Committee, was forced to issue a rule and a somewhat revised bill cleared the House on August 15, 1958, 176 to 130. The differences between the chambers were quickly resolved and Congress sent the Douglas bill to the White House. Though his secretary of labor, James P. Mitchell, urged Eisenhower to sign the measure, he killed it with a pocket veto. The Democrats then made depressed areas a major issue in the 1958 elections and that, doubtless, contributed to their great gains.

Douglas put his bill back in the Senate hopper on January 27, 1959, this time with 38 co-sponsors, including 5 progressive Republicans. A companion measure went to the House. The Senate voted favorably on March 23, 49 to 46. Eleven southern Democrats joined the opposition and five Republicans voted for the bill. But now it got stuck to Judge Smith's gluey fingers. After a bitter fight and resort to the extraordinary Calendar Wednesday procedure, on May 4, 1960, the bill was forced out of the Rules Committee and the House adopted it 223 to 162. The Senate vote on the House version took place on May 6. Senators Kennedy and Humphrey interrupted their primary fight in West Virginia, the most depressed state in the nation, to return to Washington to vote for the Douglas bill; the two other Democrats who were presidential candidates, Lyndon Johnson and Stuart Symington, joined them in support. The vote was 45 to 32; again, five Republicans joined the majority.

There was speculation over what the President would do. Eisenhower issued a stinging veto message on May 13, 1960. There was a rumor that Nixon had asked the President to sign the bill in order to remove the issue from the presidential campaign. If true, the result demonstrated again how little influence Nixon had with Eisenhower. An effort by Douglas to override fell eleven votes short of the necessary two-thirds.

In the West Virginia primary campaign Kennedy had been appalled by the poverty he observed and redoubled his commitment to the depressed areas measure. He (and Humphrey) denounced the Eisenhower veto. The Democratic platform strongly endorsed the Douglas bill. During the presidential campaign, in a major speech in West Virginia and in the third television debate, Kennedy pressed Nixon hard on the issue. His victory, of course, assured strong presidential support for the depressed areas program.

In fact, Kennedy promised to send up a bill within sixty days of his election. He announced formation of the task force on area redevelopment on December 4, 1960, and asked for its report by the new year. Senator Douglas was chairman and was seconded by Senator Joseph Clark of Pennsylvania; Myer Feldman of the Kennedy staff was secretary; and there was a very large group of industry, labor, university, and government members. The task force held meetings in Charleston, West Virginia, and Washington and drew on more than fifty subcommittee and local reports. Douglas, who thought his bill needed no further airing, interpreted the mandate broadly and dealt with far more than the bill. Considering the unwieldiness of the group and the breadth of its agenda, it is remarkable that the chairman delivered a unanimous report to the President-elect on December 27, 1960, hardly three weeks after the formation of the task force.

A distressed area, the report defined, was one in which the unemployment rate much exceeded the national average. At the end of 1960, when the latter was over 6 percent, "the distressed area rate is over 10 percent, and in some instances as high as 20 percent." Nearly a hundred labor markets suffered from such joblessness and 300 to 400 low-income rural areas were "plagued by underemployment." The causes and characteristics of distress were so diverse that there must be "an arsenal of weapons" to combat particular problems. Private industry and government at all levels must join in drawing on these ideas if distress was to be overcome.

The task force recommended policies both to relieve personal hardship and to provide jobs. The former included surplus food distribution, which Kennedy adopted in the first executive order of his presidency, improvement of the unemployment compensation and welfare systems, and emergency public works. The last would become a major interest of the new Council of Economic Advisers. The core of the job opportunity program, of course, was area redevelopment. The task force also urged that more military contracts be awarded in labor markets with high unemployment, that human resources be improved with federal assistance for education and training, and that physical resources—highways, forests, parks, and so on—be developed in the distressed areas.

Douglas noted that his bill had already received extended hearings and congressional debate and had twice been passed by Congress. Thus, there was no need to retrace those steps. To avoid controversy and delay he made the bill adopted in 1960 the model and reintroduced it. The Senate leadership assigned it the coveted designation as S.1. Douglas was joined by 43 cosponsors: 38 Democrats and 5 Republicans.

But there were snarls in the thread. Douglas had insisted that the Area Redevelopment Administration be an independent agency to preserve its autonomy. The Eisenhower administration had talked of putting it in the Department of Commerce, which did not sympathize with the program. Douglas assumed that Kennedy would support his position and was upset when the President moved the other way. The official reasons were that Commerce Secretary Hodges was supposedly eager to get the program, that Budget Director Bell thought the government would be tidier without another independent agency, and that Commerce was an especially stodgy department that needed "some yeast." The real reason, William Batt said, was "to get some conservative support for a contentious issue and try to get some business support." It never came.

On January 25, 1961, the President strongly urged Congress to enact the Douglas bill promptly, but added: "In my judgment, the department best equipped to supervise and coordinate the program is the Department of Commerce." "Bill," Douglas told Batt, "I have to give up on the location of the agency. I agreed to put it in the Commerce Department. I could fight the southerners and I could fight the Republicans but I couldn't fight the southerners and the Republicans and the White House too." Kennedy, Batt said, made "a terrible mistake."

Another problem arose in the task force. Douglas and Batt argued that big rich firms would not be attracted to depressed areas by ARA loans. As an alternative, they urged a tax incentive. Otherwise, only small and poor businesses would participate in the program. As Batt put it, "In effect, you're putting the weakest companies in the weakest areas." But the new Treasury officials were "just dead set against it." Myer Feldman, who represented Kennedy on the task force, vetoed the proposal because "we weren't going to fool around with the tax . . . system for any of those purposes."

While there was vigorous debate in Congress on S.1 over several issues—long-term versus annual appropriations, a prohibition on assisting a depressed area for "pirating" a firm from another location, indefinite or a four-year life for the program—passage came swiftly. The Senate approved the bill by voice vote on April 20, 1961, and the House acted six days later by a margin of 223 to 193. The President signed the Area Redevelopment Act on May 1, 1961, almost six years after Douglas had introduced the bill. In order to achieve that gratifying victory, however, the senator over the years had made many political concessions that would seriously impair its effectiveness.

With customary blunt honesty, Douglas had given his original bill the title "Depressed Areas Act." But he soon discovered that those in distress did not want to call attention to that fact. They preferred a name that was upbeat and euphonious. It became the Area Redevelopment Act.

A crucial problem was the definition of the "redevelopment" areas that would be eligible for assistance. In his original bill Douglas had primarily addressed urban centers suffering from chronic unemployment. His paradigm was the one-industry town in which the big plant shut down. But so narrow a definition was politically impossible because there were not enough states and congressional districts with such problems. Members of Congress without depressed areas did not want their constituents taxed for the benefit of others. The main pressure came from the South, where agriculture was depressed, particularly in amendments introduced by Senator Fulbright of Arkansas in 1956. Douglas, without a political option, loosened the definition to include rural counties.

The statute directed the administrator to certify for eligibility three types of areas. The first were large and medium-sized urban labor markets for which statistics were collected in which unemployment had averaged at least 6 percent and had been at least 50 percent above the national average for three of the four preceding years, or 75 percent over two of the three prior years, or 100 percent over one of the two last years. The House and Senate conferees instructed ARA to declare smaller areas eligible if it could find statistics. Second, the southerners got broad language directing the agency to certify rural counties with low farm incomes, the overwhelming majority in their region. Under the 1960 bill 662 rural counties would have qualified, all but 13 in the South; Wisconsin would have had none. Senator Proxmire of that state inserted an amendment in the final bill instructing ARA to "distribute the projects widely among the several States." Finally, at the urging of the Montanans—Senators James Murray and Mike Mansfield and Representative Lee Metcalf—Douglas included Indian reservations where desperate poverty left no room for argument.

The act created seven programs: (1) a $100 million loan fund to encourage the building or expansion of private plants in distressed areas; (2) a similar program with an equal amount for rural areas; (3) a $100 million loan fund for communities to improve public facilities to attract new industries; (4) $75 million for grants to communities so hard up that they could not assume loans; (5) $4.5 million for retraining workers for the new jobs; (6) $10 million for training subsistence payable to the states to be made at the unemployment insurance rate for no more than 16 weeks; and (7) $4.5 million for technical studies to determine the development potential of distressed areas. The rate of interest on loans would be that currently charged by the Treasury plus 1/2 of 1 percent to cover administrative expenses and losses. A loan could not exceed 65 percent of the cost of the project. The total outlay came to $394 million. But this was deceptive because $300 million (the three loan funds) would be borrowed from the Treasury, would revolve, and, if successful, would produce income and be repaid. The probable real cost, therefore, was only $94 million.

Thus, this was an impecunious program from the outset that could not possibly fund the very large number of projects it was authorized to sponsor. Douglas himself was the main culprit. Growing up in the Maine woods in what some would call rural poverty, he learned early on to make do with very little and to admire Yankee frugality. In the Senate he was unusual among liberal Democrats by working to root out waste and to promote savings. An unreconstructed professor, he wrote a book on the subject—*Economy in the National Government.* Now he applied the same standards to his own program. Some members of the task force had wisely urged higher expenditures. But Douglas insisted on using the amounts that Congress had voted in the 1960 bill and he prevailed.

Finally, the act created the Area Redevelopment Administration within the Department of Commerce, headed by an administrator. The life of the program was fixed at four years, to expire on June 30, 1965.[2]

Bill Batt was invited to the White House ceremony at which the President would sign the Area Redevelopment Act. He missed the flight in Harrisburg and "broke all the speed laws" driving to Washington, arriving just in time. Kennedy moved around the group shaking hands. When he came to Batt, he said, "I hope you're going to run this for us." Batt said, "I'd love to do it." Since he was the best qualified person, the choice surprised no one. "I was kind of a logical person for it, I suppose." He had been recommended by Douglas, Galbraith, and his boss, Pennsylvania's Governor David L. Lawrence. Hodges was delighted to get him. Batt brought along his top assistant in the state's department of labor and industry, Harold W. Williams, who would become deputy administrator.

At the outset Batt made several important decisions. ARA would be a small coordinating agency of 325 employees, concentrated in Washington. This worked quite well in the first year when the number of applications was small, but became a problem in the latter part of 1962 when there was a shortage of field people. It was severe in states where many areas were eligible, like Kentucky, with only one man, and in the West where distances were vast, such as Colorado and Wyoming, both covered by one person. Batt needed employees who were self-starters, highly motivated, and devoted. He was amazed at how many turned up. "Really, people found us." ARA had something of the appeal of the Peace Corps. It was, he said, "a wonderful time to come to work for the government."

The other structural decision, urged by Douglas and in part required by the law, was to farm out work to delegate agencies. In effect, ARA became the prime contractor and others subcontractors. This led to an immediate jurisdictional dispute with the Department of Agriculture and its lobbyists. The department, faced with a high overhead and a shrinking constituency, sought to obtain authority over rural areas and even over smaller urban districts. Batt, supported by Hodges, resisted and won. Agriculture would process rural applications, but ARA would make the final decisions. Similar arrangements were made with Housing and Home Finance for community facilities and with the Small Business Administration for industrial loans. Manpower training was jointly administered by the Labor Department and HEW. Labor also advised on area designation. The Office of Indian Affairs handled the reservation program.

Subcontracting was at best awkward. Batt was more blunt: "This is a miserable way to run a railroad." Where missions were parallel or supplementary, he pointed out, the system worked well, as with Labor, Indian Affairs, and Vocational Education at HEW. There was trouble with Agriculture, state vocational education agencies, and SBA. From ARA's point of view, Sar A. Levitan wrote, "it would be advantageous for most if not all its programs to be administered directly by its own staff."

ARA's immediate substantive task was to designate the local areas that were eligible for assistance. The statute and the legislative history made this hopelessly complicated. While the original intent was the redevelopment of areas that had declined, the jurisdiction included rural areas that had never developed. The two did not mix well and this was evident in the statistics needed for designation. The government had reasonably good data for major urban labor markets as well as for smaller ones with a labor force over 15,000, including 8000 nonagricultural workers. But it had none for small markets and Agriculture collected no statistics on rural underemployment. Census had a decennial series on county income, but the 1960 results were not yet available and ARA had to rely on 1949 data, which were obsolete.

The politicization of area designation had begun long before the act was passed and it soon reached a crescendo. Members of Congress brought formidable pressure on ARA to select areas in their states and districts which Batt found impossible to resist. This took place, of course, with a severely restricted budget which could not possibly accommodate all chosen areas.

The designation problem created debilitating results. In the organizational phase ARA was compelled to give primary attention to selecting areas, time it could have used to administer programs. By early 1962 it had chosen 129 large cities, 657 small urban-rural areas, and almost 900 rural counties. "More than a sixth of the total U.S. population," Levitan wrote, "lived in these presumably depressed areas." Once loose criteria were established, they could not be abandoned in the face of close congressional scrutiny. Even worse, in 1962 and early 1963, ARA added 170 areas on the basis of flimsy unemployment insurance statistics. Thus, the agency raised expectations which, in many cases, it could not satisfy.

From his experience in Pennsylvania and with his eye primarily on redevelopment, Batt was convinced that credit was the "root" of the depressed areas problem for smaller firms. Banks would not lend for the entry of new plants or the expansion of old ones in these distressed communities. For Batt, therefore, the industrial and commercial loan program was the central ARA policy.

The SBA processed these loan applications and made recommendations to ARA. The law authorized a loan of up to 65 percent of the cost of land, buildings, machinery, and equipment; the local community must provide 10 percent; and the remaining 25 percent would come from entrepreneurs, banks, or, in some cases, SBA. The standard for an ARA loan, looser than that applied by lending institutions and SBA, was that there should be "a reasonable assurance of repayment." The applicant had to show that he had been unable to obtain the credit from conventional sources and that the community was committed. SBA, which was fussy about financial security, took three to four months to process applications.

Each deal was unique and required its own financing, as a couple of illustrations will show: A furniture factory in Mingo County, West Virginia, was financed by $200,000 from the owners, over $650,000 from ARA, $288,000 supplied by SBA and three local banks, and, since the county was broke, 10 percent from the state. Early in 1962 American Car and Foundry announced that it would close its plant in Berwick, Pennsylvania, idling 1900 workers. Several executives and community leaders formed the Berwick Industrial Development Association to continue a defense contract at ACF. ARA loaned $504,000, ACF and local funds covered the remaining obligations under the law, and SBA provided $350,000 for working capital. This saved 300 jobs.

As of June 30, 1964, ARA had approved 354 industrial and commercial loans with a combined value of over $150 million, which created almost 40,000 jobs. Measured by size of investment, the leading states in descending order were Michigan, Maine, Arkansas, Louisiana, Pennsylvania, Puerto Rico, Kentucky, Oklahoma, and West Virginia. These loans were made to small firms in a wide range of industries. For example: Blue Flame Coal of Price, Utah; Northwest Oyster Farms of Nahcotta, Washington; Minnesota Wild Rice Harvesters of Deerwood, Minnesota; Greater Hazleton Can-Do of Hazleton, Pennsylvania; Goodluck Glove of Metropolis, Illinois; National Seating of Gilbert, West Virginia; Iron Mountain Stoneware of Southerland Valley, Tennessee; Gault Tool of Ada, Oklahoma; American Boatbuilding of Providence, Rhode Island. Batt said the ARA worked harder on West Virginia than on any other state. "Jobs are life in West Virginia. It's like water in the Middle East or food in Bihar, India." He thought the program there was a great success, as it was in the anthracite region of Pennsylvania. ARA also had a big impact in the South, particularly in rural Georgia and Tennessee. The joke in the office was, "Keep your Confederate money because the South will rise again."

The criteria for loans to improve public facilities were: the project must improve the opportunity for industrial or commercial development; funds were not otherwise available; and the project must conform to the community's development program. The standards for grants for public facilities were more stringent: the community must contribute "in proportion to its ability" to do so; there is "little probability" that the project would be undertaken without the grant; and the amount would be limited to the difference between financing obtainable from other sources and the cost of the project.

By June 30, 1964, ARA had made 153 loans/grants totaling almost $90 million, which created over 30,000 jobs. This investment was in community infrastructure, particularly sewage and water systems, but also in port facilities, natural gas, railroad spurs, airports, hospitals, industrial parks, and courthouses. Examples: Dora, Alabama, got water and gas systems and an access road; Blairsville, Georgia, built an airport; Harlan and Middleboro, Kentucky, purchased hospitals; Wilkes-Barre, Pennsylvania, obtained a research center; LaFollette, Tennessee, built a dam and a water and sewage system; Elkins, West Virginia, got a hospital and a water-sewage facility. The stress in these programs, Batt pointed out, was "to help create public works which would contribute to private investment and jobs." The U.S. had a huge backlog of areas requiring water purification and waste disposal systems. ARA took a big step toward meeting it.

Two of the tourism projects provoked sharp criticism—$10.3 million to the Oklahoma Lake Redevelopment Authority for two luxury hotels on Lake Eufala and $3 million to Duluth for a convention center. ARA justified its heavy stake in both on the grounds that they would stimulate satellite investment, resulting in more jobs. Several of the research programs were imaginative: a $400,000 grant to Wilkes-Barre for construction of a graduate training laboratory to attract engineers and scientists to work for high-tech firms in that depressed anthracite town; a $900,000 grant for a marine research center at Yaquina Bay, Oregon, to help the fishing industry, to search underseas for oil and minerals, and to help solve the pollution of local waters by the timber industry; a $642,000 grant for a timber conservation and development center at the University of Kentucky, also to help local industries.

The statute established a job training and retraining program for the unemployed and the underemployed in the designated areas. This was limited to institutional (classroom) programs; there was no provision for on-the-job training. An enrollee received subsistence at the level of the average unemployment insurance benefit of his state for a maximum of sixteen weeks. While some courses were longer, this restriction tended to compress them into sixteen weeks.

The administrative structure could hardly have been more complicated. It was imposed on the existing federal-state vocational education program, which traced its origin back to the Smith-Hughes Act of 1917. ARA hardly participated and employed only one part-time coordinator. The Labor Department was more or less in charge and determined training needs and potential job vacancies. Once need was established, HEW's Vocational Education Division fixed the course content and contracted with the state to produce it. Agriculture was consulted for rural programs and the Bureau of Indian Affairs for Indian programs.

In practice the system usually worked from the bottom up. The local employment service office spotted a skill shortage that met ARA standards and confirmed it with its local advisory committee. It then worked out a course with the local vocational education people, which they presented to their state office. If approved, the proposal was forwarded to Washington where it worked its way through the bureaucratic maze, hopefully to final approval and the grant of funds from the Office of Manpower, Automation and Training in the Department of Labor. The locals, of course, complained with justification of federal delay, red tape, and added cost. Nevertheless, this cumbersome system seems to have produced results.

By June 30, 1965, 44,975 unemployed or underemployed workers had enrolled in 1,060 training programs at a cost of $25 million. They were from 48 states, Puerto Rico, and American Samoa, with most from Michigan, Pennsylvania, and West Virginia. According to 1963 data, the average cost to train an individual was $567, about $250 for instruction and $317 for subsistence. About five of six trainees completed the courses. Of those who finished, 69 percent got jobs, usually in the occupation for which they had been trained.

There were a number of interesting programs. In rural New Jersey farm laborers learned to operate, maintain, and repair agricultural machinery. The success of this course led to similar programs in Arkansas, Oklahoma, and Texas. The second time around, New Jersey, learning from experience, added literacy training. Indians from reservations in New Mexico and Arizona were instructed in electronic soldering and welding in Los Angeles.

There was, however, little fit between new enterprises financed by ARA and the training programs. Employers almost invariably found the low-skilled people they needed in the depressed local labor market.

The final ARA program was for technical assistance to communities in developing long-range economic plans. The act required a local government that wanted assistance to submit an overall economic development program, and by May 1, 1963, 850 areas, including 42 Indian reservations, in 48 states had done so. Most were booster plans for attracting new manufacturing firms and very few were carefully drawn. The small ARA staff was overwhelmed by the review process and was almost helpless in improving them. ARA did finance studies to improve the efficiency of the scallop and flounder industries in New Bedford, Massachusetts, and to exploit a large salt bed in Monroe and Washington counties, Ohio. Perhaps its most important contribution was to urge communities to merge into larger areas to develop combined plans for such projects as highways, water development, and power grids. In two cases—Appalachia and the Upper Peninsula of Michigan—it recommended regional plans.

For the Keynesians ARA was a bizarre public works program which created employment in a cumbersome and halting manner. They much preferred what came to be called accelerated public works. The Samuelson

Task Force had proposed such a program and the Council of Economic Advisers tried to convince Kennedy to offer legislation. But in 1961 he declined to worsen the budget deficit. Senator Clark had introduced an accelerated public works bill in 1960 and, with AFL-CIO support, he now turned up the heat. Strongly backed by the cities, this former mayor of Philadelphia introduced a public works bill. Shortly after the turn of the year Kennedy conceded.

He sent a $2 billion stand-by capital improvements bill to the Hill on February 19, 1962. Its main feature, devised by the Council, was a "trigger" mechanism which the President would fire when the unemployment rate reached a prescribed level. It soon became painfully evident that the Congress would not delegate its control over federal expenditures to the President. The administration lowered its sights and Sorensen called a meeting of Goldberg, Heller, Gordon, Bell, Batt, and Feldman, now on the White House staff. Two issues required presidential decisions and the group moved into the Oval Office. One was the amount of money to be requested; Kennedy decided on $900 million, a third of it for rural areas. The other was whether the projects should be limited to distressed areas certified by ARA, as Batt urged, or to more broadly defined labor surplus areas, as Goldberg argued. The problem with the Batt proposal was that it would pick up only a handful of major cities—Detroit, Pittsburgh, and Providence—while Goldberg's would bring in many more. Kennedy chose the broader definition. He signed the Accelerated Public Works Act on September 14, 1962, and gave administrative responsibility to ARA.

According to Batt, the program went "like a house afire" and was extremely popular. He and Jack Conway, a former UAW official who was now deputy administrator of the Housing and Home Finance Agency, set up a streamlined approval procedure. City and county governments had a huge backlog of projects they were "aching to do," for which the engineering was completed, and they applied quickly. The contrast with ARA was stark. Further, as Batt pointed out, "It is much, much more difficult, obviously, to loan money than it is to grant money." There was no fraud and no scandal. As with ARA, Batt pushed projects that would make communities attractive to private investment and job creation. Again, there were many water purification and waste disposal facilities. As of August 31, 1965, ARA had approved 7,711 public works projects. The total cost of the program, including matching funds, was $1.7 billion, of which ARA contributed $843 million. The estimate was that these projects generated about 200,000 man-years of on- and off-site employment.

Appalachia was the nation's largest and most notable depressed area. While highly diverse, parts of the region, particularly the Cumberland plateau in eastern Kentucky, were among the most backward in the U.S. In 1960 the nine Appalachian governors called for a regional development program and conferred with the Douglas task force. After the creation of ARA, they descended upon Batt for help. He had little enthusiasm for complex regional programs and was put off by the refusal of the governors to put up any state money.

Early in 1963 the governors complained to the White House. In March there were enormous floods in the Cumberlands and Governor Bert T. Combs of Kentucky went to Washington for federal help. He brought a list of items he wanted, the last of which was for the President to create a multistate Appalachian Regional Commission. This aroused Kennedy's interest. Sorensen and Feldman met with Batt and Williams, which led to a draft program put together by the ARA administrators. The President then called the Appalachian governors and top federal officials to a big White House conference on April 9, 1963, which created the President's Appalachian Regional Commission.

Franklin D. Roosevelt, Jr., had just become under secretary of commerce. Batt advised Kennedy to put the commission in the Commerce Department and to place Roosevelt, who was much interested, in charge. He did so and Roosevelt, Batt said, did "an impressive job." As it worked out, the Appalachian program was overwhelmingly highway construction, building roads with federal dollars to facilitate access to the region. According to Batt, the governors loved highways and that was the one project on which Democrats and Republicans could agree. "They both drive cars." Roosevelt put together "a curious hydra-headed affair," Batt observed, "but the darn thing seems to work."

In the spring of 1963 Batt concluded that the law must be amended because funds would soon be exhausted and to allow for greater flexibility. The Budget Bureau, the Housing Subcommittee of the House Banking and Currency Committee, and Senator Douglas agreed. But there were dangers in going to Congress. Hodges offered no support. There was resentment in many areas because they had gotten no assistance or had to wait interminably. In a period of slow sales some industries complained that ARA was financing new competitors. Organized labor, particularly the Ladies Garment Workers, which had earlier supported the Douglas bill, now worried that ARA was funding runaway nonunion shops. In 1961 the Pennsylvania House Republicans, led by William Scranton, had strongly backed ARA; now that Scranton was governor, they drifted away. *Reader's Digest* and *Human Events* attacked the program relentlessly. Despite these discouragements, Batt thought he could make it—if he was lucky. He was not.

The measure went to the House first and reached a floor vote on June 12, 1963. On the preceding day Governor Wallace had been evicted from the schoolhouse door at the University of Alabama by federal forces. That evening President Kennedy had gone on television to call for passage of a civil rights act. Batt, who figured on a twenty-vote margin, "saw my support melt away." Thirty-nine congressmen who had voted for ARA in 1961 switched: 19 Democrats and 20 Republicans. Eighteen of the Democrats were from the South, four from Alabama. As Charlie Halleck saw the southern Democrats vote against ARA, he realized that he could lick the bill by turning the screws on his Republicans. The bill was defeated by five votes, 209 to 204. Two weeks later the Senate, with vigorous leadership from Douglas, passed it 65 to 30, which at least made Batt feel better.

Despite significant achievements, particularly in assisting depressed areas by expanding their industries, improving their public facilities, creating jobs, and training workers, ARA was ill-starred. Neither the statute nor the agency distinguished between area redevelopment (industrial) and regional development (rural). ARA's responsibilities much exceeded its financial resources. The labor movement became disenchanted with assistance to low-wage, antiunion firms. Members of Congress lobbied for their districts and states and politicized the ARA program. Neither Hodges nor Kennedy offered protection. The Douglas bill could only have been passed in a period of severe unemployment. Yet, as the German experience demonstrated, "an area redevelopment program," William H. Miernyk wrote, "can be fully successful only if there is 'full' employment in the economy as a whole." Industries which were unable to sell their products did not want competition from new plants in depressed areas. The result was that ARA became politically and bureaucratically expendable.

With the passage of the Manpower Development and Training Act (MDTA) in 1962, ARA's training program was overshadowed. In the same year, however, ARA assumed responsibility for administering the Accelerated Public Works Act. When the Area Redevelopment Act expired in 1965, ARA itself died, but passed its programs on to three successor agencies. The Economic Development Administration created by the Public Works and Economic Development Act of 1965 continued with area redevelopment and public works. The Appalachian Regional Commission under a 1965 statute put into effect the program ARA had helped launch. ARA's training program was absorbed by MDTA.

It is difficult to evaluate ARA. Levitan, the only scholar to study the program intensively, concluded lamely that "a diluted program to aid the unemployed in chronic labor surplus areas is better than no action at all." ARA's National Public Advisory Committee, chaired by Frank P. Graham, reported in 1964 that "ARA has made substantial progress. . . . The job of generating employment in depressed areas is a long-term undertaking. . . . It is a mistake . . . to expect too rapid results."

Senator Douglas pointed out sensibly that no serious evaluation could be made except in the field by examining particular projects. In the area he knew best, Illinois's Egypt, the record was strong, with successful projects in Carbondale, Metropolis, and Sparta. His impression of other areas was that failures "have been far outweighed by successes." Batt agreed that it was "exceedingly" difficult to put a yardstick to ARA. His hunch was that the program was "enormously successful." Mount Union, Pennsylvania, had been a slagheap that dumped into the formerly beautiful Juniata River. When ARA was through, Mount Union had a playground, a big pavilion, a new sewage disposal plant which stopped river dumping, and a factory with 500 jobs. Scranton had an industrial park and new water and sewage systems. The Food Machinery Corporation plant in Charleston, West Virginia, had been rescued and employed ARA-trained people who came out of the patches. "They've got a new life."[3]

The Manpower Development and Training act of 1962 was the Kennedy administration's primary program to overcome structural unemployment. Like ARA, MDTA originated not in the administration but in the Congress, again in the Senate. The leading spokesman for this legislation and, more broadly, for what some called the "Manpower Revolution" was Senator Joseph S. Clark, Jr., of Pennsylvania.

Clark's father had been a prominent Philadelphia lawyer, a pillar of the Republican party, and a former national tennis champion. Young Clark was educated at the Middlesex School and Harvard, where he shone both academically and in sports. He was an editor of the law review at the University of Pennsylvania Law School. Clark became a Democrat to support Al Smith for President in 1928 and FDR converted him into a dedicated New Dealer. He had an outstanding record as a colonel in the Army Air Force in the India-Burma-China theater, where he was impressed with the military's ability to deploy manpower.

In 1951, Clark and his friend Richardson Dilworth led a liberal-reform coalition to wrest control of Philadelphia from the corrupt Republican machine that had dominated the city for sixty-seven years. They succeeded and Clark became mayor in 1952. Again, he was much interested in the manpower problems he confronted in staffing his administration.

In 1956, Clark was elected to the Senate from Pennsylvania. Shortly he received the Philadelphia Award, $10,000 for outstanding civic services. He gave the money to a venerable local institution, the American Academy of

Political and Social Science, to finance a conference on the prospective need for leadership. In his keynote address on the topic he would later call "staffing freedom," Clark puzzled over an effective manpower policy, that is, matching skills to jobs in the most efficient manner possible in a free society. He was particularly interested in two aspects of this problem. The first was chronic and persistent unemployment. The other was the recruitment of leaders, devising a scheme of rewards and punishments that would move the most qualified people into the top positions.

In the late fifties Clark's outlook and role were transformed. His state, which earlier had been a leader in industrialization, now suffered from severe economic decline. Its basic industries—steel, coal (anthracite and bituminous), railroads, metal fabrication, and textiles—lost markets, shut down operations, and eliminated jobs. Even more than West Virginia, Pennsylvania became the classic case of structural unemployment. Further, when Clark took his seat in the back row of the Senate in January 1957, the northern liberals could accomplish little. There were a few powerful voices, notably Douglas and Humphrey, but the three progressive freshmen—Clark, John Carroll of Colorado, and Frank Church of Idaho—were lonely in the rear of the chamber. "The Class of 1958" changed the composition and mood of the Senate; the three in the last row were now joined by fifteen new Democratic senators. As Clark saw it, the time had come to push forward.

In 1959 the senator's legislative assistant, James L. Sundquist, met with Batt and William Cooper, who administered Pennsylvania's underfinanced training program for jobless adults in its vocational schools. Sundquist asked whether there was anything besides ARA that Clark could back regarding unemployment. Batt suggested making Cooper's program national, and Sundquist asked Cooper to draft a bill. He passed the idea on to the American Vocational Association, which proposed adding to the hoary Smith-Hughes complex another grant-in-aid category for training the adult unemployed. A rather prosaic proposal, it was dressed up in the form of the Employment Act of 1946 with a council of manpower advisers and an annual manpower report of the President. Senator Jennings Randolph of West Virginia, who was up for re-election, introduced this first Clark training bill.

In 1959–60 Clark sat on the Special Committee on Unemployment Problems of which Senator Eugene J. McCarthy of Minnesota was chairman. After extended hearings, the committee proposed a comprehensive program to deal with joblessness. Clark inserted a recommendation for training and retraining. Since private industry was incapable of providing a national program, the federal government must step in. The committee cited with approval a proposal by the American Vocational Association for area vocational education centers. It recommended that a federal program should give preference to school-age youngsters and to workers over thirty-five who had become structurally unemployed. For the latter a subsidy related to unemployment insurance was required.

The Pennsylvanian and Secretary of Labor Arthur J. Goldberg meshed smoothly. "Goldberg," Clark later said, "thought exactly the way I did." The secretary, sensitive to structural unemployment, viewed manpower training as a critical national priority and considered it the central Labor Department program. Under Secretary Willard Wirtz shared this view and would succeed Goldberg when the latter was named to the Supreme Court on August 29, 1962.

When Kennedy had appointed Goldberg as Secretary of Labor it seemed inevitable and proved popular. Barry Goldwater, a rather bizarre authority, considered him the best man in the cabinet and he may have been on the mark.

Goldberg's life was a saga of the New Immigration. His parents had left Russia in the 1890s to settle in the Jewish ghetto on Chicago's West Side. He was the youngest of eight children and his father died when he was three. He was a delivery boy at twelve. He attended Crane Junior College and the Northwestern Law School, putting himself through by working nights in the post office and during vacations on a construction gang. He was editor of the *Illinois Law Review* and graduated *summa cum laude*. A special dispensation allowed him to take the Illinois bar exam before he was 21, which, of course, he passed. He practiced law in Chicago.

In 1936 Goldberg was active in the FDR re-election campaign and met a number of labor leaders. He began to represent the new CIO unions. In 1942 General William J. Donovan of the Office of Strategic Services engaged Goldberg to establish an underground intelligence network of transport workers behind the German lines in Europe. His performance was said to have been notable.

Philip Murray, who was president of both the Steelworkers and the CIO, brought Goldberg to Washington in 1948 as general counsel of both organizations. But Goldberg maintained his own law firm. He argued many of the important Taft-Hartley cases during the postwar period and Murray relied on him heavily in the troubled steel negotiations. Goldberg was the author of the procedure by which the CIO dealt with its Communist-dominated affiliates. At the time of his death in 1952, Murray considered his recruitment of Goldberg to have been his major contribution to the CIO.

Goldberg was the leading architect of the merger of the AFL and the CIO in 1955 and wrote a book about it, *AFL-CIO: Labor United*. The AFL leaders were so impressed with him that they kept him on as special counsel. Since

David J. McDonald, who succeeded Murray as president of the Steelworkers, was a lightweight, among other defects, Goldberg handled the steel negotiations for the union during the fifties. He won a notable victory at the close of the 116-day strike in 1959. When the McClellan Committee exposed corruption among some unions in the late fifties, Goldberg handled the formation of the AFL-CIO ethical practices committee and was the author of its code. He worked closely with Senator Kennedy in 1958–59 on labor reform legislation, which influenced the Landrum-Griffin law of 1959.

In 1960, Goldberg was an early Kennedy supporter and spearheaded the campaign for labor backing. George Meany, the president of the labor federation, did not put Goldberg's name on the list of AFL-CIO candidates for Secretary of Labor (there was some building trades opposition), but spoke highly of him. Goldberg had some misgivings about taking the job, believing that the secretary should come out of the labor movement. Aside from his standing with the unions, Goldberg was much respected in the intellectual community. A brilliant speaker, he appeared often on college campuses at a time when the labor movement was under attack. His wife was an abstract painter and they were modest collectors of Picasso, Matisse, and Ben Shahn. But Goldberg was more than an intellectual lawyer; he was a mover and shaker, the kind of can-do guy that Kennedy wanted. As Nicholas Katzenbach observed, "He was active in everything and full of life and zest and all the things that President Kennedy loved and full of imagination and ability to achieve, to deliver."

Some experts in the industrial relations community who knew both had an equally high regard for Wirtz. He had been born in DeKalb, Illinois, in 1912, where his father taught at the normal school. He attended that school, spent a year at Berkeley, and finished at Beloit College in Wisconsin. He was big man on campus—football, president of the student body, an outstanding academic record. He taught high school history for two years and then went to the Harvard Law School, where he became editor of the law review. He then turned to law school teaching, briefly at Iowa and then at Northwestern.

Wirtz moved to Washington in 1942 to become assistant general counsel of the Board of Economic Warfare and then shifted to the National War Labor Board, which, incidental to its main functions, served as a master class for training experts in labor relations. Wirtz was a star pupil. He became general counsel and a member of the board. When it phased out at the end of 1945, he became chairman of its successor, the National Wage Stabilization Board. In 1947 he returned to the Northwestern Law School.

In 1950 the new governor of Illinois, Adlai Stevenson, scoured the state for able and honest people and named Wirtz to the Liquor Control Commission, a part-time job. He and Stevenson became warm friends. In 1956 he was a top strategist in Stevenson's second unsuccessful run for the presidency. They then formed a law firm in Chicago. In the late forties and fifties Wirtz was one of the nation's leading and most skilled mediators, fact-finders, and arbitrators, called upon repeatedly by the government as well as by labor and management.

Aside from an outstanding intellect and a certain boldness of mind, Wirtz was exceptionally articulate and was much in demand as a public speaker. A. H. Raskin of the *New York Times* noted his "abhorrence of cant and a clinical precison in the use of words." His wit was legendary. Wirtz was modest and made little of his talents and accomplishments. Rather unusual among "the best and the brightest" of the Kennedy-Johnson era, he did not make a fetish of "toughness." While as tough as any, Wirtz did not hesitate to show that he cared—for particular individuals and for the whole society.

During the late fifties Seymour Wolfbein, the deputy assistant secretary of labor, had begun work on manpower training and developed a legislative proposal. Secretary Mitchell and most of Eisenhower's cabinet approved, but the President denied his support. When Goldberg took over, Wolfbein held the same position and was strongly encouraged to continue. The secretary put him to work with Clark's staff drafting an alternative to the first Clark bill.

The idea of a council of manpower advisers was scrapped. If the Council of Economic Advisers was concerned about unemployment, as it certainly was, that was deemed sufficient. More important, the Labor Department and the Bureau of the Budget thought the Smith-Hughes system "archaic," and the Office of Education, which administered that program, had no interest in manpower training. But the Vocational Education Association, which wielded considerable power in Congress, could not be brushed aside. Thus, in his education message of February 21, 1961, the President called for a re-evaluation of Smith-Hughes and related statutes and instructed HEW to name a panel of experts. The resulting study, *Education for a Changing World of Work,* indeed found the system out-of-date and ill-suited for dealing with labor displacement due to modern technological change.

The President submitted the Manpower Development and Training bill to Congress on May 25, 1961. It would create two programs—institutional (classroom) and on-the-job (OJT) training. The former would become part of the Smith-Hughes system of vocational education and would be administered by HEW, which would develop training programs jointly with the states. The Labor Department would handle OJT by contracting with public agencies, employers, and unions.

These programs would be directed primarily at unemployed individuals with inadequate or obsolete skills, though some employed persons in need of additional training would also be admitted. Those in institutional programs would receive a training allowance at the rate of their state's unemployment insurance benefit for not more than 52 weeks. This policy under MDTA, also the case with ARA, pegged allowances at a low level since unemployment payments averaged only 35 percent of earnings. OJT trainees would be paid by their employers, but the government would provide half their compensation or half the retraining allowance, whichever was less, in any case not over $46 a week. A person out of work more than six months who found a job in another town would be eligible for a relocation allowance, half the cost of moving.

The Labor Department would have primary responsibility for administration. Over four years, 800,000 workers would be trained at a cost of $700 million. The bill would also require the Labor Department to mount an extensive research program covering the impact of automation, the practices of employers and unions that impeded mobility, and the adequacy of the nation's manpower development efforts. The Secretary of Labor would submit an annual manpower report to the President, who would transmit it to Congress. For the first two years the federal government would pay the full cost of training allowances; after June 30, 1964, the states would pay half of this amount.

The hearings before the Senate Subcommittee on Employment and Manpower, with Clark in the chair, and the House Subcommittee on Unemployment and the Impact of Automation in early June were brief and uneventful. Goldberg forcefully carried the load for the administration. The AFL-CIO supported the bill. The American Vocational Association urged the transfer of administrative responsibility to HEW, but Secretary Abraham Ribicoff and Assistant Secretary Wilbur Cohen said that they preferred to keep it with Labor. Significantly, industry did not oppose MDTA; in fact, it was silent. Goldberg noted that the seven prominent business members of the President's Labor-Management Advisory Committee had joined in the committee's unanimous support for the training bill in principle. The full Senate Committee on Labor and Public Welfare reported favorably and more or less unanimously on July 31, 1961. Even Senator Goldwater said, "I support the basic aims of this legislation," though he reserved the right to propose amendments on the floor. The House Committee on Education and Labor endorsed the bill on August 10 with four of the 11 Republicans in opposition. The Senate passed the bill 60 to 31 on August 23. This smooth sailing came to an abrupt halt on September 6 when the House Rules Committee by a tie vote failed to report a rule. This was resolved in February 1962 by substituting the Senate for the House bill, thereby gaining a rule by an 8 to 7 vote. The House passed the new version on February 29, 354 to 62. President Kennedy signed the Manpower Development and Training Act on March 15, 1962.[4]

There would have been a certain administrative symmetry in assigning the new programs to the Labor Department's old-line bureaus—institutional training to the Bureau of Employment Security and its component, the U.S. Employment Service (USES), OJT to the Bureau of Apprenticeship Training (BAT), and research to the Bureau of Labor Statistics (BLS). But Goldberg was concerned about their stodginess and thought he might do better with an innovative new agency.

Sometime before MDTA was passed, according to Wolfbein, Goldberg called him in on Friday afternoon to talk about training and automation. The secretary asked for a memorandum on what the department should do. In the document he submitted on Monday morning Wolfbein noted "all the eternal verities" and recommended the establishment of an office of manpower and automation. "It was very dangerous to write memos in those days," he said later. Goldberg told him he liked the proposal very much and was appointing him the director of the new office. When the law was passed, the name became the Office of Manpower, Automation and Training (OMAT). This action did nothing to smooth inter-agency friction.

MDTA got off to a very slow start. While the act was passed on March 15, Congress did not appropriate funds until late August 1962. A large and complex administrative apparatus (USES had 2000 offices) had to be put in place—writing regulations; negotiating, approving, and funding projects; obtaining facilities, equipment, and instructors; and recruiting trainees. The bureaucratic problems were formidable. USES, responsible for local recruitment, administered the unemployment compensation system and was accustomed to referring the best qualified people for vacant jobs, a process called "creaming," which conflicted with picking the less qualified for training. HEW's Office of Education, which, along with the state vocational training agencies, was responsible for institutional training, was little interested and unorganized. BAT, which administered OJT, was experienced in running apprenticeship programs for the AFL-CIO skilled trades, had little concern for the unskilled, and was understaffed. OMAT, which recruited a number of bright young people, did launch a few imaginative "education and demonstration" programs. But by December 31, 1962, only 430 projects had been approved with 16,157 trainees, of whom only 6,315 were actually receiving instruction.

During 1963 the institutional training programs took hold. In February 1963, Secretary Wirtz named Under Secretary of Labor John L. Henning manpower administrator in order to get the departmental horses to pull together as a team. This may have helped. More important, the administrative hurdles were gradually surmounted. Facilities, equipment, and instructors moved into place in many communities and the Employment Service tested, counseled, and selected trainees. At the outset public schools were used during off hours. A shortage of school space and a preference for offering the trainee a variety of training choices led to the emergence of the skill center. In Detroit, for example, twenty-two scattered courses were consolidated into a surplus federal building known as the Detroit Skill Center. Ideally a center would provide counseling, basic education, and skill training in many occupations. As time went on an increasing number approached the ideal. Between August 1962 and December 1964, 4,985 institutional training projects with 284,449 trainees were approved. The latter were primarily white male heads of households in their prime working years with a high school education and a good deal of job experience. The Employment Service was creaming, selecting from the unemployed those with the best chance of completing a program and landing a job.

OJT lagged badly. Congress had expected it to constitute one-third of the training program. In fact, it was 6 percent in 1963 and 12 percent in 1964. BAT was part of the problem; it was reluctant to participate and distrusted by other agencies in the department. More important was the fact that OJT programs were much harder to launch than institutional ones because they required employer participation and few employers volunteered. In a period of high unemployment it was cheaper to hire off the street than to train; many employers lacked the personnel and facilities; there was the risk that the worker the employer had trained would leave for a competitor; and the $25 reimbursement seldom met the costs of training. Unions, too, were hesitant about OJT because it might reduce job opportunities for their members. By December 1964 only 1,571 OJT projects had been approved involving a mere 35,262 trainees. Even the very limited experimental and demonstration programs covered more trainees. The inadequacy of OJT was a serious failing because on-the-job learning had many advantages over school training.

During 1963 several pressures compelled the administration to return to Congress for amendments to MDTA. Early in the year it was clear that only a handful of the states would be ready to pick up half the cost of training on July 1, 1964. Without a change in the law the system would collapse. Experience during the first year showed that the number of recently unemployed who needed retraining was relatively small because they tended to be re-employed in similar jobs. Workers who most needed training were those who had never learned skills, the long-term unemployed, and untrained young people.

By mid-1963, MDTA, like so much else, had gotten caught up in the civil rights crisis. Lyndon Johnson, Kennedy wrote Wirtz and HEW Secretary Anthony Celebrezze, "feels . . . that the Federal Government should and could be doing much more to relieve Negro unemployment by additional and intensive job training programs for the unskilled, the illiterate and those on public welfare." He raised a number of policy options.

Wirtz replied on June 10 that about 13,000 nonwhites were being trained under MDTA and ARA, and there would be around 30,000 in 1964. These were dishearteningly small numbers. "The basic problem today is a shortage of jobs. We could do a great deal more than is being done in the training of Negroes. In the present labor market, however, the danger would be that many of them would be trained, but unemployed." Nevertheless, Wirtz strongly urged that the upcoming civil rights message stress the elimination of job discrimination and increased education and training to qualify blacks for the jobs that exist.

On June 19 the President submitted a special message to Congress on civil rights and job opportunities. He proposed amendments to MDTA to provide education and training for blacks, particularly teenagers. The age for admission to training would be cut from 19 to 16, a higher proportion of training payments would be permitted to out-of-school jobless youths, and basic education would be expanded. Experience demonstrated that many who applied for MDTA programs were functionally illiterate and were deficient in arithmetic and the ability to communicate.

Congress responded promptly and generously. The life of MDTA was extended a year to 1967. Full federal funding was provided until June 30, 1965, two-thirds in fiscal 1966, and one-half in fiscal 1967. Training allowances were stretched to 72 weeks, twenty of them for basic literacy education. The minimum age for admission was reduced to sixteen. Allowances were liberalized by adding $10 a week to the state unemployment compensation benefit rates and an institutional trainee would be allowed to work 20 hours a week without suffering a loss of income. Despite MDTA's many problems, the program enjoyed greater bipartisan support in 1963 than it had the previous year. The amendments passed the Senate 76 to 8 and the House 392 to 0.

As a result of these amendments, MDTA increasingly focused on the problems of blacks in the labor market. According to data compiled in 1967, all states exceeded their nonwhite proportion of the population in institutional training, usually by multiples of two to five. In southern states it was almost twice the population share. In OJT programs the percentage of nonwhites was much lower but still well above the black ratio of the 1960 population. Nonwhites, however, were underrepresented in training for professional, technical, and skilled occupations. MDTA's impact on blacks in the labor market was mainly to move them out of service jobs into semi-skilled and clerical occupations. Nonwhites were less likely to be placed after completing their courses in the jobs for which they had trained, particularly in the South. Clearly, there was bias against blacks in the skilled trades in OJT.

Wolfbein, who insisted on running integrated programs in the South, was called a four-letter word on television by George Wallace at the governors' conference in Miami Beach in 1963. Wallace then took him aside and said, according to Wolfbein, that he had "nothing against bureaucrats, white people, Jews, and all the other things that I was, and that, if I didn't make too much noise about it, I could go ahead and get some programs going [in Alabama] which were integrated." There was no such luck in Mississippi. Wolfbein pointed out that the stress on training blacks and, later, the poor made MDTA's task much more complicated. "Setting up a piece of legislation which was supposed to be responsive to these other dimensions is quite a different dish of tea from a simple economics-based business of training and retraining unemployed people."

The relationship of MDTA to unemployment was complex. Some conservatives, like Fed Chairman Martin, were convinced that raising the level of skills in the labor force would create new jobs. After Tobin and Heller had discussed the economy with him, Tobin wrote, "Martin *does* believe that a large part of the growth in unemployment is technological and structural and is not susceptible of remedy by expansion of money and credit. He believes that appropriate policy should aim at labor mobility and price reductions to pass on technological gains." To be sure, training increased the mobility of labor, but, as Wirtz had stressed to Kennedy, it created no jobs; it merely raised the educational level of the unemployed. Thus, a training program would be most effective when the demand for labor was high and there were skill shortages. The year 1962 was not a good one to launch MDTA; as unemployment declined later in the decade, its effectiveness increased. Because the numbers were so small, however, these theoretical considerations had little practical impact. In a labor force of almost 70 million, over 4 million were out of work in the early sixties. The sponsors of MDTA had hoped that close to a million people would be trained by mid-1965. The actual number who completed training by that time was only a little over 200,000. This small number could be folded into the labor market with relative ease.

Between September 1962 and September 1967, 346,700 persons completed institutional training. Based on three-, six-, and twelve-month USES follow-up surveys, 90 percent got a job during the first year after finishing training and 77 percent were employed when last reached. Three-fourths of those employed at last contact considered their jobs training-related. Of those out of work at last contact, one-third, mainly women, had dropped out of the labor force. A sample study of those who completed institutional training between June 1, 1964, and February 28, 1965, was compared with a sample of their friends, neighbors, and relatives who were also jobless and untrained. Those who had taken training were significantly more successful in getting and retaining jobs. A study of institutional training dropouts showed that one-third quit the program to take jobs, many for jobs for which they had been trained. The Ford Foundation Project on Retraining reached a similar conclusion. The director, Gerald G. Somers, wrote, "In almost all of the surveys at least 75 percent of the trainees were employed after their training." A 1967 study of MDTA OJT completers showed that the great majority had been kept on by the contractor and a few had moved on to another job.

The administration, Congress, and the public at large were reassured by these results. By 1966, according to Wolfbein, MDTA had joined mother and the flag in popularity. He was amazed by the overwhelming votes that came from the Congress. Others were cheered by MDTA and ARA, as well, because they represented the first serious efforts to launch an American manpower policy. Margaret S. Gordon of the University of California, Berkeley, studied the European experience, which was much more developed, and urged "a permanent government retraining program in the United States. Such a step would be consistent with the growing realization that we should move toward a permanent and continuous program of manpower policies."[5]

6

Updating the New Deal

EVELYNE TWILLEY worked in the S. S. Kress store in Gadsden, Alabama, and was a member of Local 506 of the Retail, Wholesale, and Department Store Union. She had been hired in 1956 at 67½ cents an hour. Now, because of the new union contract, her wage had risen to 90 cents. The "girls" at the store were paid between 70 and 98 cents (only one got the top rate). Evelyne considered herself lucky to work in a union shop. The unorganized stores in Gadsden, like Elmore's, F. W. Woolworth, and McClellan's, paid much less.

She was also fortunate in that, unlike other Kress employees, she did not have to contribute to her family's support. In fact, they helped her from their farm at Fort Payne by providing meat, fresh vegetables, butter, and eggs. After taxes and Social Security her take-home pay was $30.18 a week. She paid $25 a month rent for her apartment and had to buy some food, like bread at 22 cents a loaf and milk at 28 cents a quart. She could buy some of her clothes at the store, where Kress gave employees a 10 percent discount. Evelyne made out, but she had to watch the pennies.

In 1960 the federal minimum wage was $1 an hour and retail trade was excluded from coverage under the Fair Labor Standards Act. "I hope you gentlemen will do whatever you can to raise the minimum wage," Evelyne told the House Subcommittee on Labor on March 31, 1960. "We have been watching and waiting for news from Congress for a long time and it would certainly make a great difference in the way we live if we could get the same minimum wage that other people do." She had a lot of company. According to the Gallup Poll, more than 75 percent of Americans favored an increase in the minimum wage.

The time had come, many thought, to bring the New Deal social legislation up to date.[1]

The Fair Labor Standards Act of 1938 was a monument of FDR's New Deal. Its enactment had come in the face of bristling opposition from business and the South and more cautious antagonism from the American Federation of Labor. The President had to call Congress into special session and take his case to the country. While Roosevelt got a law, he needed to make very significant concessions in the amount of the minimum wage and especially in the statute's coverage. The minimum was fixed at 25 cents in 1938, rising to 30 in 1939, and, depending on the industry, to between 30 and 40 cents thereafter until 1945, when it was expected to reach 40 cents. Fewer than 11 million workers were covered, less than one-fourth of the employed labor force, including agriculture. The exemptions were legion: intrastate industries, farming, fishing, seamen, many transportation industries, executive, administrative, and professional employees, retail trade, many service trades, most of construction, and many newspapers.

There were two major amendments between 1938 and 1961. In 1949, in response to the tremendous increase in the cost of living and average wages following the war, the minimum wage was raised to 75 cents. But coverage was actually reduced. In 1955, this time because of the inflation that accompanied the Korean War, Congress raised the minimum to $1 an hour effective March 1, 1956, but made no change in coverage.

By the late fifties there was widespread recognition that FLSA needed an overhaul. Rising prices and wages had eroded the minimum wage and many thought that coverage was a disgrace. Secretary of Labor Mitchell certainly felt this way and even his boss, President Eisenhower, while hardly willing to lead the charge, had called for a modest increase in coverage since 1955. For the AFL-CIO improvements in the minimum wage were the highest priority. This was not because their members would benefit directly (virtually all were paid above the minimum), but because the unions felt responsible to give the unorganized workers a voice. The big Democratic gains in the 1958 elections, particularly in the Senate, seemed to open the door. Senator Kennedy, his eye now fixed on the presidency, was courting labor and the liberals. As chairman of the Subcommittee on Labor, he became the spokesman for improvements in the minimum wage.

Kennedy introduced a bill approved by AFL-CIO on February 16, 1959, and James Roosevelt of California brought it into the House on the same day. It would raise the wage to $1.15 three months after passage and to $1.25 a year later, and it would expand coverage very significantly to 9.5 million additional workers. Kennedy's subcommittee held hearings between May 7 and June 4. The Eisenhower administration countered by introducing a bill on May 15 that would keep the minimum at $1 and expand coverage modestly. The subcommittee approved Kennedy's bill on July 10 but there the matter rested for the remainder of the year.

The House Subcommittee on Labor Standards, with Phil M. Landrum, the anti-union Georgia Democrat, in the chair, held hearings between March 16 and May 19, 1960. While some witnesses supported the Kennedy bill, the great majority were from industries that would be covered for the first time who opposed it strongly. Examples: J. T. Meek, president of the Illinois Retail Merchants Association, said that "retailing has no proper place in this act." R. A. Wright, owner of the Burke Motor Inn in Carroll, Iowa, told the committee that he would have to raise his wages 20 to 50 percent and would lose "most of my clientele." R. A. Wagner, executive director of the Cambria County, Pennsylvania, gasoline dealers association, stated that coverage would "cripple gasoline retailing." Such testimony took its toll.

Roosevelt could not muster the votes in the House. On June 25, 1960, Representative W. H. Ayres, an Ohio Republican, offered a bill that would raise the minimum for those already covered to $1.15 and would extend coverage to some interstate retail chains at $1.00. While he said this would bring in 1.4 million new workers, the Labor Department estimate was 500,000 to 700,000. And there was a big loophole: Macy's, Gimbel's, and Marshall Field, all major interstate chains, would escape coverage. Representative A. P. Kitchin, a North Carolina Democrat, substituted the Ayres for the Roosevelt bill and it passed the House on June 30, 1960, by a vote of 341 to 72.

While the Senate was much more favorable, there was some erosion. One problem in the spring and summer of 1960 was that Kennedy was so busy campaigning for the presidency that he had little time for the minimum wage. During the West Virginia primary Senator Clark said, and the *New York Times* reported, "If sonny boy gets back from the cricks and hollers long enough to have a subcommittee meeting, we will have a bill this year." He did come back and the subcommittee adopted a diluted version of his bill by a vote of 12 to 2, with only Goldwater and Dirksen voting against it. The wage would rise in three steps to $1.25 and coverage extension was reduced to 5 million. On August 18 the Senate passed a still paler version. The wage would move to $1.15 in 1961, $1.20 in 1962, and $1.25 in 1963; coverage was pared to 4 million.

A conference committee sought to reconcile the substantial differences between the houses in late August. Kennedy offered to reduce the wage to $1.15 but refused another concession on coverage. The House members declined his offer, deadlocking the conference. Kennedy said, "We'd rather come back and try to do it in January."

He meant it. When he became President, Kennedy moved with determination and speed to get his bill enacted. On February 2, 1961, less than two weeks after the inauguration, he wrote, "I urge the Congress to raise the minimum wage immediately. . . ." Goldberg was at his heels. He sent over the draft bill on February 6 and the President submitted it to Congress the next day, urging "prompt consideration."

It was much like the 1960 Senate version. For those presently covered the minimum would rise to $1.15 the first year, $1.20 the second, and $1.25 thereafter. Coverage would be extended to local transit, construction, gasoline service, and large enterprises in retail trade. Employees of newly covered firms would have a minimum of $1.00 the first year, $1.05 the second, $1.15 the third, and $1.25 thereafter. FLSA had always required that covered employees be paid time and one-half for hours worked in excess of 40 weekly. In industries not covered millions worked longer hours with either no premium pay at all or with some less favorable provision than time and one-half. The bill proposed gradualism for those newly covered: no overtime pay the first year, time and one-half after 44 hours in the second, 42 in the third, and 40 hours beginning with the fourth year.

Responding to Kennedy's wishes, both houses moved rapidly. The House subcommittee (Roosevelt had replaced Landrum as chairman) held hearings in the latter part of February, the Senate subcommittee at the end of that month and in early March. While the opponents were again present, the proponents were now out in full force from the administration, led by Goldberg, the labor movement, outside experts, and a star witness, Senator Douglas.

By 1961 the arguments over the minimum wage were well over a century old. For anyone with some exposure to labor history they were well worn, if not fatigued. While the spokesmen on each side tried to give them a topical twist, the fundamental reasoning did not change. The case for a legal minimum was basically humanitarian/moral and ran this way: Workers in low-skilled jobs at the bottom of the income scale are, virtually by definition, unorganized and powerless. They include heavy concentrations of blacks, Hispanics, women, and, prior to FLSA, children. Since there is an oversupply, they compete against each other and drive wages down. Employers pay them at the lowest wage needed to obtain their services. This is less, usually far less, income than is necessary to provide a minimum standard of living of health and decency. The only way to break this cycle is for the state to fix a minimum wage.

The proponents restated this argument. Msgr. George G. Higgins, director of the Social Welfare Department of the National Catholic Welfare Conference, favored the bill "on the basis of simple humanity." The minimum wage rested on a "moral basis," the "ethical demand for a living wage." The National Consumers League had supported the minimum wage for over 60 years. Mary Dublin Keyserling, a member of its board, testified for the bill as "an expression both of humanitarian interest and also of the conviction that inadequate wage levels are not only a burden on those directly affected, but are a heavy charge on the community." Goldberg cited President Kennedy, who had said that very low wages "promote the spread of slums, of crime, of disease, of all ills that grow from hopeless poverty."

The opposing argument is at bottom self-interest. Many employers, particularly smaller firms in competitive industries, do not want to share their income with employees any more than necessary. Since they can make better wage deals themselves when the state does not intervene, they oppose the minimum wage. But self-interest, or, as some would put it, greed, does not have an inspiring ring in a public forum. Here neo-classical economic theory provided a sanitized rationale by substituting the worker's interest for the employer's. An employer, the argument goes, does not pay a low wage to exploit his worker, but, rather, because this is his true worth—in economic lingo his marginal value product. If the state intervenes to raise the wage with a legal minimum, the employer will eliminate the now overpriced worker's job, replace him with a machine, or raise prices. The minimum wage, stress those who oppose it, creates unemployment.

This argument, like so much else in neo-classical theory, is correct in extreme cases and at most only partly so in ordinary circumstances. If, for example, Congress had increased the minimum wage to $2 in 1961, there certainly would have been a significant amount of resulting joblessness. But if it was raised in stages to $1.25 over several years, as was the actual case in 1961, the displacement effect would be hardly noticed.

Again, the opponents of the bill restated these arguments. Eugene B. Sydnor, Jr., the president of Southern Department Stores of Richmond, forecast dire consequences if his firm was covered under FLSA. Southern operated 32 "junior" department stores, mainly in small towns in Virginia, but also in the Carolinas. It had 739 employees, overwhelmingly women in their middle years, of whom 324 received less than $1 an hour and who worked an average 42-hour week. If he had to pay $1, Sydnor said, he would have to raise prices, increase imports, lay off between 30 and 40 employees, and cut the hours of about 74 workers to the 23 peak selling hours weekly. If the minimum was raised to $1.25, he would have to at least double these figures.

The Chamber of Commerce of the U.S., which represented many employers exempted from FLSA, like Southern, was the historic adversary of the minimum wage. Its spokesman for many years had been Emerson P.

Schmidt, the director of research, a man not inclined to restraint. The increase in wage costs due to the administration bill could not possibly be taken out of profits and could not be absorbed by rising productivity. It must, Dr. Schmidt predicted, significantly increase unemployment and inflation. The former would aggravate the problems of distressed areas, the latter the balance of payments. Schmidt attached no numbers to these assertions and admitted that he had not examined the problems statistically.

In his academic career at the University of Chicago, Douglas had been the nation's leading authority on wages and had played a key role in both the 1949 and 1955 increases in the minimum wage. He now responded to the labor displacement theory, relying on Labor Department studies. In 1942, when the wage was raised to 32½ cents in four low-wage industries, there was no disemployment effect. The 1949 jump was big, from 40 to 75 cents, 87 percent. A study of five low-paying industries showed no employment declines in three and drops of 4 and 2 percent in the other two, which could have had other causes. The 1956 increase to $1 from 75 cents was 33 percent. Of more than 24 million workers who might have been affected, no more than 1800 were laid off. Since the 1961 proposal was for a 25 percent increase over three years, Douglas expected even less displacement this time. He anticipated virtually no impact on U.S. exports. There had been none after the last round and almost all exporters paid wages above the minimum. As the author of the Area Redevelopment Act, he would have been much concerned over an adverse impact on depressed areas. He anticipated none.

The House subcommittee sent the bill to the full Committee on Education and Labor on February 28, 1961, with only one change: the three-year phase-in was shortened to two in two steps: $1.15 and $1.25. The committee accepted the bill with this change on March 13 by a vote of 19 to 12. But there was trouble on the floor. On March 24 the House again substituted Ayres-Kitchin for the Kennedy bill, this time by a vote of 216 to 203. This probably was a political maneuver by certain representatives, many from the South, who expected the administration bill to pass but hoped to get additional industries exempted from coverage in conference.

The Senate committee adopted the bill on April 11 by a vote of 13 to 2 (Goldwater and Dirksen again dissenting). The Senate adopted it 65 to 28 on April 20. The wage advance would be in two steps—$1.15 for two years and $1.25 in the third. Coverage would be extended to 4.1 million new workers.

The bargaining in conference was over coverage. As the price for votes on the floor, the House conferees succeeded in excluding cotton ginning and other farm-related groups, laundries, and auto and farm equipment dealers. On May 3 both houses acted favorably, the Senate 64 to 28, the House 230 to 196. In the latter a number of southern Democrats and a handful of

Republicans switched to support. Kennedy signed the law on May 5, 1961.

The basic 1961 amendments to FLSA were as follows: For currently covered workers the minimum wage would become $1.15 on September 3, 1961, for two years and $1.25 thereafter. For those newly covered the minimum would be $1 for three years, $1.15 in the fourth, and $1.25 in the fifth. There would be no weekly overtime requirement for the latter group for two years. In the third, time and one-half would be paid after 44 hours, in the fourth after 42, and in the fifth year after 40. Coverage was extended to 3,624,000 new workers, 2,182,000 in retail trade, 1 million in construction, and lesser numbers in gasoline service stations, local transit, fish processing, and seamen and telephone operators.

This was, as the President noted, by far the most significant change in FLSA since 1938. Since March 1956, when the minimum went to $1, Senator Douglas pointed out, an increase of 11 cents was needed merely to compensate for the rise in consumer prices. If this was multiplied by the gains in productivity, a wage of $1.254 was justified, almost exactly the 1961 amount. The 1949 and 1956 adjustments gave low-wage workers 53 percent of average hourly earnings in American industry. An increase to $1.25 now would restore this precise relationship. The gains in coverage were long overdue and significant, but were barely a start—only 27.5 million workers would now be covered out of 70.8 in the labor force. Kennedy was particularly annoyed over the exclusion of laundry employees. *Congressional Quarterly* observed that Congress had concluded "a bitter minimum wage battle by passing landmark legislation," an encounter from which the President had emerged with "a major victory."[2]

With the exception of highly placed women such as Eleanor Roosevelt and Frances Perkins, who served as role models, the New Deal did virtually nothing to promote equal rights for women. The issue was simply not on the political agenda. The women's organizations that had fought successfully for the right to vote and for protective labor legislation during the Progressive Era were politically exhausted. They shunned the use of the word "feminist." The Equal Rights Amendment, introduced in the twenties, split these organizations by drawing support from business and professional women and opposition from those who feared that ERA would undermine the protective laws. In any case, ERA was a non-issue. This public indifference continued for another generation. The question of equal rights for women finally emerged in the sixties, but in a peculiar way which involved the Fair Labor Standards Act.

When Kennedy was elected President he was essentially neutral (perhaps uninvolved is a better word) regarding women's rights. This was not because

The Council of Economic Advisors strongly urged a tax cut to spur the economy. FROM LEFT: Chairman Walter W. Heller and members Kermit Gordon and James Tobin. *Kennedy Library*

Secretary of the Treasury C. Douglas Dillon, here with Kennedy, was reluctant to go along with tax reduction. *Kennedy Library*

In a notable address at the Yale commencement on June 11, 1962, Kennedy strongly endorsed the "New Economics." *Kennedy Library*

Here Kennedy on September 10, 1963, addresses a businessmen's organization that backed tax reduction. *Kennedy Library*

Congressional leaders at the White House. STANDING FROM LEFT: Speaker John McCormack; Senate Majority Leader Mike Mansfield; Senator Tom Kuchel of California; Kennedy; Johnson; House Ways and Means Committee chairman Wilbur Mills; Goldberg; House Minority Leader Charlie Halleck. Mills played a central role in tax and Medicare policies. *Kennedy Library*

Signing the Area Redevelopment Act on May 1, 1961. Its author, Senator Paul H. Douglas of Illinois, looks over the President's shoulder. *Kennedy Library*

The President with Secretary of Labor Willard Wirtz, who succeeded Goldberg when he was appointed to the Supreme Court in 1962. *Kennedy Library*

Kennedy signing the minimum wage law, May 5, 1961. FRONT ROW: (*first from left*) David Dubinsky of the Ladies' Garment Workers; (*fourth*) Senator Robert Kerr of Oklahoma; (*fifth*) Representative Adam Clayton Powell; (*seventh*) Representative James Roosevelt; (*eighth*) Secretary Goldberg; (*ninth*) George Meany, president of AFL-CIO; (*tenth*) House Whip Carl Albert. *Kennedy Library*

Kennedy giving the pen with which he signed the Manpower Development and Training Act on March 15, 1962, to its author, Senator Joseph S. Clark, Jr., of Pennsylvania. *Kennedy Library*

Modern women's rights policy emerged in the Kennedy administration with Esther Peterson, head of the Women's Bureau, leading the way. Here (*in white*) she stands behind the President as he signs the Equal Pay Act on June 10, 1963. Peterson was also responsible for the report of the Commission on the Status of Women. *Kennedy Library*

Kennedy campaigning for Medicare. *Kennedy Library*

Senator Robert Kerr showing the President his prize cattle at his baronial spread in Oklahoma. Kerr masterminded the Senate defeat of Medicare in 1962. *Kennedy Library*

The Health, Education and Welfare team meeting with the President at Palm Beach on December 28, 1962, to map out the 1963 strategy for federal aid for education. FROM LEFT: Secretary Anthony Celebrezze, Kennedy, Commissioner of Education Francis Keppel, and Assistant Secretary Wilbur Cohen. *Kennedy Library*

Senator Wayne Morse of Oregon steered the 1963 education legislation through the Senate. Here with the President. *Kennedy Library*

The President giving the pen with which he signed the Peace Corps Act on September 22, 1961, to his brother-in-law Sargent Shriver. Shriver headed the team that drafted the bill, pushed it through Congress, and became director of the Peace Corps. Senator Humphrey, who sponsored the bill, is to the left of the President. *Kennedy Library*

Peace Corps volunteer M. L. Corwin with Filipino children. *Kennedy Library*

President Kennedy at his final press conference on November 14, 1963. Here he predicted that his entire legislative program would be enacted within eighteen months. Had he lived, it would almost certainly have occurred. But eight days later he was assassinated in Dallas.
Kennedy Library

John F. Kennedy's body leaving the White House for the Capitol on November 23, 1963. Note the riderless horse. *Kennedy Library*

RIGHT: President Kennedy's casket lay in state on the Lincoln catafalque in the Capitol Rotunda on November 24–25, 1963. Many thousands of citizens paid their respects. © *Daniel Budnik/Woodfin Camp & Associates*

Among those grieving at the Kennedy funeral: LEFT: His friend Senate Minority Leader Everett Dirksen of Illinois. BELOW: Former Presidents Harry Truman and Dwight Eisenhower. *Kennedy Library*

he denigrated the competence of women. Esther Peterson, who worked closely with him, said that he always treated her as a person without regard to gender and that he valued her opinions. Rather, he did not consider women's rights a significant issue at the time. This was evident in his talent search for "the best and the brightest," which was confined to individuals already at the top in their fields, ignoring the fact that women, hobbled by historic discrimination, were unable to get that far. He did ask Margaret Price, director of women's activities at the Democratic National Committee, for a memorandum on the performance of recent administrations in the appointment of women. She reported in December 1960 that Wilson, Franklin Roosevelt, Truman, and Eisenhower had named a substantial number. She appended a list of prominent female Democrats that she thought he should consider. In March 1961 Mrs. Roosevelt came to the White House to give him another list, three pages of names, of qualified women. But Kennedy named no woman to the cabinet or to any other top job, though the total number he appointed did not compare unfavorably with Truman and Eisenhower. This, obviously, was not an important question for him.

One woman he did bring into his administration was Esther Peterson, who was on Price's list. Of Danish extraction and Mormon heritage, she had been raised in Utah and had attended Brigham Young University. During the depression she worked for the Amalgamated Clothing Workers and after the war she became the union's legislative representative in Washington. She developed a good working relationship with a new member of the House from Massachusetts, John F. Kennedy, who sat on the Labor Committee. Between 1948 and 1957 Peterson was in Europe where her husband was the American labor attaché in Stockholm and Brussels. She returned to Washington in 1958 as legislative representative for the Industrial Union Department of AFL-CIO. She again dealt with Kennedy, now chairman of the Senate Labor Subcommittee. In 1960, Peterson went to work full time on his presidential campaign.

After the election her old friend Arthur Goldberg offered her a choice of jobs in the new administration. Hardly ambitious, she asked to be named director of the Women's Bureau. She thought she might be able to help "the disadvantaged women, those with the low wages, and who have a lot of kids, who don't make enough money, the people in the laundries and the industries that have just not kept up with our times." Kennedy wondered why she wanted to head so obscure an agency, but granted her wish. At Goldberg's urging, he also appointed her assistant secretary of labor in charge of women's affairs, which made her the highest ranking woman in the government.

Esther Peterson knew where she was going. She was politically savvy, had a direct line to the White House, and had a clear idea of what she wanted to accomplish. She had little use for ERA and the kind of women who backed it. Peterson thought it was a threat to the protective laws and, if adopted, would benefit women in business and the professions, who least needed help. She was concerned about working women, most of whom needed this state protection because they were not covered by FLSA. Further, having battled the National Association of Manufacturers and the Chamber of Commerce for years on labor issues, she thought it suspicious that they should endorse ERA. Peterson was not upset that Kennedy had not named women to high positions because she did not consider it an important question.

Peterson's strategy was to shift the debate away from ERA toward a broad re-examination of the position of women in American society through a high-level commission. She wanted to give the new President "a substantive program he could put his teeth into." Until 1961 this proposal was assumed to require legislation. Goldberg suggested doing it by executive order and Kennedy readily agreed. There was some fuss over the chair. Peterson wanted Mrs. Roosevelt, whose life-long commitment to women's rights and unchallenged stature made her the obvious choice. The White House staff was concerned because she had supported Stevenson for the nomination and her relationship with Kennedy was hardly warm. Peterson insisted, "You've got to have *the* best chairman or this won't amount to anything." Kennedy conceded on condition that Peterson must persuade her to accept. Mrs. Roosevelt was reluctant because of poor health, but Peterson convinced her of the importance of the undertaking for a cause dear to her. She said, "This may be the last thing I really do." In fact, she died in 1962 before the commission reported. Peterson and Roosevelt carefully selected the other members to be certain that a majority would not endorse ERA. Actually, only one of the 26 members was a firm supporter of the amendment.

Kennedy issued Executive Order 10980 establishing the President's Commission on the Status of Women on December 14, 1961. He gave it broad authority to investigate and make recommendations for action on the role of women in many areas of the nation's life. Roosevelt became chairman, Peterson executive vice-chairman, and Richard A. Lester of Princeton, a noted labor economist, vice-chairman. The other members included five cabinet officers, the chairman of the Civil Service Commission, two senators, two representatives, and an assortment of educators, union leaders, representatives of church organizations, and prominent women.

Over the preceding generation there had been enormous changes in the status of women in U.S. society. "The time was right," Margaret Mead, the noted anthropologist, wrote in the introduction to the report, "for a new stocktaking. . . . *American Women* marks where we now stand." The commission delivered its conclusions and recommendations to President Kennedy on October 11, 1963.

"This report," the commission wrote, "is an invitation to action." The more important recommendations were the following: Discrimination based on sex should be challenged not by constitutional amendment but rather by suits to establish "the principle of equality" under the Fifth and Fourteenth amendments. Child-care facilities should be provided for working mothers with the costs borne by the families, voluntary agencies, and public funds. Paid maternity leave should be available to women who work. A widow's benefit under Social Security should be raised to the amount her husband would have received had he lived. Discrimination should be eliminated from state laws affecting the family and the property rights of married women. FLSA coverage should be broadened to include industries which employed many women—hotels, restaurants, laundries, retail trade, agriculture, nonprofit organizations. The President should issue an executive order providing equal opportunity in employment for women in private firms working under federal contract. The Civil Service Commission should facilitate wider use of part-time employment in the federal service.

Finally, the commission recommendation on equal pay read: "State laws should establish the principle of equal pay for comparable work." Two aspects of this statement require comment. The first is the use of the word "comparable" rather than "equal" to modify work. The former, of course, was much broader. The other is that the commmission was silent about federal law. This was because the President had signed the Equal Pay Act on June 10, 1963, four months before the commission reported.

"Equal pay for equal work" for women was an old principle that had made repeated policy appearances. Both the Industrial Commission of 1898 and the Commission on Industrial Relations of 1915 had endorsed it. The National War Labor Board of World War I had applied it in over 50 cases. The Classification Act of 1923 had eliminated salary differentials based on sex in the federal service. During World War II the second NWLB had made equal pay for equal work General Order No. 16, which introduced the principle widely in American industry. In 1946 the women's equal pay bill passed the Senate, though it failed in the House. While many bills were introduced and there were desultory hearings in 1950, nothing happened at the federal level between 1946 and 1962. Meantime, by 1962, 20 states had enacted equal pay laws. Virtually all were in the North and they varied significantly in effectiveness. In addition, a study of 510 union contracts in 1956 showed that 195, 38 percent, had equal pay clauses.

Peterson thought the time was ripe for federal legislation. She studied the state laws and "concluded very definitely that we had to have a federal law." She put her troops to work in the Women's Bureau collecting data and framing and answering arguments. She persuaded Goldberg to approve and the solicitor's office to draft a bill. The White House, however, did not help. Peterson said, "We were on our own."

The bill was introduced early in 1962. At its heart was a prohibition on wage differentials because of sex. An employer in commerce would be forbidden to "discriminate . . . on the basis of sex by paying wages to any employee at a rate less than the rate he pays to any employee of the opposite sex for work of comparable character on jobs the performance of which requires comparable skills. . . ." There was this exception: unequal wages paid "pursuant to a seniority or merit increase system which does not discriminate on the basis of sex."

The bill raised the semantic question noted above: Would the principle of nondiscrimination apply to "equal" work or to "comparable" work? The answer was critical in determining the reach of the bill.

Prior to World War II many employers, perhaps a large majority, paid women less than men for the same work. Wartime demand for female labor, the NWLB policy, and union pressure combined to turn this around in the postwar era. Walter Fogel, who made the most careful analysis of this issue, concluded, "There existed relatively little unequal pay for equal work when the EPA was passed in 1963." The real differential between women and men in the labor market was not over hourly wage rates, but, rather, because of annual incomes. In 1960, a typical year, women's median yearly incomes were $3,293, compared with $5,417 for men, 61 percent. This was primarily because employment was sharply differentiated by sex and because "women's" jobs paid much lower wages than "men's" jobs. Thus, a law that required the same pay for equal work would have little impact. If the statute used the criterion of "comparable" work, it would apply more widely, perhaps much more so, depending on the legislative history, the enforcement policy, and the court interpretations. Peterson and Goldberg, of course, understood this and offered the comparable work standard in the hope of substantially raising women's incomes.

Another provision of the bill that was certain to arouse controversy was its coverage. It would apply to every employer with employees who were engaged in interstate or foreign commerce or who produced goods for commerce. There were no exemptions. Thus, coverage would much exceed that under FLSA.

The legislative history of the Equal Pay Act in 1962–63 was a steady retreat from the Labor Department positions on comparable work and coverage. EPA, therefore, repeated the experience of FLSA in 1938. While business did not mount a major open campaign against the bill, it worked effectively behind the scenes with Republicans and southern Democrats to limit its impact severely.

The high point for the administration came in the report of the House Committee on Labor on May 17, 1962. Comparable work was preserved; employers with federal contracts over $10,000 must pay both sexes equally; and the employer was forbidden to reduce male wages in order to establish equality. But the committee made the first concession on coverage by exempting employers with fewer than 25 employees.

When the bill moved to the floor of the House the concessions increased. The law would be limited to a single plant or establishment. That is, it would be legal for a multi-plant firm to pay a higher rate to a man in plant A than to a woman in plant B for the same work. More important, at the urging of Katharine St. George, the New York Republican, the comparable work standard was eliminated and replaced by equal work. This was in the face of a strong letter from Goldberg arguing that it would defeat the purpose of the bill. St. George, Peterson said, "just makes me furious because here's this woman who was supposedly in our corner who speaks for the opposition." This "ERA type . . . really wrecked us." The House passed the measure with these changes on July 25, 1962, by voice vote.

There was more trouble in the Senate. In order to get something passed, the sponsors tied a bill resembling the House version of EPA to a foreign service building bill on October 3. The maneuver failed to win unanimous consent a week later.

In 1963 the Labor Department started over again, this time in the Senate. Earlier Representative Charles Goodell, the progressive New York Republican, had urged the FLSA precedent—that is, make EPA part of FLSA, accept the same coverage, and combine enforcement of both in the Wage and Hour Division of the Labor Department. This was an enormous concession on coverage, exempting women in agriculture, hotels, restaurants, laundries, smaller retail establishments, and administrative and managerial positions. Peterson was pleased to have enforcement in Wage and Hour because it was the logical location. The Labor Department, eager to get anything, agreed to these changes. The bill then sailed through the Senate on May 17 with no opposition.

The House bill much resembled the Senate's, including Goodell's proposals on FLSA. On May 23 the House weighed its version against the Senate's and adopted its own. On May 28 the Senate went along. The President signed the Equal Pay Act on June 10, 1963. Peterson, bitterly disappointed, bit her lip and told the press that "it was a marvellous beginning." Goldberg commended her for her skill in dealing with the media.

While the new law was known as the Equal Pay Act, it actually became Section 6(d) of the Fair Labor Standards Act. The critical provision read as follows:

> No employer having employees subject to any provisions of this section shall discriminate, within any establishment in which such employees are employed, between employees on the basis of sex by paying wages to employees in such establishment at a rate less than the rate at which he pays employees of the opposite sex in such establishment for equal work on jobs the performance of which requires equal skill, effort, and responsibility, and which are performed under similar working conditions, except where such payment is made pursuant to (i) a seniority system; (ii) a merit system; (iii) a system which measures earnings by quantity or quality of production; or (iv) a differential based on any other factor other than sex: *Provided,* That an employer who is paying a wage rate differential in violation of this subsection shall not, in order to comply with the provisions of this subsection, reduce the wage rate of any employee.

Historically speaking, the Equal Pay Act of 1963 was an anomaly. The civil rights issue, including the end of discrimination in the labor market, was an important fact of American life in 1962–63. But the civil rights movement was fully engaged in achieving this goal for blacks without assuming the added responsibility for women. Black women were far more concerned with race than with sex. The feminist organizations—the National Federation of Business and Professional Women's Clubs, the American Association of University Women, the YMCA, and others—testified, but it is probably an exaggeration to say that they had even slight influence. (The first of the "modern" feminist groups, the National Organization of Women, was not formed until 1966, three years *after* passage of EPA.) The EPA was a one-woman accomplishment for which Esther Peterson deserves the credit. It was a symbolic triumph but, if one accepts Fogel's analysis, a victory with little substance. Virtually everyone in Congress voted for it, even Senator Goldwater, because almost no one opposed the principle of equal pay for equal work. Members of Congress could support it with impunity because they felt certain that it would have little or no practical impact.

The President's Commission on the Status of Women was far more important. In the executive order Kennedy had called for overcoming "prejudice and outmoded customs" in order to achieve the "full realization of women's rights." Cynthia Harrison, who has studied the commission most diligently, thought that it did much to accomplish this aim. It became, she wrote, "the starting point for governmental discussion of women's status that continued for at least two decades." For the Kennedy administration it was "the jewel in its crown for women."[3]

One of the New Deal's most imposing achievements was a national collective bargaining policy. This was the Wagner Act of 1935, the National Labor

Relations Act (NLRA), along with amendments to the Railway Labor Act, which applied similar principles to the railroads and the airlines. While the Wagner Act was altered significantly by the Taft-Hartley amendments of 1947 and the Landrum-Griffin amendments of 1959, they had only slight impact upon the law's representation system. Workers went about selecting representatives for collective bargaining in 1961 pretty much the way they had in 1937. This was because the procedure had been hammered out carefully from tough experience and because it was eminently democratic.

This representation procedure, set forth in Section 9 of the act, consisted of these elements: First, there must be a determination of the "unit appropriate," which would settle both who was eligible to vote in an election and the precise boundary of the bargaining unit if the union won. Second, the National Labor Relations Board (NLRB) would conduct a secret ballot election or its equivalent, a check of membership cards. In an election those eligible to vote would cast ballots either for the union or for no union. Third, the result would be determined by the majority rule. If no union won, that was the end of it. If the union won, the NLRB would certify it as the lawful bargaining representative of the employees in the unit appropriate. Fourth, the victorious union would be the exclusive representative of all the employees with both the right and the duty to represent all of them. Finally, the employer would have the duty to bargain with the certified union (with Taft-Hartley the union also would have the duty to bargain with the employer).

But according to Section 2 of NLRA, employers subject to the statute "shall not include the United States, or any State or political subdivision thereof." Thus, employees in the public sector—federal, state, and local—were excluded from coverage. This did not forbid public workers to engage in collective bargaining, but it prevented them from doing so within the carefully crafted NLRB system.

Outside the Board's procedure a major roadblock stood in their way—the concept of sovereignty. Derived from English common law, it held that the sovereign is the supreme and exclusive source of power, which cannot be shared. In a democracy like the United States the people are sovereign and they delegate their power to their elected officials alone. In the employment relations area, Kurt L. Hanslowe pointed out,

> governmental power includes the power, through law, to fix the terms and conditions of government employment, that this power reposes in the sovereign's hand, that this is a unique power which cannot be given or taken away or shared, and that any organized effort to interfere with this power through a process such as collective bargaining is irreconcilable with the idea of sovereignty and is hence unlawful.

Many, probably most, jurisdictions held to this theory and their courts sustained it. In particular, they insisted that a strike against the government was abhorrent to public policy.

In fact, however, unionism and some form of bargaining in the federal service went back more than a century. By the 1950s there were many labor organizations active among federal government employees, with three notable areas of concentration—the Post Office, the Tennessee Valley Authority, and the Department of the Interior, particularly the Bonneville Power Administration.

In 1961, 762,000 federal workers belonged to employee organizations, making up one-third of federal employment. Over one-half, 489,224, worked in the Post Office, which was 84 percent organized. They were members of a dozen different unions or associations based on craft (letter carriers, clerks, and so on) and race (white and black letter carriers). But the Post Office Department never adopted a formal labor relations policy and never fully accepted collective bargaining, though it did so sporadically in local agreements. Policy seemed to reflect the views of whoever happened to be Postmaster General at the time.

The Tennessee Valley Authority adopted a contrary policy, virtually full acceptance of the NLRA collective bargaining model. The TVA's "mission" had multiple purposes—power generation, fertilizer production, navigability and flood control on the Tennessee River and its tributaries, reforestation and conservation, agricultural and industrial development, and, as the founding statute in 1933 put it, "other purposes." Thus, TVA was created as an autonomous public corporation run by a board of directors with exceptionally broad authority, including exemption from the federal civil service.

In 1933 the board adopted a "short-range goal" of twelve years for the building of nine dams and related facilities to provide electric power, navigation, and flood control. TVA, therefore, immediately became an enormous construction enterprise. There were two ways to go: force account (doing the building itself) or low-bid contracting (hiring the least expensive private construction firm to perform the work). The latter had worked poorly at Boulder Dam. Further, the board preferred force account in order to devise a sound and enduring system of labor-management relations, to provide employment in the appallingly depressed Tennessee Valley, and to train manpower for similar projects in the future. Two roughly similar major programs that emerged later—atomic energy and space—would adopt the opposite policy of contracting out, and neither would develop a collective bargaining policy.

As a construction operation, TVA was required automatically to deal with the building trades unions. This meant accepting union organization and bargaining based on craft and the principle of the exclusive jurisdiction of each union.

While both the TVA board and staff were convinced that the agency should adopt a policy approximating private sector collective bargaining, there were serious preliminary problems: the absence of a federal policy, the concept of sovereignty, and hostility to unions in the South. This led to two years of study and consultation with several of the nation's leading authorities on collective bargaining, Dr. William M. Leiserson, then chairman of the National Mediation Board, and Otto Beyer, at the time director of labor relations for the Federal Coordinator of Transportation.

In 1935, TVA published its Employee Relationship Policy, which was a little Wagner Act adapted to its situation. Employees received the right to organize and designate representatives of their own choosing free from the interference of management. Bargaining units would be based on craft and TVA would encourage the unions to form central bodies. Elections would be held under the majority rule for the selection of representatives. This policy led to the selection of fourteen unions, which in 1937 joined in the Tennessee Valley Trades and Labor Council. Three years later TVA and the Council, now representing fifteen unions, signed the first general agreement covering blue-collar employees. TVA sought a similar agreement with its seven white-collar unions, two representing professional employees, but ran into many difficulties, including the inability of the unions to join in the Salary Policy Employee Panel. In this area, therefore, the pattern of individual union agreements emerged. In any case, the entire TVA system came to operate under collective bargaining pretty much in conformity with the NLRA model.

The Interior Department was an old-line cabinet-level department which differed markedly in structure from TVA's public corporation. During the thirties craftsmen employed by the Bonneville Power Authority (BPA), accustomed to collective bargaining with private utilities in the Northwest, pressed for union representation. They formed the Columbia Power Trades Council. In 1944, BPA recognized that the problem must be faced and proceeded to negotiate the first agreement the next year. Interior approved and in 1948 the department issued a policy statement allowing its subordinate agencies to bargain with their employees. The Reclamation Bureau made an agreement with craft workers at the Grand Coulee Dam, and other bureaus followed. In 1959, BPA took the lead in developing a dual system: a Basic Labor Agreement setting forth broad policies, and Supplementary Labor Agreements prescribing in detail wages, hours, and working conditions.

The postwar economic expansion, urbanization, and baby boom created an enormous need for public services, particularly at the state and local levels. Between 1946 and 1960 public employment as a whole grew from 5.6 million to 8.5 million or 16 percent of nonagricultural employment. The federal payroll increased quite modestly. By contrast, the states almost doubled their employment, from 804,000 in 1946 to 1,592,000 in 1960, about one-third of the latter in education. Local government employment leapt upwards from 2.8 million in 1946 to 4.8 million, with half in education.

This rapid growth created many tensions in employment relationships, primarily because public wages and salaries lagged behind those in the private sector. Many government employees formed and joined unions and demanded collective bargaining. But the sovereignty concept and court decisions reinforcing it prevented governments from developing policies to accommodate these demands. Thus, political pressure built up during the fifties to junk the sovereignty idea and to create a movement for collective bargaining in the public sector. This historic process emerged first in the cities, barely surfaced in the states, and blossomed in the federal government.

Unionization of municipal employees grew rapidly during the fifties. By 1955, 874 of the 1,347 cities with a population over 10,000 including all 18 of those with over half a million people, had such organizations. The American Federation of State, County and Municipal Employees (AFSCME), which organized on an "industrial" basis, was the largest union with members in 365 cities. The Fire Fighters were active in 614 municipalities, police unions in 60. Many engaged in what could loosely be called collective bargaining. The most dramatic breakthroughs occured in Philadelphia and New York City, both already highly unionized in the private sector.

Locals of the Fraternal Order of Police and the Fire Fighters had been founded in Philadelphia early in the century. A special city ordinance in 1939 authorized the first written agreement with Local 222 of AFSCME covering a few hundred workers in the Department of Public Works on a limited number of issues. In 1944 the city recognized District Council 33 as the successor to Local 222. The 1951 Home Rule Charter assigned authority over pay, classifications, and other matters to the Personnel Department and the Civil Service Commission, subject to approval of the administrative board (the mayor, director of finance, and managing director).

In 1952, Joseph S. Clark, Jr., became the reform mayor of Philadelphia. Despite the absence of a state law empowering cities to engage in collective bargaining, Clark was much interested in introducing a system in the city that would approximate the private sector model. He brought in Eli Rock, a labor relations expert and prominent arbitrator, as consultant and relied heavily on his advice.

For Rock the critical issue was the appropriate unit. While unit determination was important in the private economy, it was crucial in the public sector. It affected the scope of bargaining, the role played by various branches of government, the peaceful settlement of disputes, the orderliness of the bargaining process, and "ultimately," Rock wrote, "perhaps the success of the whole idea of collective bargaining for public employees." His primary concern was proliferation, "excessive fragmentation." There was no choice about uniformed police and firemen, whose "community of interest," as the NLRB put it, was very sharply defined. For others, though small units were not necessarily bad, Rock thought they should be avoided if at all possible.

Thus, Clark and Rock opted for the broadest possible bargaining unit in the interests of administrative efficiency and harmony. Their pragmatic philosophy carried them the rest of the way to the Wagner Act model: recognition of the union once its majority was demonstrated and then good faith bargaining. Over the next few years under Mayors Clark and Richardson Dilworth Philadelphia developed three collective bargaining systems—for police, for firemen, and for general employees. The unions of uniformed employees quickly demonstrated their majorities and negotiated comprehensive agreements. At the outset AFSCME, because it had not gained majorities everywhere, made limited agreements. By 1958, however, District Council 33 had surmounted this hurdle and entered into an agreement which gave it exclusive bargaining rights for all city employees who were nonuniformed, nonsupervisory, and nonprofessional. Except for absence of the right to strike and arbitration as the terminal step in the grievance procedure, this agreement was no different from a sophisticated contract in the private sector. In 1960 union security in the form of maintenance of memberhsip was added. In that year the police and fire unions had 8,112 members among the 8,519 employed; AFSCME had 10,158 of the 19,000 general employees.

New York City achieved results similar to Philadelphia's for its more than 200,000 employees, but with an entirely different style. Here, too, a new mayor launched the program. One did not need to search far for the source of his interest in collective bargaining policy. Robert F. Wagner, Jr.'s father was also the father of the Wagner Act.

Wagner became mayor in 1953 and created a new department of labor the next year with the express responsibility for developing a collective bargaining policy for city employees. Ida Klaus, a labor lawyer who had done legal work for the NLRB, became the key figure as counsel to the department. Her philosophy, she wrote, was for the city to formulate "a definitive and systematic program enunciating fundamental rights and procedures." Klaus drafted and in 1954 the mayor issued the Interim Order on the Conduct of Labor Relations. It was certainly "interim." Unlike Clark, Wagner approached problems in easy steps. The order merely guaranteed the right of employees to organize, ordered departments to establish grievance procedures, and provided for union-management committees to discuss working conditions in agencies that were substantially unionized. During 1954–55 the Department of Labor conducted extensive research into public sector collective bargaining, which led to the publication of nine monographs. The latter, in turn, became the subject of public hearings. After three years of study, Klaus wrote and the department issued a 110-page statement in June 1957, *Report on a Program of Labor Relations for New York City Employees,* culminating in a proposed executive order along with rules and regulations.

The mayor, heavily engaged in his 1957 re-election campaign, hardly mentioned the program. He won overwhelmingly and on March 31, 1958, issued Executive Order 49.

In effect, Klaus had written a little Wagner Act for the city, even borrowing generously from language of the statute. The more important provisions were the following: Employees would have the right to self-organization for collective bargaining and the equal right to refrain from such activity. A union designated by a majority in the unit appropriate would become the exclusive representative. The commissioner of labor would decide unit questions. The standards to be applied, largely taken from NLRA, were vague. The Department of Labor would conduct elections and card checks and would certify the majority representative. The commissioner of labor would also have the reponsibility to resolve substantive disputes.

Since Wagner was uncertain of his authority, he covered only the agencies whose heads he had appointed and even here excluded the police. Thus, only half the city's employees came under the order at the outset. The exclusions were important—education, higher education, transit, housing, the Triborough Bridge and the tunnels, cultural institutions, the courts, and a number of city and county offices, as well as the police. Over time many of these agencies came into the program.

The commissioner of labor got into trouble at once over unit. Instead of devising a rational scheme of representation, as Philadelphia had done, he simply waited for the unions to file claims. They did so in large numbers and often in peculiar ways. Over the first decade of the program the commissioner issued almost 800 certifications, at least one for a unit of only two employees. "The result," Raymond D. Horton wrote, "was a crazy patchwork of bargaining units, excessive in number and highly complicated."

Because of these problems with unit, no one knows just how much unions and bargaining grew after the issuance of the order, though there is no doubt they increased rapidly. It was not until 1970, after several important strikes, that the system stabilized. By that year the city had over 400,000 employees. While there were many unions, the big six had 80 percent of the membership. AFSCME, District Council 37, which had 20,000 in 1961, had over 90,000 in 1970. The United Federation of Teachers, formed in 1961, had 70,000 members in 1970. By that time the Transport Workers and the Patrolmen's Benevolent Association were both over 25,000. The Fire Fighters had 11,000 and the Sanitationmen had enrolled 10,000 members. By 1970 certainly, and probably several years earlier, virtually all of New York City's eligible employees were union members covered by collective bargaining agreements.

"The states," Jean T. McKelvey wrote, "have been laggards. . . . The cities have been innovators and the states . . . have been slow to erect constructive, as opposed to restrictive, legislation governing public employees." In fact, prior to the launching of the federal program, only Wisconsin had acted and its legislation was puzzlingly truncated. In 1959 that state passed an enabling statute: Public employees received the right to organize and join unions for the purpose of collective bargaining as well as the right to refrain from doing so. Public employers were forbidden to restrain employees in the exercise of these rights. That was all. In 1962 the legislature took another step by authorizing the Wisconsin Employment Relations Board to conduct representation elections and to engage in mediation and fact finding in labor disputes.[4]

The movement for public sector collective bargaining, so marked in the cities and beginning in the states, emerged inevitably in the federal service. For more than a decade bills had been introduced in Congress. The most important, Rhodes-Johnston, had been authored by Representative G. M. Rhodes, a Pennsylvania Democrat, and was sponsored by Senator Olin D. Johnston, a South Carolina Democrat. It had the backing of the government unions, AFL-CIO, and, in a general way, of Senator Kennedy when he had been chairman of the labor subcommittee. The bill had gotten nowhere during the Eisenhower years, but Kennedy's election opened the door. According to a Department of Defense count in March 1961, there were nineteen federal collective bargaining bills in the congressional hopper.

Rhodes-Johnston was an anomaly and bore little resemblance to the Wagner Act model. It provided no means of determining appropriate unit, of selecting a representative by secret ballot election, or of granting a union exclusive representation. Thus, there could be multiple unionism in the same unit, an invitation to disorder, if not chaos. The bill merely provided that an agency would have the obligation to bargain with a union when it "represents" the employees in a unit. Supervisors would be punished, even dismissed, if they failed to comply. Disputes would be subject to mediation by the Federal Mediation and Conciliation Service and to final arbitration. Unions would be entitled to the checkoff by payroll deduction of their dues and fees.

Early in 1961 the House Committee on Post Office and Civil Service asked the Department of Defense for its views on this bill. The department had many reactions, virtually all negative. On March 22, Cyrus R. Vance, Secretary of the Army, set forth these objections to Budget Director Bell. He was "acutely aware" that pressure for legislation was building because of "the lack of a definitive executive policy." Vance thought it could be deflated if he could tell Congress that the President was considering an executive order "embodying those principles of the proposed legislation which are desirable and in the public interest." He therefore submitted the draft of such an order. Though it differed in almost every way from Rhodes-Johnston, it came no closer to the NLRA model. The Budget Bureau brought Myer Feldman of the White House staff into the discussion and solicited the views of the Civil Service Commission, the Labor Department, and the Post Office. The upshot was that in April 1961, the President asked Goldberg to head "an informal study group to make recommendations." It would also consist of a designee of the Secretary of Defense, the chairman of the Civil Service Commission, the director of the Budget Bureau, and the Postmaster General.

The study group quickly concluded that Defense was correct in urging an executive order. The alternative of proposing legislation to Congress was likely to be complex, time-consuming, and uncertain. If the adminstration took the legislative route, it would have to confront Rhodes-Johnston with no satisfactory substitute. But drafting an order was also a complex process which would require careful study.

On June 22, 1961, therefore, the President issued a memorandum on employee-management relations in the federal service to the heads of all departments and agencies. He enunciated several broad policies: the right of federal workers to form employee organizations to deal with labor issues, "consultation" by management with these organizations, membership without discrimination based on race, color, religion, or national origin, and refusal by federal officials to deal with organizations "which assert the right to strike against or advocate the overthrow of the government of the United States."

The President also created the Task Force on Employee-Management Relations in the Federal Service to study the problem and make recommendations to him. Goldberg was again named chairman, Civil Service Commission chairman John W. Macy, Jr., became vice-chairman, and the other members were Defense Secretary Robert McNamara, Postmaster General J. Edward Day, Budget Director David Bell, and Theodore Sorensen, special counsel to the President. Aside from alternates, Daniel P. Moynihan, Goldberg's special assistant, was staff director, and Ida Klaus, brought down from New York, was the consultant. According to Wilson R. Hart, Goldberg was "the guiding genius and prime mover" and Klaus was "the working genius." Kennedy asked for the report by November 30, 1961, and he received it on that date.

The Task Force obtained information on personnel practices from fifty-seven departments and agencies and held hearings in Washington and six other cities. Despite the fact that one-third of its employees were members of organizations, "the Federal Government has little in the way of formal policy to guide collective dealings. . . ." The only statute was the Lloyd-LaFollette Act of 1912 which had taken on the common-law meaning that a government worker could join or refrain from joining an organization, the former only if it "does not assert the right to strike or advocate the overthrow of the Government." Otherwise, departments and agencies were on their own. Twenty-two had no policies at all. Eleven merely restated Lloyd-LaFollette. Twenty-one had adopted a 1952 guide prepared by the Federal Personnel Council which encompassed the right to join lawful organizations, management's desire to encourage discussion, criteria for matters to be discussed, and the services to be provided organizations, such as access to bulletin boards. The Department of the Interior was unique in having "a comprehensive code of labor relations procedures." The extent of organization was lopsided. Craftsmen and other blue-collar workers were much more organized than white-collar employees. TVA was almost completely unionized; the Atomic Energy Commission appeared to have no members. In the Post Office almost half a million workers belonged to organizations; in the State Department there were eleven.

The time was ripe, the Task Force urged, for the federal government "to come forth with a positive and comprehensive policy." But there were provisos. First, while the Task Force sought to approach the Wagner Act model, it was impossible to impose it in full.

> Despite the obvious similarities in many respects between the conditions of public and private employment, the Task Force feels that the equally obvious dissimilarities are such that it would be neither desirable, nor possible, to fashion a Federal system of employee-management relations directly upon the system which has grown up in the private economy. Nor is it necessary. The needs of the present

> can be fully met by adopting elsewhere in the Government the best features of employee-management systems which have been operating in some areas *within* the Federal structure.

Second, "many of the most important matters affecting Federal employees are determined by Congress and are not subject to unfettered negotiation by officials of the Executive Branch." The head of a department or agency in most cases has no authority to bargain over wages, hours of work, vacations, and holidays, among other things. Thus, "the benefits to be obtained for employees by employee organizations, while real and substantial, are limited."

Finally, collective bargaining must function side by side with civil service, "the essential basis of the personnel policy of the Federal Government." Civil service must remain the "indispensable method of selecting government employees and rewarding their achievements." While many opponents of public sector collective bargaining thought there was an irreconcilable clash with civil service, the Task Force foresaw "no conflict."

The remainder of the Task Force report consisted of its policy recommendations incorporated into Executive Order 10988, which the President issued on January 17, 1962. Federal employees would have the right freely and without reprisal "to form, join and assist any employee organization or to refrain from any such activity." For the purpose of the order, an employee organization could be either a labor organization of federal employees or a union which included both federal and private sector workers. But such an organization could not be one that asserted the right to strike against the government, that advocated the overthrow of the government, or that discriminated in membership on the basis of race, color, creed, or national origin.

Employee organizations could qualify for three levels of recognition. This was an innovation with no precedent in the Wagner Act or otherwise. The first, informal recognition, required no showing of membership. The organization was authorized only to present its views on matters of concern to its members, but imposed no duty on the agency to "consult" with it.

The next step, formal recognition, allowed an organization to speak "as the representative of its members in a unit as defined by the agency." It would be granted only when no other organization was the exclusive representative in the unit and it had gained "a substantial and stable membership" of at least 10 percent of the employees. When these tests were met, the formally recognized organization would have the right to present its views in writing, and the agency would "consult" with it in formulating personnel policies and rules affecting working conditions.

At the third stage, the NLRA model, "an agency shall recognize an employee organization as the exclusive representative of the employees in an appropriate unit when such organization is eligible for formal recognition . . . and has been designated or selected by a majority of the employees of such unit. . . ." While the unit could take a variety of forms, there must be "a clear and identifiable community of interest among the employees concerned." Managerial, executive, supervisory, and personnel employees were excluded. Professionals could not be put into a unit of nonprofessionals unless a majority of the former voted for it. Either the agency or the organization seeking recognition could request the Secretary of Labor to nominate an arbitrator from the national panel maintained by the Federal Mediation and Conciliation Service to determine the appropriate unit and/or hold a representation election.

An exclusive representative would have the right to negotiate an agreement culminating in a written memorandum of understanding. The agency would have the duty to "meet at reasonable times and confer with respect to personnel policy and practices and matters affecting working conditions." But an agency would have no obligation to bargain over its mission, budget, the organization or assignment of personnel, or the technology of its operation. Agreements must conform to federal law, regulations, and policies. Management would reserve the right to direct the work force. Grievance procedures would terminate in advisory arbitration subject to the approval of the head of the agency.

Agreements made prior to January 17, 1962, were immune from the terms of the executive order. Thus, ongoing relationships with more favorable terms than 10988, such as those at TVA and the Department of the Interior, were protected. The FBI, the CIA, and other investigative and intelligence agencies were excluded from coverage under the order.

The Task Force was entirely negative on union security. "The union shop and the closed shop are inappropriate to the Federal Service." This is because they are "contrary to the civil service concept." This seems justified in the case of the closed shop, which requires that an employee be a member of the union prior to hire. But it makes no sense with the union shop, which allows an employer freedom of choice in hiring and only requires membership thereafter.

The Task Force was baffled by the problem of the checkoff of union dues. It was persuaded by some court decisions that this device was unlawful in most federal operations. On the other hand, the checkoff was easy to administer, encouraged stability in bargaining relationships, and was widely used in the public sector in the cities and states, as well as by TVA and the Bonneville Power Authority. The Task Force, therefore, blessed dues withholding in its report provided that it was voluntary and that the employee organization bore its costs. But the executive order was silent on the issue and the Task Force recommended that the President ask Congress for a law authorizing the checkoff. In January 1963 the Comptroller General ruled that the President had the authority to institute withholding without legislation, and in November the Civil Service Commission established the checkoff on the terms laid down by the Task Force.

The executive order did not provide for an enforcement agency like the NLRB. The Labor Department and the Civil Service Commission would share that responsibility. The Task Force recommended that these agencies address the problem of adapting to the federal sector standards for fair conduct for unions and employers as provided for the private sector under NLRA. They did so and on May 21, 1963, the President issued the Standards of Conduct for Employee Organizations and the Code of Fair Labor Practices in the Federal Service.

The issuance of Executive Order 10988 on January 17, 1962, transformed labor relations in the federal government. It was followed by a burst of union organizing, unit determination arbitrations, and the negotiation of collective bargaining agreements.

Prior to the order there were only 26 units with exclusive representation covering 19,000 employees, all at TVA and the Interior Department. By 1963 there were 470 exclusive units with 670,000 employees. The latter figure reflected the dominance of the Post Office. The postal unions quickly won exclusive representation at 24,000 local units which were grouped into seven national bargaining units. On March 20, 1963, Postmaster General Day signed a unified department-wide agreement with all the unions. With a coverage of over half a million workers, it was the largest agreement, public or private, with a single employer in the United States.

In 1967, President Johnson appointed the Review Committee on Employee-Management Relations in the Federal Service, chaired by Secretary Wirtz, to assess the five-year experience under 10988. There were then 1,238,748 employees in 1,813 exclusive units. In addition, there were 1,172 grants of formal and 1,031 of informal recognition. More than a million employees belonged to labor organizations. The Review Committee wrote:

> The benefits from the program have been many. There has been a marked improvement in the communication between agencies and their employees. Employees now actively participate in the determination of the conditions of their work. This participation has contributed significantly to the conduct of public business. The collective bargaining agreements that have been negotiated have given continuity and stability to the labor-management relationship.

The impact of 10988 was not confined to the federal sector. The executive order also served as a model for the states and local governments, which during the sixties enacted a great number of public sector collective bargaining laws.

In an "evolutionary analysis" of public sector labor legislation B. V. H. Schneider placed the order at the opening of the stage she called "beginning collective bargaining." It was marked by "general acceptance of the concept of the legal duty on the part of the employer to bargain in good faith." While the scope of bargaining was severely restricted, 10988 had "the effect of legitimizing formal negotiating procedures, unit determination, exclusive recognition, and unfair practice procedures in the public sector. . . . It wove together the objectives of the civil-service cooperation principle with procedural elements of the NLRA." Of crucial significance, Schneider noted, was "the adoption of exclusive representation and written agreements—steps toward the conventional private-sector model and away from the 'government is different' syndrome, all of which would have seemed inconceivable 10 years earlier."[5]

7

Federal Aid for Education

FEDERAL AID for education was a political snake pit. Worse, there were many kinds of vipers, who all attacked relentlessly. Their venom was invariably fatal to legislation and sometimes inflicted severe political injury upon congressmen, senators, and even Presidents. Hugh Douglas Price varied the images, writing of "legislative mayhem" and bills that have become "politically accident-prone."

While public education in the U.S., as the conservative rhetoric insisted, was the primary responsibility of the states and localities, there was a long and notable history of federal assistance. In both the Survey Ordinance of 1785 and the Northwest Ordinance of 1787 the Continental Congress reserved public lands to support schools. The Morrill Act of 1862 used the same means to assist the states in founding the land-grant colleges to educate in the "agricultural and mechanical arts." In 1867, during Reconstruction, Congress established the "Department" of Education, which was soon stuffed into the Interior Department and in 1929 came to be known more appropriately as the Office of Education. The Smith-Hughes Act of 1917 provided federal grants to the states to support vocational education in agriculture, home economics, and the trades. During the Great Depression, the Works Progress Administration put unemployed teachers to work and the Public Works Administration built and repaired school and college structures. The Lanham Act of 1940 provided aid to local communities whose school systems had been "impacted" by swollen untaxed military and defense plant installations. The G.I. Bill of Rights, enacted in 1944, gave financial support for the education of veterans of World War II and was later extended to those who served in Korea. In 1946 the National School

Lunch Act supplied food to the states for hot lunches for school children. In 1950, Congress authorized long-term low-interest loans to colleges for the construction of dormitories. In 1958, following the shock of the Soviet *Sputnik*, the National Defense Education Act authorized $1 billion for loans and grants to colleges to improve the teaching of science, mathematics, engineering, and foreign languages.

Between the Great Depression and Kennedy's election in 1960 the nation's schools deteriorated physically. During the slump and the war there was little new construction of schoolhouses and the existing structures wore out. There was also a severe shortage of qualified teachers. Wartime and postwar inflation eroded the real value of their salaries. The baby boom enormously increased the number of pupils who, by 1961, had moved into the elementary and secondary schools and were now advancing inexorably upon the colleges. As a result, classrooms were jammed, students were on shifts, and buildings were obsolete and shabby. In many states teachers' salaries were so low that it was difficult to recruit. Teachers, as part of the public sector collective bargaining movement, were now joining unions and demanding increases in pay. Most local governments and the states, relying primarily on property taxes, had strained their financial resources to the limit.

If one viewed the educational system this way, the solution to the problem was obvious: The federal government, with much greater financial resources based on the income tax, should assist the states and localities financially to build classrooms, to provide teachers with decent salaries, and, by an equalization formula, to transfer funds from the rich to the poor educational systems. This analysis was eminently straightforward, sensible, and convincing. If one doubts this conclusion, he must examine the transformation of Senator Robert A. Taft of Ohio, "Mr. Republican," the Senate minority leader. He had been a diehard conservative on aid to education, believing the federal government should keep its hands off a state and local matter. He had led the Senate to defeat the 1943 school bill. But Taft had listened to the witnesses in the 1945 and 1947 hearings of the Committee on Education and Labor. During the 1948 education debate in the Senate, Taft said:

> I changed my mind. Fundamentally, Mr. President, I think we have a tremendous obligation to provide equality of opportunity to the children of the United States. . . . No child can have equality of opportunity in my opinion unless to start with he has a minimum education. . . . Because of the way wealth is distributed in the United States . . . I do not believe we are able to do this without a federal-aid system.

But there was little that was straightforward, sensible, or convincing about the politics of federal assistance. In the postwar period two cold embers once again burst into flame: religion and race. They guaranteed that the debates would be uncivil, bitter, divisive, and destructive.

"Nothing," legal scholar Paul G. Kauper wrote, "is better calculated to stimulate argument, arouse controversy, excite the emotions and even produce internal visceral reactions than a discussion of church-state relations." By 1960, 5 million children attended Roman Catholic primary and secondary schools in almost 10,000 parish systems, about 85 percent of those enrolled in religious day schools. The Catholic population and, thus, the parochial school system, was highly concentrated—New Hampshire, Massachusetts, Rhode Island, Connecticut, New York, New Jersey, Pennsylvania, Ohio, Illinois, and Wisconsin. Excepting Louisiana, there were insignificant numbers in the South. Other religious systems were run by the Lutheran Church–Missouri Synod, the Protestant Episcopal Church, Orthodox Jewish congregations, the Evangelical Lutheran Joint Synod of Wisconsin, the Seventh-Day Adventists, and the Christian Reformed Church.

Politically, of course, the Catholics dominated the debate. On educational questions the church spoke through the National Catholic Welfare Conference—made up of the five cardinals of the U.S. and the ten bishops and archbishops who headed the conference's departments. Their position was fundamentally reactive. As members of a minority religion, Catholics allowed others to take the lead; they would respond. Before the war, when no one raised a basic issue, the church remained silent. After the war, when federal aid to education became an important issue, Catholics announced their stands. They took no position on the adoption of a broad national policy. But if Congress enacted federal aid for public schools, the church wanted equal treatment for parochial schools. It favored long-term low-interest loans for school construction (direct grants may have been unconstitutional). But it was not interested in government support for teachers' salaries. By 1959, Catholic schools employed over 22,000 lay teachers. If public school salaries rose, the church schools would have to increase their pay scales to remain competitive. The Catholics were much interested in categorical support for particular programs—transportation, nonreligious textbooks, health services, counseling, programs for the handicapped, and income tax relief for parents whose taxes helped support the public schools and who also paid tuition at parochial schools. By 1960 the Catholic schools and colleges were already receiving some federal aid under Smith-Hughes, the National School Lunch Act, and the National Defense Education Act.

Federal support for religious schools raised a basic constitutional issue. Opponents relied on the doctrine of the separation of church and state. But the Constitution had no such provision. President Jefferson had first endorsed separation in 1802, but he later retreated from it. The First

Amendment, the only part of the Constitution that deals with religion, reads in relevant part, "Congress shall make no law respecting an establishment of religion, or prohibiting the free exercise thereof. . . ." These prohibitions were extended to the states in *Cantwell v. Connecticut* in 1940 under the due process clause of the Fourteenth Amendment.

Query: Did federal or state assistance to parochial schools constitute the establishment of a religion or prevent the free exercise of faith? The *Society of Sisters* case gave some comfort to churches which provided schools for general education as well as teaching religion. The Society of Sisters was chartered by the state of Oregon, empowering it to establish orphanages and schools. But Oregon later enacted a law which required parents and guardians to send children between eight and sixteen to public schools. The Society, of course, challenged the statute. The Supreme Court in 1925 set the law aside for interfering unreasonably with the liberty of parents and guardians to control the education of children for whom they were responsible. The court, some thought, leaned the other way in 1947 in the *Everson* case. It sustained 5 to 4 the right of local authorities in New Jersey to reimburse parents for the cost of transporting their children to both public and parochial schools. But Justice Black, speaking for the majority, added a famous dictum which included this sentence: "No tax in any amount, large or small, can be levied to support any religious activities or institutions, whatever they may be called, or whatever form they may adopt to teach or practice religion."

Two authorities, the first Lutheran, the second Roman Catholic, reached essentially the same conclusion after a careful review of the cases. Kauper wrote, "Congress may grant some assistance to these [parochial] schools as part of a program of spending for the general welfare, so long as the funds are so limited and their expenditure so directed as not to be a direct subsidy for religious teaching." Robert F. Drinan, S.J., was in substantial agreement:

> To conclude . . . that Justice Black and the Supreme Court in *Everson* ruled out all aid for the *secular* aspects of sectarian schools is simply going beyond the evidence contained in that opinion. In fact, such a conclusion is in conflict with the basic idea underlying the decision, the concept that it is "obviously not the purpose of the First Amendment" to cut off church schools from services which are "separate and . . . indisputably marked off from the religious function. . . ."

The conclusion one might derive from these analyses is that there was room for constitutional compromise provided that there was the political will to do so.

In the conflict over parochial schools political will withered. Normal alignments were stood on their heads. Liberal northern Democrats of the Catholic faith who represented heavily Catholic districts were under pressure to desert their commitment to federal aid if parochial schools were excluded. Moderate Republicans, concentrated in the heavily Catholic Northeast, were under the same constraints. Liberal southern Democrats who were eager to upgrade their backward public school systems were compelled by their Protestant constituencies to vote against bills that included church schools. Conservatives, both Republican and southern Democratic, who came from predominantly Protestant areas had another reason to oppose school bills: certain of victory, they chewed pleasurably on the discomforts of the other side.

Race was equally divisive. Federal support for the education of black children went back to the Freedmen's Bureau during Reconstruction. In the latter part of the nineteenth century the southern states established segregated school systems with the blessing of the Supreme Court. The founding of the NAACP shortly after the turn of the century gave blacks their own voice on educational policy. NAACP always supported federal aid for education, but for forty years shifted positions from assisting segregated black schools to eliminating segregation. In 1950 it came out firmly for the end of separate schools based on race and brought the cases that led in 1954 to the *Brown* decision. "The raising of the issue of segregation as such," Frank J. Munger and Richard F. Fenno, Jr., wrote, ". . . transformed a money issue into a moral one."

It also became a political bludgeon. Congressman Adam Clayton Powell, the flamboyant and capricious preacher from Harlem's Abyssinian Baptist Church, was chairman of the House Committee on Education and Labor. Following the court's 1954 decision, he introduced the Powell Amendment, which would forbid financial assistance to racially segregated schools. Characteristically, Powell sometimes offered his amendment and sometimes withheld it. While *Brown* alone might have doomed passage of a school bill, the Powell Amendment removed any remaining shred of doubt. Southern and border state Democrats voted solidly against it. Some northern liberal Democrats cast ballots against a bill with the amendment because they thought its purpose was to torpedo the legislation. Conservative Republicans who opposed federal aid had it both ways. They voted for the Powell Amendment to curry favor with blacks and then voted against the bill with the amendment.

The combined negative forces of religion and race tore apart the politics of federal aid for education for almost a generation. One can start at almost any point in time to demonstrate this. The bill for 1949–50 is as good as any and also provides a small political dividend. In 1949 the Senate passed a school aid bill 58 to 15. The North Carolina Democrat Graham A. Barden, a brilliant parliamentary tactician who was chairman of the House committee and strongly opposed to the legislation, introduced an amendment that would restrict assistance to public schools, denying parochial schools even help for categorical services. Francis Cardinal Spellman of New York denounced Barden's move as "discrimination [as] shocking as it is incomprehensible." In her newspaper column the following week Eleanor Roosevelt, without mentioning the legislation, endorsed the separation of church and state and opposed the use of tax funds to support parochial schools. Spellman, who had no use for temperate language, denounced her record of "anti-Catholicism" and "discrimination," which, he said, was "unworthy of an American mother." The resulting firestorm became what Price called "one of the ugliest and most bitter disputes to hit Capitol Hill in this century." Barden's committee voted 13 to 12 to defeat the Senate bill. Two promising young members of the House committee joined the majority: John F. Kennedy, because he represented a heavily Catholic district, Richard M. Nixon, because he was a conservative Republican.

Between 1951 and 1953, Democratic liberals revived the land-grant feature of the Morrill Act by urging that income from the lease of tidelands oil fields be used to support education. Bills proposed in the House by Mike Mansfield of Montana and in the Senate by Lister Hill of Alabama, Paul Douglas, and Herbert Lehman of New York went down in defeat. In 1955 the Eisenhower administration offered an extremely limited school construction bill, mainly for loans. It simply died. In 1956 a more generous construction measure, again backed by Eisenhower, got stuck with the Powell Amendment. The House defeated it 224 to 194, with conservative Republicans voting for the amendment, which passed, and then against the bill.

Price called the 1957 experience a "near miss." Pro-aid Democrats, convinced that they must make peace with the President to get anything, lowered their sights to the 1956 construction bill and, working with the administration, added a few sweeteners. But Eisenhower, worried about a budget deficit, seemed in a misty way to back off. The ludicrous debate in the House consisted mainly of speculation among Republicans over whether Eisenhower would sign or veto the bill his administration had joined in proposing. But Howard Smith, the chairman of the Rules Committee and a very astute parliamentarian, did not want the President to bear the heavy burden of decision. He moved to strike the enabling clause of the bill and the motion passed 208 to 203.

In 1959 the Senate again acted favorably. Amazingly the House Rules Committee allowed a construction bill to reach the floor. For the first time the lower chamber passed an education bill, but with the Powell Amendment! Judge Smith immediately shifted into reverse. For a House committee to get to conference, it was necessary to have a rule. None could be jimmied out of Smith's committee. "Clearly," Price wrote, "if Kennedy won the 1960 election, one of the first pieces of business would be deciding on a strategy to deal with the impasse in the House, and especially with the strategic Rules Committee."

Just prior to adjournment for the presidential campaign on September 1, 1960, Senator Clark offered an amendment to provide $1.1 billion for construction and teachers' salaries. It was defeated 44 to 44. There was then a motion to reconsider, immediately followed by a motion to table. Again, the vote on the latter was 44 to 44. The motion to table, of course, required a majority to pass. Vice President Nixon, about to start his campaign, was cornered and had to cast the deciding vote to table.

Thus, federal aid for education became a major issue in the 1960 presidential campaign. Nixon, as his vote to table indicated, seems to have regarded it as a no-win political problem, more an annoyance than an issue. Kennedy, by contrast, was wholly committed to federal aid by 1960. This was, Sorensen wrote, "the one domestic issue that mattered most to John Kennedy." "The human mind," he never tired of saying, "is our fundamental resource." The nation's military, scientific, and economic strength rested on the educational system, and the beggared public schools were crippling the nation. He rattled off the discouraging statistics, which he had memorized.

The Democratic platform, written by northern liberals, strongly backed assistance for both school construction and teachers' salaries; the Republicans supported help for construction and were silent on pay for teachers. Kennedy hammered at the issue throughout the campaign, culminating in a major speech in Los Angeles. It was an important question in the first television debate. In a speech in Houston to the Protestant Ministerial Association on September 12, Kennedy said, "I believe in an America where the separation of church and state is absolute." He thereby locked himself into a position of supporting only public education at the primary and secondary levels. (Kennedy later told Nicholas Katzenbach with feeling: "Eisenhower could have dealt with this whole problem, but I can't.") He repeatedly attacked Nixon for refusing to help teachers and for his tie-breaking vote in the Senate. Kennedy also urged an equalization formula to transfer funds from the rich to the poor states. With his victory, Munger and Fenno wrote, he became "the first President to make aid a major element in his domestic program and to give it vigorous support. By the time he took office, President Kennedy's position was clear-cut."[1]

During the interregnum Kennedy appointed a task force on education. Frederick L. Hovde, the president of Purdue, was chairman, and the other members were John Gardner, president of the Carnegie Corporation, which had a special interest in education, Alvin Eurich, vice president of the Ford Foundation, Francis Keppel, dean of the Harvard School of Education, Russell Thakery, executive secretary of the American Land-Grant Colleges Association, and Benjamin Willis, superintendent of schools in Chicago and soon to be president of the American Association of School Administrators. The task force was certainly distinguished, but it had several shortcomings. Its strength was in higher education, while it was expected to address mainly primary and secondary education. The other deficiency was that, while race and religion were the fundamental political problems, there was no one to speak for the white South, the NAACP, or the Catholic and Protestant churches.

The report was delivered to the President-elect at the Carlyle Hotel in New York on January 7, 1961, and was released to the press that day. The task force members were deeply committed to federal aid for education and their report was a shocker. They had three major recommendations.

The first basic recommendation was for federal support for the nation's troubled public schools. A demanding "priority," they wrote, "should be . . . a vigorous program to lift the schools." The President should urge three legislative changes to achieve this aim.

The first proposal was for a grant of $30 per pupil per year based on average daily attendance at all the public schools in the U.S. The states would receive the grants and then transmit them to local boards of education. They would have discretion to employ the money for "construction, salaries, or other purposes related to the improvement of education." The states and the boards must not be allowed to substitute federal for state and local dollars.

The second proposal addressed the problem of low-income states with poor public education systems. They were defined as states with personal incomes below 70 percent of the national average, estimated at one-fourth of the total, mainly in the South, with about 7 million pupils. These states would be the beneficiaries of an equalization formula providing an additional $20 per child.

The third proposal dealt with the "unique and grave educational problems" in the major cities, those with over 300,000 population. They, too, would receive equalization grants of $20 per pupil. The funds would be used by local boards for research and experimental programs in the special problems of urban schools, for construction, for improving community service by the schools, and for guidance and job placement for students over sixteen. Eligibility would be based on density of population, the quality of housing, and the percent of students finishing high school. This was, Keppel said, the first time anyone had addressed the problem of education in the slums. "We did poke a finger in there" and it immediately caught Kennedy's eye.

The second basic recommendation was federal support for colleges and universities whose enrollments "are now at an all-time high" with "the period of greatest increase *immediately* ahead." The task force asked for a fund of at least $500 million in the first year, $350 million for matching grants and $150 million for loans, to rise in succeeding years as the number of students increased. The money would be used for the construction of academic facilities. In addition, the college housing loan program, which had been "outstandingly successful" in providing dormitories during the preceding decade, would be dramatically expanded—$150 million in the current year and $350 million in each of the following four years.

The final basic recommendation was to strengthen the National Defense Education Act. Since the priority was "the critical shortage of teachers at all levels," the fellowship program should be enlarged to attract able new people and to allow those already in service to improve their skills. In addition, college loan funds should be increased and the construction of television networks encouraged. Here the task force provided no price tag.

The combined cost of these recommendations was estimated to be at least $9.39 billion over four years, around $2.31 billion a year. This far exceeded the most generous authorization in the Senate's 1960 debate, which had failed to pass. Kennedy was jolted by the cost. Keppel recalled later that the report, if adopted, "would probably have broken the federal government's bank in no time at all." The conservative press had a field day.

Even more serious politically was the fact that all private elementary and secondary schools (not colleges) were excluded from the program, including, of course, parochial schools. The Catholic hierarchy, already concerned about Kennedy's promise to the Houston Ministerial Association, was aroused by the task force report. Cardinal Spellman three days before the inauguration attacked it vitriolically in a major speech at a Catholic high school in the Bronx. "It is unthinkable that any American child be denied the Federal funds allotted to other children which are necessary for his mental development because his parents chose for him a God-centered education." As a Catholic President in a predominantly Protestant country, Kennedy was politically strapped into a position of opposing aid to parochial schools. If he earlier had any doubt, now, on the eve of his presidency and before he even had a school bill, he learned with certainty that he would be forced into battle with his own church.

As if this were not bad enough, Kennedy's appointments to the Department of Health, Education and Welfare insofar as they affected education may be charitably described as weak. Wilbur Cohen thought this was because of the Catholic issue, on which Kennedy was extremely sensitive. That is, he wanted to keep control of educational policy in the White House in his own hands, assisted by Sorensen, Feldman, and Cohen.

Ribicoff became secretary of HEW. But he regarded the department as unmanageable, was so indiscreet as to say so publicly, and made little effort to run it. According to Cohen, he was "an extremely poor administrator." Rather, he was "very egotistical" and "a political animal." He really wanted to be a senator and soon quit to make the run in Connecticut, successfully.

The search for a commissioner of education was an ordeal. The Hovde task force was asked to suggest a name and came up with James Allen, the commissioner of education in New York State. Keppel thought him the best commissioner in the nation, someone who was aware of the educational problems in the central cities, and almost unique in being acceptable to both the lower school and higher education communities. Keppel spent an afternoon trying to persuade Allen to take the job, but his personal problems were insurmountable. Someone finally came up with the name of Dr. Sterling McMurrin, a professor of Philosophy at the University of Utah. He must have been the most obscure person in the administration. Nobody seems to have heard of or from him. Kennedy, evidently, did not know of his existence until he resigned dramatically through his congressman in 1962. Ribicoff told McMurrin he would handle the politics of education himself and that the commissioner should be the in-house educator. Keppel said, "This didn't work worth a damn."

There was, as Price pointed out, another formidable roadblock—the Rules Committee. In the late fifties the northern Democrats became incensed over Judge Smith's chokehold—his personal control over procedure and the Republican–southern Democratic majority that dictated substance—which prevented a liberal bill from reaching the floor. Rules had eight Democrats and four Republicans, but Smith and William Colmer of Mississippi, both extreme reactionaries, invariably joined the Republicans on important issues. A 6 to 6 tie defeated a rule. In 1959 the liberals exacted a promise from Speaker Rayburn that their bills would reach the House. He was counting on his old buddy, Joseph Martin, the Republican leader, with whom he had been cutting deals for years. But almost immediately after Rayburn made the pledge the House Republicans voted 74 to 70 to replace Martin with Charlie Halleck, a hard-nosed partisan who liked the Rules Committee exactly as it was. Mr. Sam had to go back on his promise, and that put his personal integrity on the line. During the 1960 campaign Kennedy called for change in the committee and Nixon opposed it. A showdown was inescapable and both sides gathered their forces.

Rayburn had hardly known Kennedy prior to the 1960 Democratic convention and, as Lyndon Johnson's close friend and key adviser, had viewed him with suspicion. During the campaign, D. B. Hardeman and Donald C. Brown wrote, Mr. Sam's "attitude toward the young presidential nominee shifted dramatically." After the second television debate he exclaimed, "My God, the things that boy knows!" A southern Baptist himself, he was enormously impressed with Kennedy's speech to the Houston Ministerial Association. He campaigned vigorously for the ticket in several states. When it won, Hardeman and Brown wrote, "the New Frontier had no more enthusiastic supporter than Rayburn. . . ."

For a month the Speaker waged a masterful campaign to undermine Smith and Halleck, which Neil MacNeil has set out in detail. The climactic debate to enlarge the Rules Committee to 15 members with the addition of two Democrats and one Republican took place on January 31, 1961. It was a historic moment for the House of Representatives. As the speeches ground to a close, Rayburn left the rostrum to take his seat on the floor. Mr. Sam was seventy-nine, had sat in his beloved House since 1913, and had been Speaker when the Democrats were in control since 1940. When he stood, the House and the galleries rose with him to a thunderous ovation. Always taciturn, he spoke briefly to explain his plan, concluding, "Let us move this program." He won 217 to 212. The House erupted into pandemonium.

Rayburn named Carl Elliott of Alabama and B. J. Sisk of California, both liberals, and Halleck picked William Avery of Kansas, a moderate conservative, for the now open seats on Rules. But the new 8 to 7 majority did not mean that a liberal bill would automatically get a rule; rather, it signified that there was no longer a virtual certainty that it would be denied one. An education measure, guaranteed to arouse bitter controversy, might still face trouble.

During January and early February 1961 the administration—Sorensen, HEW, the Bureau of the Budget—and the congressional leadership scrambled to prepare the legislation. There were three options—the bill that passed the Senate in 1960, the Hovde task force report, and some big critical thinking at Budget. This last, anticipating a leap upward in federal support for education from its current 9 percent level, called for using the federal financial lever to push up the state and private contributions. The Bureau also wanted to consolidate the federal agencies dealing with education. These ideas, while interesting, were not carefully thought through and, more important, did not address political realities. The Hovde proposals, while implausibly expensive, were instructive, particularly by raising the issue of schools in the central cities. The 1960 Senate bill was much the most attractive politically. The upper house had voted 51 to 34 for the measure and would almost certainly do so again, which was half the battle. Further, as Democratic Senator Wayne Morse of Oregon, who would steer the bill through the Senate, liked to say, it was a "clean bill." That is, it made no reference whatever to either religion or race. This, of course, coincided exactly with Kennedy's views and the bill became the basis for the 1961 legislation.

The President sent a special education message to Congress on February 20. He recited the litany of needs:

> Too many state and local governments lack the resources to assure an adequate education for every child. Too many classrooms are overcrowded. Too many teachers are underpaid. Too many talented individuals cannot afford the benefits of higher education. Too many academic institutions cannot afford the cost of, or find room for, the growing numbers of students seeking admission in the 60's.

Kennedy recommended a three-year program of federal assistance for public elementary and secondary classroom construction and teachers' salaries based upon the Senate's bill. The expenditure would be $666 million in the first year, later rising to $866 million. He adopted the Senate's equalization formula, which had been borrowed from the Hill-Burton Hospital Construction Act. As it worked out, the formula produced marked differentials between states. For fiscal 1963, when the full program was expected to be in operation, the range per student per year would be from a high of $33.80 in Mississippi to a low of $15 in Alaska, Connecticut, Delaware, Illinois, Massachusetts, New Jersey, New York, and Rhode Island. Then, adopting the principle of the Hovde proposal in a general way, Kennedy asked that 10 percent of each state's funds be reserved for "'areas of special educational need'—depressed communities, slum neighborhoods, and others." "In accordance with the clear prohibition of the Constitution," he stated, "no elementary or secondary school funds are allocated for constructing church schools or paying church school teachers' salaries. . . ."

For higher education the President recommended several programs that would assist both public and private, including church, institutions. The current College Housing Loan Program would be extended and expanded for five years to provide loans for residential housing for students and faculty at a rate of $250 million a year. A similar five-year program, costing $300 million annually, would provide loans for the construction and renovation of classrooms, laboratories, libraries, and related structures. Finally, a five-year state-administered scholarship program would be established at a cost of $26,250,000 in the first year to assist "talented and needy young people." The scholarships would average $700 a year with a $1000 maximum. There would be 25,000 the first year, 37,500 the second, and 50,000 thereafter. These benefits would go to students, not colleges, and they would be free to select their own programs and institutions.

The National Catholic Welfare Conference hardly waited for the ink to dry on the President's message. It met on March 1, 1961, and Archbishop Karl J. Alter of Cincinnati issued its four-point statement the next day: (l) The question of whether there ought to be federal aid to education was a matter on which "Catholics are free to take a position in accordance with the facts." (2) If there is federal aid, "in justice Catholic school children should be given the right to participate." (3) Long-term low-interest loans to private schools were clearly constitutional and the Kennedy bill should be amended to include them. (4) "In the event that a federal aid program is enacted which excludes children in private schools these children will be the victims of discriminating legislation. There will be no alternative but to oppose such discrimination."

With this statement and a massive press and letter-writing campaign that followed, the church put intense pressure on Catholic members of Congress, 88 in the House and 12 in the Senate, to say nothing of the President. Aid to parochial schools, of course, was a leading question at the White House press conference on March 8. Support through NDEA, Kennedy said, raised no constitutional issue because it was "tied very closely to national defense." Loans to universities and scholarships for college students also caused no difficulty because the decision to attend was the voluntary choice of an individual. Assistance to elementary and secondary education, however, posed a critical constitutional question. This was because this form of "education is compulsory. . . . Every citizen must attend school." Kennedy continued, "I do not think that anyone can read the *Everson* case without recognizing that the position which the court took, minority and majority, in regard to the use of tax funds for nonpublic schools raises a serious constitutional question."

The Senate Subcommittee on Education held hearings between March 8 and 20, 1961. The outcome was never in doubt. All six Democrats, Chairman Morse, Lister Hill of Alabama (chairman of the full committee), Pat McNamara of Michigan, Ralph Yarborough of Texas, Joseph Clark of Pennsylvania, and Jennings Randolph of West Virginia, along with two of the three Republicans, Clifford Case of New Jersey and Jacob Javits of New York, strongly supported federal aid. Barry Goldwater of Arizona was the lone opponent. Morse was in full command and his performance was masterful.

This eccentric, egotistical, and brilliant man was a bundle of contradictions. He was the Senate's most notorious maverick, was known as that body's hair shirt, and his biographer called him the "tiger in the Senate." Yet he could listen patiently to views he detested, could conduct impeccably fair hearings, and described himself as a "private" in President Kennedy's army. While anyone who dealt with Morse had to be wary, Kennedy and Morse respected each other and got along very well. Kennedy told Sorensen admiringly that Morse was "the only man in the Senate who speaks in precise paragraphs without a text." He was also close to Lister Hill, the courtly southern gentleman whose gentle manner was the exact opposite of his. Morse had started as a Republican, became an Independent, and, finally, turned into a Democrat. Never at a loss for words, he held the record for the longest speech ever delivered in the Senate.

Morse had been a professor of law and dean of the University of Oregon Law School. During the war he had been a public member of the National War Labor Board. He had also been a distinguished labor arbitrator. An incurable academic, Morse conducted hearings like college seminars and, when it seemed appropriate, would ask witnesses to write "papers." He did exactly this with Ribicoff. The HEW General Counsel then submitted a brief on the constitutionality of federal aid for education that had actually been written in the Department of Justice. It became an important document in the debate. The brief, of course, reinforced the position the President had already taken.

When Congress had opened in January 1961, Morse, under the Senate's rules, was first in line to become chairman of the subcommittee. He was eager to get the job for several reasons: he had a deep commitment to education and to federal assistance; the presence of prickly constitutional questions aroused his favorite intellectual interests; his parliamentary skills were honed by exposure to collective bargaining; he was attracted to the public prominence that would inevitably accrue to the individual who steered this important legislation through the Senate; and the fact that, since he had been elected to the Senate in 1944, no consequential piece of legislation had carried his name. Some of his colleagues thought him too irresponsible for this task. But Kennedy did not hesitate. He took it for granted that Morse would get the job and proceeded on the assumption that they must work together. Morse became chairman and Kennedy's confidence was fully rewarded. "Some people don't seem to realize that when I am given a responsibility, I do a job," Morse said. "I'm going to bleed for my President."

Morse had long been committed on principle to Kennedy's position on parochial schools. In pursuit of his goal of a "clean bill," he had gotten Adam Clayton Powell to promise not to introduce his amendment. Thus, the only roadblock that remained was religion and nobody knew how to clear that one from the highway.

In anticipation of the testimony of Monsignor Frederick G. Hochwalt, the spokesman for the National Catholic Welfare Conference, Morse said it would be "presumptuous for a Senator to be giving gratuitous advice," but he would risk it. He asked the church to join first in supporting a public school bill and then trying later with a separate bill to determine whether loans to parochial schools were constitutional. Hochwalt refused to play this game because aid for public schools would pass and "a second measure, which would provide for our schools, wouldn't have much of a chance." He stressed the needs of Catholic parents who would "feel the double burden of supporting two school systems." Protestant spokesmen, notably the National Council of Churches, adamantly opposed federal aid for parochial schools. Morse could not make the issue go away.

During April the administration tried its hand. The President, of course, could not be directly involved. Sorensen and Ribicoff met secretly with a Washington priest with ties to the NCWC. The scheme was to broaden NDEA, which already provided loans to Catholic schools for defense-related instruction. The act would be amended to include construction of facilities not only for science, mathematics, and foreign languages, but also for English, physical fitness, and school lunches. "This, in effect," historian Hugh Davis Graham wrote, "would enable church schools to obtain low-interest, government-guaranteed loans to construct virtually any structure short of the chapel." Morse and Clark prepared an amended National Defense Education bill.

Morse's subcommittee overwhelmingly reported the administration's school bill. The full Committee on Labor and Public Welfare voted for it 12 to 2 on May 11, 1961, with only Goldwater and Dirksen opposed. The only change was inclusion of the already noted equalization formula. On May 25 the Senate passed the bill 49 to 34. As in 1960, the Senate had endorsed federal aid for school construction and teachers' salaries and by virtually the same vote. Thus far the administration's strategy had worked.

But the situation in the House was entirely different. Price has suggested the reason. "The typical Senator . . . is a living compromise, representing large and generaly diverse constituencies. . . . Most House members, by contrast, represent small districts, which are often relatively homogeneous. . . ." Excepting much of the South and a handful of mountain states, Senators had substantial numbers of both Protestant and Catholic constituents. Many congressmen, however, represented districts which were overwhelmingly one or the other.

The House Education Subcommittee hearings were held in March and pretty much followed the Senate pattern. The subcommittee adopted a somewhat different bill on a straight party-line vote of 4 to 3 on May 9, 1961.

Soon thereafter the full committee reported it out favorably 18 to 13, with all the Republicans and one Democrat in opposition.

But by early June the House had become a battlefield. The flood of mail from both Catholics and Protestants was enormous. Hearings before the Education and Labor Committee on amending NDEA to allow loans for construction at private schools had gotten off to a rocky start. Neither Kennedy nor Ribicoff would support the proposal publicly, and poor McMurrin, who testified for the administration, was unable to take a stand. John McCormack, the House majority leader, himself a Catholic and speaking for an almost totally Catholic district in Boston, emerged as the strategist for the parochial schools. The issue was whether the Rules Committee would allow the NDEA amendments or the administration bill to go to the floor first, the question Morse and Hochwalt had debated.

The newly constituted Rules Committee withered under the religious crossfire. The opponents of both bills could count on five Republicans, along with Smith and Colmer. Thus, they needed only one vote to kill. If the NDEA amendments came up first, three southern Democrats, all Methodists, who normally supported a school bill, were virtually certain to vote against a rule on the parochial school issue. If the administration bill received priority, three northern liberal Catholic Democrats, who supported aid to education, would have their necks on the block—Ray Madden of Gary, Indiana, Thomas P. ("Tip") O'Neill, Jr., who represented the President's old Boston-Cambridge district in Massachusetts, and James Delaney of a heavily Catholic Queens district in New York City. There were rumors of a conflict between Rayburn and McCormack. On June 15 it was revealed that O'Neill and Delaney would not vote to allow the public schools bill to reach the floor unless the NDEA amendments were also allowed to do so. The tension in the Rules Committee attracted great public attention.

On July 18 the committee exploded. Colmer moved to table all three education bills before the House—public schools, NDEA amendments, and also the higher education measure. The motion passed 8 to 7. The majority consisted of Smith, Colmer, the five Republicans, and Delaney. Madden and O'Neill, tortured, nevertheless joined the minority. Larry O'Brien had worked over an adamant Delaney to no avail. "He didn't want a thing," O'Brien said later. "I wish he had." The Washington joke was that "the unholy alliance" of Republicans and southern Democrats had now got religion.

The administration struggled to rescue something from this political debacle. On July 20, Ribicoff sent four alternative education packages to the White House for the President to pick one that might have a chance of passage. He came up with what was called the Emergency Educational Aid bill, which consisted of one-year extensions of the impacted areas and National Defense Education statutes, funds for classroom construction in school districts with "a proven classroom shortage," and scholarships for college students. Congress showed no interest. On August 14, 1961, Sorensen wrote Kennedy: "The compromise measure advanced by Secretary Ribicoff last week failed to win a commitment from key House leaders, apparently because it went too far—and from key Senate leaders, apparently because it did not go far enough." The year 1961 was a total loss.

Recrimination became the order of the day and there was plenty to go around. "Morse (and to a lesser extent Humphrey)," Sorensen wrote Kennedy, "have expressed their bitter opposition to letting the House, the House Rules Committee and the Catholic Church exercise a veto over the Senate passed Administration bill." Most blamed the President for lack of leadership. But Price, the most astute of the observers, thought that Kennedy, given the vulnerabilities of school legislation and his personal disability as a Catholic, did about as well as could be expected. "If politics is viewed as the art of the possible, the President's inaction does not appear so strange." As usual, the President dealt with the disaster with wry humor. He reminded his audience at the Gridiron Club dinner that when Al Smith had been defeated in 1928, he sent the Pope, who was allegedly planning to take over the country, a one-word telegram: "Unpack!" "After my stand on the school bill," Kennedy said, "I received a one-word wire from the Pope myself. It said, Pack!"[2]

If the theme of federal aid for education in 1961 was defeat, in 1962 it became fragmentation. There were so many proposals floating around the Hill that it was difficult to tell them apart. The Kennedy administration's leadership quit. Wholly aside from the religious issue, the supporters of federal aid fought among themselves. All in all, it was a very bad year.

On October 6, 1961, Ribicoff had sent the President a perceptive and very discouraging political assessment of the recent disaster.

> The passage of any broad-scale education legislation will be a most difficult task. From my trips around the country, discussions with members of Congress, and review of editorials and news columns . . . , I am convinced that there is not a full commitment to education in this Nation. People are concerned about the education of their own children, but there is very little realization . . . that increased financing for education, shared by the Federal Government, is urgently needed to improve education . . . and, indeed, insure this Nation's survival.
>
> A broad program of grants to States for public school construction and teachers' salaries is virtually impossible to pass. There is substantial Southern opposition to any bill for elementary and secondary schools, even though aid for teachers' salaries has some Southern support. Republican opposition to any general aid bill

> is strong, and is overwhelmingly against teachers' salaries. Northern Democratic support cannot be assured unless some satisfactory resolution of the religious issue is achieved.

Ribicoff offered a narrow window of hope for 1962. A higher education bill, though it faces "some problems with the religious controversy and some with racial issues," might pass if brought up "by itself." There was more support for aid to the medical professions. A few narrow categorical programs might be glued to NDEA extension. But a schools bill "cannot be enacted by this Congress." It might make a good issue in the 1962 elections.

The President may have agreed with this savvy political advice, but he did not take it. He was too deeply committed to education and was determined to turn the country around. Thus, in his State of Union message on January 11, 1962, he declared:

> I sent to the Congress last year a proposal for Federal aid to public school construction and teachers' salaries. I believe that bill, which passed the Senate and received House Committee approval, offered the minimum amount required by our needs and—in terms of across-the-board aid—the maximum scope permitted by our Constitution. I therefore see no reason to weaken or withdraw that bill; and I urge its passage at this session.

The special message on education that the President sent to the Congress on February 6, 1962, set forth a comprehensive program for overhauling the nation's educational system. He renewed the proposals for elementary and secondary as well as higher education. In addition, he urged support to improve teacher quality, to expand education in the health sciences, to increase the numbers of scientists and engineers, to reduce adult illiteracy, to provide schooling for the children of migratory farm workers, to expand educational TV, to increase special education for the handicapped, and to provide assistance to the arts. This snowstorm of proposals was blinding, and Congress, already in a negative mood, made no effort to clear a pathway.

Worse, the administration fell into disarray. By the summer of 1962 progressive Republicans who supported federal aid were chiding Kennedy for abandoning ship. On July 13, Ribicoff resigned to run for the Senate. The President replaced him promptly with Anthony J. Celebrezze, the mayor of Cleveland. According to Cohen, while he turned out to be a very good secretary, he was picked because he was Italian to help young Edward Kennedy garner Italian votes in his race for the Senate in Massachusetts. In any case, Celebrezze was of no help on education legislation in 1962. On July 27, McMurrin resigned, to the President's outrage, through his congressman. Lister Hill, the Senate stalwart on federal aid, was up for re-election in Alabama against a staunch segregationist. HEW was beginning to tighten the screws on NDEA grants to institutions that segregated, and Alabama, with George Wallace waving the bloody shirt, was seething with white support for its racially separated school system, including the university. Hill could hardly help the President.

As Ribicoff had pointed out, there was a slim possibility for assistance to higher education based on bipartisan support. In the House this would require administration deference to historic Republican opposition to the President's proposal for federal funding of scholarships for students. Conservatives did not believe that young people should be subsidized to go to college. On January 30, 1962, the lower chamber passed a bipartisan higher education bill by a vote of 319 to 79. It provided a five-year construction program for college facilities, $180 million annually for matching grants, of which the federal share would be one-third, and $120 million a year for loans. On February 6 the Senate passed a contradictory bill by a vote of 69 to 17 with strong bipartisan support. It provided only loans, not grants, for construction, as well as undergraduate scholarships.

The National Education Association, whose members worked mainly in primary and secondary schools with a few in teachers' colleges, nevertheless immediately entered the fray. Representing about three-quarters of a million public school teachers and a much smaller number of administrators mainly in the West and South, NEA had been the historic leader in asking for federal aid for the public schools. Its main concern was for teachers' salaries, though it also supported assistance for construction. NEA now announced that the House version conflicted with its policy against grants to private institutions.

It was necessary for the House to have a rule authorizing its committee to go to conference to adjust the differences with the Senate bill. Judge Smith took three months to extract a promise from Adam Clayton Powell that, as the latter put it, "under no circumstances when we go to conference will we recede from the House position." Smith's committee then issued the rule on May 9, 1962.

During June other school organizations joined the NEA in opposing grants to private colleges—the American Association of School Administrators, the American Vocational Association, the Council of Chief State School Officers, the National Congress of Parents and Teachers, and the National School Boards Association. The conference met on June 19 and 22 and neither side would give ground. On June 25 the Supreme court struck down a classroom prayer prescribed by the New York Regents in *Engel v. Vitale*, raising a question over the constitutionality of even loans to church colleges. Many southern Democrats were outraged by the court's decision. Representative George Andrews of Alabama said, "They put Negroes in the schools and now they've driven God out." On June 26, NEA, exploiting the decision, wrote all members of Congress of its opposition to aid to sectarian institutions. On June 27 the conference committee adjourned to study the court's decision. In July the NEA annual convention reiterated its stand and Ribicoff and McMurrin resigned. It seemed as though a higher education bill had stopped breathing.

But two stubborn conferees, Representative Edith Green of Oregon and Senator Clark, searched for a "resuscitation" formula. They thought they found one: limiting construction grants to buildings to be used for science, engineering, and libraries, and only loans, not scholarships, for students, though 20 percent of the fund could be set aside for "non-reimbursable loans" to "exceptionally needy" and promising students (the "non-reimbursable loan" set the science of semantics back a generation). The conference committee filed its report based on this compromise on September 17, 1962.

The process of "resuscitation" was only momentary. NEA denounced the grant provision and launched a lobbying campaign. More important, on September 19, NEA sent a telegram to all members of the House urging them to vote to return the bill to conference in order to eliminate grants to private institutions. Other public school organizations and the Southern Baptist Convention joined the attack. Mainline Republicans turned on their more progressive colleagues for breaking faith on scholarships. Halleck smelled a partisan opening. As one Democrat put it, "Charlie Halleck wasn't going to let the Democrats have a college aid bill in September of an election year." Southern Democrats worried that "exceptionally needy" was a code phrase for Negro. Senator Hill crossed the Capitol to the House cloakroom and lobbied the entire Alabama delegation to vote for recommittal.

On September 20, 1962, the House voted 214 to 186 to recommit. The reversal from the January roll call was stunning. Those who switched consisted of 78 Republicans and 33 southern Democrats. It is unlikely that the NEA caused a significant part of this shift, but it received virtually all the blame, particularly for its telegram. Kennedy was enraged and House conferees who supported the compromise attacked the NEA. A number of leading newspapers added their denunciations. Most important, as Keppel put it, there was now "a bitter relationship . . . between the school people and the college people." When this was added to the hostility between "the public school people and the whole Catholic world," education legislation was in a heap of trouble.[3]

In round one (1961) Kennedy had taken a severe pounding and had dropped briefly to his knees. Round two (1962) was worse and he had gone down for an eight-count. But in round three (1963) the tide of the fight turned. He not only kept his feet but also won going away. There were a number of reasons: The financial needs of the nation's educational systems remained enormous and, in fact, became more pressing each year, particularly for higher education. The Kennedy administration pulled itself together both in leadership and in legislative strategy. Finally, as Sundquist stressed, "people *do* learn from experience. . . . The NEA and its public school allies now knew that an all-or-nothing attitude would mean, for the public schools, nothing. Likewise, Catholic leaders now understood that an equal-treatment-or-nothing position would mean, for the Catholic schools, nothing."

After McMurrin resigned Keppel returned to the search for a commissioner of education. It was a fruitless endeavor. Then, as he said later, "I finally got a bad mixture of bad conscience that somebody ought to do this job and a little restlessness of my own at Harvard, so I sent word . . . that if the White House wanted to ask me, on a personal basis, I'd be willing to think about it hard." Ralph Dungan of the White House staff did some checking and the President was soon on the phone to ask, "Will you do it?" That was it.

Kennedy, still furious over the McMurrin resignation, decided to make a show of bringing Keppel on board and was present when he was sworn in at the White House. Logan Wilson, the president of the American Council on Education, which spoke for the universities, came to the ceremony, but the President, still seething over the NEA's conduct, refused to ask anyone from that organization. Keppel, who suffered from no shortage of political savvy, said, "He had to be calmed down." They got the NEA president-elect, Robert H. Wyatt, who was, "thank God," a good Indiana Democrat. The President invited him as a fellow-Democrat, not as a spokesman for NEA!

Kennedy was extraordinarily lucky to get Keppel. He must have been the most qualified individual in the nation for commissioner of education; in fact, judged by the standards set by his predecessors, he was grotesquely over-qualified. Dungan's calls produced impressive results: John Gardner of the Carnegie Corporation thought Keppel was "terrific," admired by both the public school and the higher education people. Benjamin Willis considered him "brilliant." Alvin Eurich of the Ford Fund for the Advancement of Education called him "the best candidate." Payson Wild, who had known him at Harvard and was now dean of the faculties at Northwestern, said he was "a wonderful choice," but "too big" for the job. Celebrezze, Dungan wrote the President, could hardly wait to get him behind a desk in Washington.

Keppel had been born in New York City in 1916, the first of five sons. His father was a distinguished educator who had been dean of Columbia College and, for many years, president of the Carnegie Corporation, where he had launched a number of valuable projects, including the American Association for Adult Education and Gunnar Myrdal's monumental study of race, *The American Dilemma.* Francis Keppel was educated at Groton and Harvard, meeting young Joe Kennedy at the latter. After graduation in 1938, he studied sculpture for a year at the American Academy in Rome, an experience that persuaded him that his talents lay elsewhere.

Keppel returned to Harvard as dean of freshmen, served in the Army during the war, and came back to Cambridge as assistant to the provost. In 1948 the dean's job at the Harvard School of Education became open. James Bryant Conant, the president of the university, was enormously interested in the condition of the public schools and was eager to revamp the graduate school completely in order to serve them better. He appointed a search committee to look for an outstanding dean, but its nominees turned the job down. Conant, much impressed by Keppel despite his inexperience in the field, then named him. His performance over the next fourteen years became legendary and he emerged as one of the nation's outstanding educators. He seemed to know everyone in that enormous field. Keppel was a superb mediator who deliberately avoided taking controversial positions in order to protect his role. As his membership on the Hovde task force attested, he believed strongly in federal aid, particularly to bring up the quality of weak school systems. He was a Democrat and an Episcopalian and got along fine with Catholics. Keppel had a wry sense of humor, much like Kennedy's. When he left Harvard he said he was doing so in order to ease the parking problem.

Kennedy, Keppel immediately realized, was convinced that "something ought to be done about this education business—it wasn't working well in the government." He wanted "to get a bill passed." Keppel was "thrust" at once into legislative sessions with Sorensen and Cohen to work up the "education package." The notion was that Congress could be won over if everything was put together in "a package thing for all the boys, the lower school fellows, the higher education people, people who were worried about handicapped children, people who were worried about vocational training, and everything else, and put them all together to neutralize the fights between the various parts of the education world." Kennedy did not expect Congress to swallow this omnibus in one gulp. "It was perfectly clear," Keppel said, "that the President realized that he was sending up a package that would probably get busted up, but that if we kept the thing fluid enough, there was a chance of getting a reasonable part of it through."

Celebrezze, Cohen, and Keppel brought the package down to Palm Beach over the Christmas holiday of 1962 for the annual presidential review before Congress convened. Kennedy wanted Keppel to have a direct line to the

Catholic church. Sorensen took him to meet the bishop of Washington, who put him in contact with key people in the NCWC. Keppel, who had considerable literary skills, and Sorensen drafted the education message.

There were some two dozen education proposals in the package, and the bill, the National Education Improvement Act of 1963, came to 182 pages. Kennedy renewed his request for aid to the schools, but briefly. His stress, as the politics demanded, was on the colleges and universities.

"The long-predicted crisis in higher education," the President declared, "is now at hand." For the next fifteen years average enrollment would grow at a rate of 340,000 a year. The need for new facilities was overwhelming and he urged assistance to both public and private institutions to build classrooms, laboratories, and libraries. In order to facilitate local access for students that lived at home, he recommended a construction program to develop two-year community colleges. Kennedy asked for the expansion of graduate training facilities for engineers, scientists, and mathematicians. Unlike his preceding messages, in 1963 the President said nothing about scholarships, an obvious bow to Republican political realities.

In the months that followed the introduction of the program, little to Kennedy's surprise, Congress broke up the National Education Improvement bill. The most important part came to be known as the Higher Education Facilities Act of 1963, and its history will be traced here. A number of the others were also passed, several quite significant, but they will be noted only briefly.

Wayne Morse and Adam Clayton Powell introduced the administration's omnibus bill and dutifully promised to keep it intact. Nobody, Republicans or Democrats, believed them. When Keppel testified before the House committee, Peter H. B. Freylinghuysen, the moderate New Jersey Republican who supported aid to education, referred to "that monstrosity that you are coming up with . . . , Mr. Commissioner." The higher education community, carefully paying lip service to the package approach, was eager to pursue its own measure and that was hardly a secret. Powell and Green opposed the omnibus from the outset, though they went through the motions for several weeks of hearings before they farmed out sections to the various subcommittees. Morse was stuck because the Senate had only one education subcommittee. Disintegration set in during the summer, and, by the early fall of 1963, Morse resigned himself to what he called "the installment approach."

The House Committee on Education and Labor held hearings on the omnibus bill between February 4 and 27, 1963. The record came to 1,112 pages. Higher education then moved to Green's Special Subcommittee on Education. She insisted on bipartisan support and Albert R. Quie of Minnesota, the ranking Republican, cooperated fully. The subcommittee reported the Higher Education Facilities bill on May 8. The full committee adopted it on May 21 by a vote of 25 to 5. The House passed it on August 14, 1963, by the healthy margin of 287 to 113 (Democrats 180 to 57, Republicans 107 to 56).

Wayne Morse, who had a cast-iron behind, was exceptionally patient when his subcommittee held hearings on the omnibus bill between April 29 and June 27, 1963, filling seven fat volumes with 4,429 pages of testimony. Though it bruised his not inconsiderable ego and caused mild unrest in Oregon, he had no choice but to follow Green's lead. The House, after all, was much the tougher body and she had steered her bill through with magisterial skill. The Morse bill, therefore, was patterned after the Green bill, though there were some differences. On September 25 the subcommittee reported the measure by a vote of 10 to 2. The Labor and Public Welfare Committee adopted it on October 7. On October 15 the Senate, to Morse's surprise and over his vigorous opposition, adopted an amendment proposed by Sam Ervin, the North Carolina Democrat, and John Sherman Cooper, the Kentucky Republican, by a vote of 45 to 33. It was an indirect anti-Catholic measure which would have conferred standing to sue in the district court of the District of Columbia upon individual taxpayers in order to test the constitutionality of the statute under the establishment clause of the First amendment. On October 21, 1963, the Senate passed the higher education bill with the amendment by a vote of 60 to 19 (41 Democrats and 19 Republicans for, 11 southern Democrats and 8 Republicans against).

The Ervin-Cooper amendment aroused concern among Catholics, of course, and earlier might have caused an explosion. But in calm 1963 the matter was handled quietly and deftly. Five prominent Catholic members of the House—James O'Hara of Michigan, who was on the Committee on Education and Public Welfare, Hugh Carey of New York, Delaney and O'Neill of the Rules Committee, and majority leader McCormack—called on Green. They told her that they opposed the amendment and that it could not pass the House. She said that the House conferees also were against Ervin-Cooper and fully recognized that Delaney and O'Neill could kill the bill by voting against a rule. The Catholics accepted her statement as final. The voice vote in the Rules Committee was so overwhelmingly in favor of a conference rule that Judge Smith did not even bother to ask for a roll call.

The conference took only two meetings and the report was issued on November 4. Since Morse agreed with the House conferees, Ervin-Cooper was removed from the bill. All nine House members and seven of the Senate conferees signed on. Only Hill and Goldwater declined to do so. The House, with virtually no debate, passed the bill on November 6 by a vote of 258 to 92.

Senate action was delayed for more than a month because of a stalemate over the equalization formula for the vocational education bill. This caused Morse to work out an intricate strategy to move both bills to a vote in tandem. Since the southerners wanted the vocational bill, the effect was to prevent a southern filibuster over Ervin-Cooper. The Senate agreed to the conference report on December 10, several weeks after the Kennedy assassination. President Johnson signed the law on December 16, 1963.

The Higher Education Facilities Act had the following major provisions. The Commissioner of Education was responsible for administration. Congress appropriated $1,195,000,000 for all purposes for three years beginning with fiscal 1964. These funds were to be allocated to the states in accordance with equalization formulas.

Grants would be made for the construction of graduate schools, including libraries, in the natural or physical sciences, mathematics, modern foreign languages, and engineering, $25 million the first year and $60 million in the second and third. The federal share of construction costs could not exceed one-third. These grants would be available to both public and private institutions, including those that were church-related.

The states would receive allotments to build academic facilities for public community colleges and public technical institutes. They would aggregate 22 percent of the total appropriation.

Low-interest loans were provided to institutions of higher education in general for the construction of academic facilities. At least one-fourth of the cost of a project must come from a non-federal source. The sum available for lending was fixed at $120 million a year for each of the first three years. Both public and private institutions could receive these loans.

Certain types of college and university structures were excluded from eligibility: those employed for events for which the public was charged admission; gymnasia and athletic facilities not related to instruction; and "any facility used . . . for sectarian instruction or as a place of religious worship" or "by a school or department of divinity."

The passage of the higher education act was a display of masterful political artistry. The omnibus strategy worked almost flawlessly and brought with it not only this legislation but six other education laws. The higher education bill was shaped to guarantee victory and, as a bonus, by handsome margins. The two most controversial issues—exclusion of parochial schools and college scholarships—were simply avoided. Thus, there was neither Catholic nor significant Republican opposition. According to Cohen, the bill was structured after the Hill-Burton Hospital Construction Act, which had provided funds for Catholic hospitals. The assumption, he said, was that "nobody objects to building a building for Notre Dame University." The

National Catholic Welfare Council endorsed both the higher education and the NDEA bills. The limitations in the law imposed a heavy price: The needs of the public schools and of college students were not met. But there would be another chance.

The higher education statute was a significant postwar breakthrough. Unless related to defense, Congress had enacted no general education law since 1945. There had always been someone ready to toss sand into the political machinery. In 1963 no one did so. This time everyone either helped or went along—the Catholics, the Protestants, many southern Democrats, many northern Republicans, the NEA (which got nothing), even Adam Clayton Powell.

When President Johnson signed the Higher Education Facilities Act on December 16, he said it was "the most significant education bill passed by the Congress in the history of the Republic." That may have been Texas hype, as one must also consider the Morrill Act and the G.I. Bill of Rights. The new law, Johnson said, would aid in providing classrooms and other facilities for several hundred thousand students; it would help finance 25 to 30 new community colleges a year; it would assist in building graduate schools at between 10 and 20 universities. The President, mindful of the rest of the package, also said, "This session of Congress will go down in history as the Education Congress of 1963. . . ." While many people were responsible, including the two gritty Oregonians, Wayne Morse and Edith Green, the person who deserved the greatest credit, Johnson said, was President Kennedy, "who fought hard for this legislation. No topic was closer to his heart."

The 1963 package included six other education statutes, one of which, the Health Professions Educational Assistance Act, was actually signed by President Kennedy "with great satisfaction" on September 24, 1963. It provided $175 million over three years for matching grants for the construction of facilities for teaching physicians, pharmacists, optometrists, podiatrists, nurses, dentists, and other health professionals. It also established a loan program of up to $2000 a year for students of medicine, dentistry, and optometry.

The other five laws were signed by Johnson after Kennedy's death. The Vocational Education Act provided for a complete overhaul and significant expansion of the old Smith-Hughes program. This was based on the recommendations of the Panel of Consultants on Vocational Education that Kennedy had appointed. Similarly, the Mental Health Facilities and Community Mental Health Centers Construction Act was a response to expert recommendations to modernize the state mental hospital systems and provided grants for building mental health facilities. The Manpower Development and Training Act was broadened. Both the National Defense Education Act and the School Assistance to Federally Affected Areas Act (impacted areas) were extended and NDEA was expanded.

If President Kennedy had been alive, he would have derived no pleasure from signing the extension of the impacted areas law, though he certainly would have done so. Keppel called it a "lousy thing," and said Kennedy "used to growl at it." It had been reasonably "clean" when it was passed during World War II, but Congress by logrolling had encrusted it with "barnacles." Eisenhower had tried to remove them and failed; Kennedy wished he could try again. But he needed political support for the package, and impacted areas brought in 4000 school systems with "a very good lobby."[4]

Francis Keppel worked for John Kennedy for only eleven months, from December 1962 to November 1963. But Keppel saw and spoke with the President on a number of occasions and, with his exceptional perception, came to understand him very well. He made an oral history for what would become the Kennedy Library on September 18, 1964, less than a year after the assassination, when his memory was fresh.

Kennedy was, of course, extremely well-educated—Choate, Harvard, the London School of Economics. Thereafter he became a one-man university extension program. He was a voracious reader and, given his high positions in public life, learned constantly from the many well-informed people he came to know. When he became President he was able to surround himself with a distinguished "faculty." Thus, education was in his bones and it was automatic for him to insist that every young American should have the opportunity to tap into that rich resource.

Kennedy was not led astray from this commitment by extraneous issues. He had no emotional connection to the Catholic parochial schools. "I felt him detached," Keppel said. One could speak to him about this boiling question with complete frankness. Keppel agreed with Sorensen's assessment: "He was the most Protestant president I ever saw on this issue." Keppel also had the feeling that Kennedy would have loved to raise educational standards, particularly in the arts. "I never got the sense that the President was personally affected by the arts, but he had a kind of feeling that he ought to be. He and the rest of society should."

Keppel was much impressed with Kennedy the politician. On January 14, 1963, while Keppel was busy putting the package together, the President, a step ahead, phoned. He was worrying about a schools bill and the House Rules Committee. If they put schools in the package, Delaney would vote against it and that would kill it. If the Catholics opposed schools, he asked, would the NEA jump on higher education? Keppel thought not because NEA had been kicked around so hard for doing so in 1962. Kennedy's prime political concern was to get a bill passed. But there was more. "Like any skilled artist, he enjoyed looking at the different colors on the palette."

He phoned Keppel with ideas, "damn near frightened the office to death—the President hadn't called since before Franklin Roosevelt." Admiral Hyman Rickover had gone to the White House to raise hell about educational standards. He told the President to order Keppel to raise national standards. Kennedy passed along the idea. While the goal was eminently desirable, the method, Keppel said, would have "loused up everything in my life." He had to spend some time smoothing over the admiral. But the President's own ideas were very good. He was deeply concerned about kids dropping out of school, the unemployed youth who became "social dynamite." "Why the hell doesn't somebody do something?" he asked. So the Office of Education did something. The President gave Keppel $250,000 from the emergency fund, and he made several TV spots. He and his brother got interested in Prince Edward County, Virginia, which had closed its public schools to avoid complying with the *Brown* decision. Keppel was ashamed of himself for not addressing the problem earlier. The situation, he thought, was "scandalous"—four years for a county in the United States with no public schools! With private foundation money they set up a special schools program for the children in the county.

Keppel continued:

> One of the many reasons why I feel the way everybody else does about his assassination is that I think he would have gotten real pleasure from the fact that that [legislative] strategy of his worked like a charm. Congress has passed more legislation on education than any since the founding of the Republic, and it was, I think, this package plus his sense of how you could get moving.

There was more. By example he taught a generation that "being intellectual was OK." And he and Walter Heller taught the country "the interrelation of education with the growth of the economy. The centers of excellence for new ideas. They bring new business."[5]

8

A Battle Lost: Medicare

EVERY INDUSTRIALIZED NATION in the world except the United States had a comprehensive health care system either nationalized or structured as social insurance. The idea emerged in this country during the Progressive Era. Theodore Roosevelt's campaign for President in 1912 included a plank calling for national health insurance. The American Association for Labor Legislation (AALL) then proposed compulsory health insurance administered by the states and in 1915 offered a model bill. The House of Delegates of the American Medical Association (AMA) endorsed the idea in 1916. But the great majority of doctors soon came to oppose health insurance and in 1920 the House of Delegates reversed itself. The AMA now stood implacably against state intervention and the AALL's bill died. The Committee on Economic Security, appointed by FDR in 1934, to frame what became the Social Security Act, proposed old-age pensions, unemployment insurance, and assistance to several categories of the needy. It also recommended that the Social Security board, when it was established, should make a feasibility study of national health insurance. The outcry from the AMA was so loud and sharp that Roosevelt dropped the suggestion lest it endanger the remainder of his program.

In 1939, Senator Robert F. Wagner, the New York Democrat, introduced the National Health Program bill, which included grants to the states to encourage, but not compel, the establishment of health insurance systems. The AMA condemned it and the outbreak of war later that year snuffed out any chance of serious consideration. In 1943, looking forward to the end of hostilities, Wagner and two other Democrats, Senator James Murray of

Montana and Representative John Dingell of Michigan, introduced a comprehensive national health insurance bill. In the ensuing seven years the Wagner-Murray-Dingell bill was the focal point of rancorous political debate dominated by the AMA and its allies. Despite the endorsement of President Truman, by 1950 the proponents had to concede that there was no hope of enacting an inclusive program. They started to examine alternatives.

Dr. Thomas Parran, the head of the Public Health Service, had suggested in 1937 that the first people to be covered by health insurance should be pensioners covered by Social Security as well as the medically indigent. In 1944, evidently independently, Merrill G. Murray of the Social Security Administration proposed "beginning" with retirees. In 1950, I. S. Falk, the head of Social Security's Bureau of Research and Statistics, and his staff, again independently, began serious work on developing a program for Social Security beneficiaries.

Oscar Ewing, the Federal Security Administrator, was convinced by 1950 that Wagner-Murray-Dingell was dead and, with White House prodding, was casting about for a politically palatable alternative. Falk's idea attracted him. It was much more limited in scope and cost than national health insurance and it addressed the elderly, the group with the greatest need. In fact, the biggest defect in the old age and survivors' insurance system was that it did nothing to protect against the main cause of dependency in old age, the high cost of medical care. The number of aged citizens was growing rapidly, from 3 million in 1900 to 12 million in 1950, from 4 to 8 percent of the population. This trend was certain to continue. Two-thirds of the elderly had incomes under $1000 annually and only one in eight had private health insurance. Most commercial carriers refused to insure old people, and unions, which were then negotiating private health plans, were unable to persuade many employers to cover their retirees.

In 1951, Falk and Wilbur Cohen, "Mr. Social Security," drafted the first Medicare bill, which Senators Murray and Humphrey and Representatives Dingell and Emanuel Celler introduced the next year. The AMA did not take it seriously and Eisenhower's election in 1952, after a campaign of opposition to "socialized medicine," seemed to finish it.

During the Eisenhower years the AMA was in command legislatively, but it ignored the condition of the elderly. Their numbers grew rapidly and they and their children had a hard time paying inflating hospital and medical bills. In 1957 the AFL-CIO made Medicare its legislative priority. Nelson Cruikshank, the head of its social security department, joined Cohen, Falk, and Robert Ball, a respected official in the Social Security Administration, to write a simple bill for old age and survivors' insurance beneficiaries 65 and over—60 days of hospitalization, 60 days of nursing home coverage, and some surgical benefits.

They needed a Democratic sponsor from the House Ways and Means Committee and started with Jere Cooper of Tennessee, the chairman. He wanted no part of mandatory health insurance. Number two, Wilbur Mills, turned them down without even reading the bill. Number three, Noble Gregory of Kentucky, declined. Number four, Aimé Forand of Rhode Island, said he would think about it. After a good deal of pressure from the draftsmen, he gave in, saying that he wanted to help "the old folks." Forand on August 27, 1957, quietly dropped the bill into the House hopper. He expected nothing to happen for ten years and, since he would retire shortly, thought action would occur long after he had left Congress.

Forand was amazed and delighted by the great public interest, particularly among the elderly, sparked by his bill. An unassuming and lackluster politician, he soon found himself a national celebrity. In the spring of 1958 both the American Nurses' Association and the American Hospital Association spoke favorably of the Forand bill. The AMA now awoke to discover that it was grappling with a tiger and began denouncing Medicare. Forand, who believed that it did not matter what they said as long as they kept talking about his bill, welcomed the publicity the doctors generated. Though Mills, who became chairman of Ways and Means when Cooper died in 1957, opposed hearings, Forand insisted. Held in the summer of 1958, they helped establish a record and attracted a good deal of public attention.

The news gradually reached the Senate. Wilbur Cohen, now at the University of Michigan, was in touch with Kennedy's office. Sorensen was an old friend and he and Feldman, who had just joined the staff, were enthusiastic about Medicare. Cohen in 1958 helped them draft Kennedy's Ten Point Program for the Aged. That fall Senator Douglas, speaking in southern Illinois, was asked about the Forand bill by "a bedraggled oldster with many missing teeth." He had never heard of it, but immediately found out. In February 1959, Wayne Morse, who had not been asked to do so and had never met Forand, introduced the Rhode Islander's bill in the Senate because he thought it was the right thing to do. William Reidy, a classmate of Cohen's (Wisconsin '34), who was on the staff of the Senate Labor Committee, had the idea of a special subcommittee on aging to help the Democrats in 1960. Kennedy was his first choice for chairman, but took time making up his mind. Pat McNamara of Michigan needed a good issue for his Senate campaign in 1960 and asked Lister Hill, the chairman of the committee, for the chairmanship. Hill gave it to him and also appointed Kennedy, Clark, Randolph, Dirksen, and Goldwater to the Subcommittee on Aging. In 1959, McNamara took the show on the road—Boston, Pittsburgh, San Francisco,

Charleston, Grand Rapids, Miami, Detroit, as well as Washington—and oldsters, spurred by the AFL-CIO, came out in droves to speak their minds. The hearings got enormous coverage from the media, including television. Medicare had now become the number one demand of the elderly. The unions organized large rallies of their retired members in early 1960. In March the candidates—Kennedy, Humphrey, and Symington—strongly supported the Forand bill before 13,000 pensioned auto workers at the State Fair Grounds in Detroit.

By the spring of 1960 health care under Social Security was probably the hottest issue of the gathering presidential campaign. The Democratic candidates cashed in. Nixon, who knew that there were 16 million oldsters, all qualified to vote, implored Eisenhower to give him an alternative. Arthur Flemming, the HEW secretary, agreed. But the President refused and in March, supported by the Republican leaders in Congress, opposed health insurance under Social Security. Nixon turned to the Republicans on Ways and Means to come up with a bill that would be acceptable to him, Eisenhower, and the AMA. One later said, "We might have been able to satisfy Nixon and Eisenhower, but, hell, we knew those fellows out at AMA headquarters in Chicago wouldn't accept anything." In late June the President relented and allowed an administration bill to be introduced.

It would create a federal-state fund out of general revenues to protect the indigent elderly against catastrophic illness. Only a couple with an annual income of less than $2500 would be covered and they would have to accept a degrading means test. In effect, it would be another public assistance program. Edward Chase, an authority on the issue, wrote, "It is hard to escape the conclusion that the plan is strictly a political gesture, reluctantly taken to ease the politically untenable situation into which sheer negativism had placed the Party." No one liked it. The AFL-CIO called it hopeless; Governor Rockefeller said it was fiscally irresponsible; Goldwater condemned it as socialized medicine; and the AMA denounced it. Nixon remained silent and Eisenhower seemed to back away.

The spotlight now shifted to Wilbur Mills. It is conceivable, had he endorsed the Forand bill, that he could have carried enough southern Democrats to clear the Ways and Means Committee. But, as the leading authority in Congress on Social Security, he was torn. He struggled intellectually with the alleged financial imcompatibility between the old-age pension and Medicare systems. "Mills," Martha Derthick wrote, "feared the costs of health insurance and the effects of unanticipated costs on the basic program. Also he was reluctant to graft payments for services, which bore no necessary relation to individual wages or tax payments, onto a program in which benefit payments had been systematically related to the individual's wages."

More important, doctors and insurance companies had worked over Mills's district in northeastern Arkansas. In 1958 the prominent Little Rock congressman Brooks Hays had been beaten by an unknown doctor with write-in votes. Mills is said to have promised the state medical society that he would oppose the Forand bill in return for AMA support in the 1960 election. As one of his colleagues said, "He is a man of great intellect and great indecisiveness." His caution was legendary and he did not move until he had ironclad majorities in Ways and Means and on the floor. There was no possibility whatever that the Forand bill would pass the House in 1960. On March 31 he reluctantly allowed a vote in the committee. It went 17 to 8 against the measure.

This should have closed the matter, but it did not. Until now the two powerful Texans, Senate Majority Leader Lyndon Johnson and House Speaker Sam Rayburn, would have nothing to do with Medicare. But Johnson was running for President and needed support from the elderly. Thus, Rayburn told Mills that something must be done for them before the election. Mills, who worshipped Mr. Sam, reluctantly agreed to let his committee reconsider the Forand bill.

On June 3, Ways and Means again rejected the measure by the same vote. Mills then proposed his own bill. It would expand the existing old age assistance program by providing medical care to the elderly poor by authorizing the federal government to make matching grants to the states. The proposal sailed through the committee and three weeks later passed the House 380 to 23. Many supporters of the Forand bill voted for it in the hope that the Senate would pass Medicare and that something like it would come out of conference. Mills felt that he had discharged his obligation to Rayburn.

If the political machinations in the House were odd, those in the Senate were bizarre. McNamara wanted his subcommittee of the Labor and Public Welfare Committee to approve a beefed-up Forand bill. But the subcommittee had no authority to report a bill. Further, Cohen and the AFL-CIO experts were convinced that the Labor Committee, dominated by liberals, did not have sufficient clout in the conservative Senate to move such a bill. Thus, they argued that it must go to the more powerful and very conservative Finance Committee, which had far more influence and, in any case, had jurisdiction over Social Security. After sharp disagreement, two bills were sent to Finance—Forand's, sponsored by Morse, and a more generous version of it, introduced by Kennedy and Senator Philip Hart of Michigan.

Finding a Democratic sponsor on the Finance Committee was an ordeal. The chairman, Harry Byrd of Virginia, not only opposed Medicare but was against Social Security. Russell Long of Louisiana thought that government medical care would convert hospitals into free vacation resorts for the elderly. George Smathers of Florida had long been an ally of the AMA. Robert Kerr of Oklahoma was plotting a one-man crusade against Medicare. Paul Douglas, who strongly favored the proposal, thought that his image as an intellectural liberal was political poison in the Senate and he got no argument. This left Clinton Anderson of New Mexico. While he had been a successful insurance executive and was fairly conservative, Anderson was attracted by the prospect of offering a liberal measure. He also liked Medicare on principle because he was getting on in years and was in ill health himself.

On May 6, 1960, McNamara, Morse, Kennedy, and Hart joined to propose a new and more generous bill. It would provide 90 days of hospital coverage, 180 days in a nursing home, and 240 days of home health care. It would not be limited to pensioned old age and survivors' insurance beneficiaries. Of the 14.8 million persons expected to be covered, 11.3 million would come from Social Security, 1.7 million from old age assistance, and 1.8 million of those who were under neither system. This became the Anderson bill.

The Finance Committee held abbreviated hearings on this bill at the end of June, just before the nominating conventions. At this time Senator Leverett Saltonstall, the Massachusetts Republican, introduced an administration bill. A group of liberal Republicans, led by Senator Jacob Javits of New York, produced a measure for matching grants to the states to purchase private health insurance policies for low-income people over sixty-five. There would be a means test. After the conventions Kennedy joined Anderson in urging the Finance Committee to accept the latter's bill. But the committee voted them down 12 to 5. The administration bill and the Javits bill were also defeated. The only one who enjoyed success was Senator Kerr.

This wily Oklahoma oilman, the wealthiest and, some said, the most powerful man in the Senate, had decided to take over the Medicare issue on his own terms. While Kerr was obsessed with power, he seems to have craved the dollar even more. Robert Kennedy thought "his only interest was money" and that he was "a real bandit." Herbert S. Parmet described his spread at Poteau, Oklahoma, as "a modern barony, as big and grandiose as its owner." It rested on 52,000 acres with some 2000 Black Angus cattle. "The House, furnished in Chinese elegance . . . had fourteen thousand square feet, three kitchens, nine bedrooms, eleven baths, and, for the grandchildren, a playroom outfitted as a ship."

But now Robert Kerr had a political problem. He was for Johnson for President but expected Kennedy to win the nomination. The latter would be a hopeless Catholic candidate in Baptist Oklahoma, and Kerr, himself up for re-election, needed an issue to separate himself so that he could win while Kennedy lost. With his salesman's optimism, he also expected to get the backing of both the elderly and the doctors in Oklahoma.

Senator Douglas had quipped that an expert on Social Security was anyone with Wilbur Cohen's phone number. Thus, Kerr called Cohen in Ann Arbor. He wanted a federal-state matching grant bill aimed at the medically indigent and he wanted Cohen to write it for him. He also wanted to be sure that a lot of his Oklahoma constituents would qualify.

This put Cohen in an awkward position. Virtually all the supporters and opponents of Medicare regarded the Mills bill that had passed the House and Kerr's related idea as an *alternative*. If such a measure were enacted, they argued, Medicare would be derailed and might never get back on track. Cohen's analysis was more sophisticated. There were two groups of aged who needed assistance: those on old-age pensions who could be covered under Social Security, and the medically indigent who were outside that system. Thus, it was necessary to have two programs—Medicare and what in 1960 would be called Kerr-Mills and later became Medicaid. As a practical matter, the Anderson bill was politically dead in 1960. Why not, Cohen argued, take the lesser program first and follow up with Medicare later? He had little trouble convincing Kennedy, Sorensen, and Feldman of the soundness of the strategy. Douglas needed more persuading and McNamara never came around.

During the summer of 1960, Cohen, using the Mills bill as a model, drafted a measure for Kerr, making certain that both Oklahoma and Arkansas were treated generously. Cohen asked Kerr whether the AMA would support the bill. Kerr assured him that he would take care of the doctors. A masterful hustler, Kerr descended upon AMA headquarters in Chicago and used every trick in his almost bottomless bag. He warned the doctors that it was his bill or Medicare sooner or later, probably sooner. The stone faces did not twitch.

The Kerr bill was politically irresistible because it allowed senators to vote for health care and to evade Medicare on the eve of the election. It swept through the Finance Committee 12 to 5, with only the liberal Democrats in opposition. The vote on the floor was 91 to 2, only Goldwater and Thurmond voting no. The conference quickly resolved the differences with the House bill and in September 1960 Eisenhower signed Kerr-Mills.

The passage of this legislation, however, did nothing to stem the debate. The experts agreed that Kerr-Mills would help only a small fraction of the elderly, and people under Social Security continued to demand Medicare. The Democratic platform and Kennedy, therefore, hammered away for Medicare. His election guaranteed a major legislative effort to win enactment.[1]

But even before he took office Kennedy knew that the political prospects for Medicare ranged from dim to hopeless. There were two reasons: extremely discouraging numbers on Capitol Hill and the massive negative campaign being waged by the American Medical Association.

The bill must start in the House Ways and Means Committee. In the summer of 1960 the committee had voted 17 to 8 to defeat the Forand bill. After the election there were still 16 opposition votes. Of the 25 members, 10 were Republicans, all adhering strictly to the party line. Of the 15 Democrats, eight were from the South or the Border and six of them, including Wilbur Mills, were on record against Medicare. The nine Democratic liberals were hopelessly outnumbered. Further, the division within Ways and Means approximately reflected the split in the House. HEW in 1961 figured that it was 23 votes short on the floor. Kennedy knew this, and, more important, Mills knew it. The chairman derived no gratification whatever from batting his head against a brick wall. Further, Kennedy was eager to get important bills dealing with taxation, trade, and welfare through Ways and Means. Mills did not want to waste the committee's time when it had important work to do. He and Kennedy seem to have made a silent agreement that Medicare would go to the end of the queue. Kennedy, however, did insist that, as the Democratic membership on the committee turned over, all new appointees must be supporters of Medicare. But this strategy would take years to work itself out.

The Senate side was no more heartening. On December 15, 1960, Clark sent the President-elect his assessment of prospects for the seven major measures that would be the heart of "the Kennedy 1961 legislative program," with Medicare at the top of the list. During the summer the Senate had defeated the Anderson bill 51 to 44. The 16 members of the Finance Committee who cast ballots voted 11 to 5 against Medicare. The six Republicans were solid and the five southern and Border Democrats who joined them were almost as reliable. A 9 to 8 majority for Medicare would need a favorable vacancy appointment and three Democratic switches, a hopeless prospect. Clark's outlook for the bill: "poor."

Kennedy's victory had frightened the AMA and it stepped up its already formidable campaign against "socialized medicine," "the most deadly challenge" the medical profession had ever faced. Its dues started to climb from $35 to an eventual $70 a year. Doctors were showered with literature to display in their offices and give or send to their patients. Speeches were written and delivered. An immense letter-writing campaign was launched. Metropolitan newspapers ran full-page ads and radio and TV were filled with commercials. Taking a leaf out of labor's book, the doctors floated the American Medical Political Action Committee. The AMA put the arm on a number of organizations considering support for Medicare—the United

Presbyterian Church, the YWCA, the American Hospital Association, and the American Public Health Association—with varying success. The AMA's stridency, bullying, and resort to falsehoods created an inevitable reaction, particularly among doctors in the medical schools. This led to formation of the Physicians Committee for Health Care for the Aged through Social Security, which included a number of the nation's most eminent medical people.

During the interregnum Kennedy had asked Cohen to chair the task force on health and Social Security. Cohen picked two noted doctors, Dean Clark, director of the Massachusetts General Hospital, and James Dixon, the president of Antioch College, along with Herman A. Somers, a political scientist at Haverford, and Elizabeth Wickenden of the New York School of Social Work, who was a friend of Lyndon Johnson's. Because of the Kennedy family's concern with mental retardation among children, two distinguished medical school pediatrics authorities interested in establishing a national institute of child health were added—Robert E. Cooke of Johns Hopkins and Joshua Lederberg of Stanford, the latter a Nobel Prize winner.

Cohen conceived of this task force report as the means of establishing the health and Social Security agenda for the Kennedy presidency and later thought that this had been substantially achieved. He delivered it to Kennedy at the Carlyle Hotel in New York on January 10, 1961. The report was comprehensive, covering a dozen major topics of which Medicare was only one, though the first.

Social Security, the task force wrote, is "the only sound and practical way of meeting the health needs of most older people." Initial contributions of 0.5 percent of payroll (0.25 percent each on the employer and the employee), rising gradually to 0.8 percent, would fund "a reasonably adequate benefit program." Since hospital costs were "impossibly heavy," the report urged incentives to develop alternative nursing home and home health services, which were significantly less costly. The task force sought to reassure the doctors: nothing would be "socialized"; the patient would have free choice of physician, hospital, and nursing home; nobody would control the practice of medicine; and providers would be paid on the basis of "reasonable cost," as agreed to by the doctors, hospitals, and HEW.

Kennedy accepted the task force report and then urged it upon Congress in a special message on February 9, 1961. He sent up the Medicare bill four days later. Since Forand had retired, it was necessary to have a new sponsor in the House. Cecil R. King of California, who ranked among Democrats only behind Mills on Ways and Means, introduced the bill. Thus, it now became known as King-Anderson.

Though it was constitutionally necessary to start in the House, and Ways and Means had jurisdiction, Mills would have nothing to do with Medicare. After a struggle he reluctantly agreed to hold hearings, which he seldom attended, that took place between July 24 and August 4, 1961. Both sides brought in everyone they could think of to make a record, and a record it was—four volumes of 2,281 pages. In terms of public impact, the hearings were pretty much a standoff; if anything, the AMA may have done a bit better. As far as Mills was concerned, that was the end of it. Nothing else happened that year. On August 30, Senator McNamara, concerned, wrote the President for a promise that he would pursue Medicare vigorously the next year. Kennedy replied the next day, "I assure you that I intend to recommend that this legislation be given the highest priority at the next session of Congress."

When the new Congress convened in January 1962 Senator Anderson asked Chairman Byrd of the Finance Committee to hold hearings on his bill. Byrd ignored him. Anderson then introduced a motion for hearings that was defeated in the committee 10 to 7. Five southern and Border Democrats—Byrd, Kerr, Smathers, Talmadge, and Fulbright—joined all but one of the Republicans in opposition.

On January 3, 1962, Ribicoff and Cohen met with Mills to discuss strategy. The chairman suggested that his committee take up three important bills in this order: taxation, welfare, and trade. He omitted Medicare because he could not get a favorable vote in Ways and Means. But he made this unusual proposal: He predicted that the welfare bill would pass both the committee and the House. When it reached the Senate, Anderson could attach Medicare to it and, hopefully, the package would slide through. In the conference between the houses, Cohen wrote Feldman, "Mr. Mills . . . after protracted discussions and some compromises . . . would accept some health insurance provisions on the grounds that it was necessary for him to get the welfare amendments passed for the State of Arkansas and that he had to reluctantly accept the 'watered down' provisions of a health insurance bill." Ribicoff was enthusiastic and Cohen recommended that the President adopt the Mills strategy.

But the administration now split badly in a fight between the under secretary of HEW, Ivan Nestingen, the former mayor of Madison, Wisconsin, supported by Walter Reuther, and Ribicoff and Cohen, assistant secretary of HEW responsible for legislation, backed by the White House staff and AFL-CIO. Nestingen wanted to start a prairie fire of public support for King-Anderson; his opponents hoped to turn around three or four members of Ways and Means to give Mills a majority by making compromises in the bill in order to win their votes. In fact, the administration, which seemed to be losing its grip, adopted both strategies.

During the spring of 1962 AFL-CIO organized large rallies of old people to support Medicare, held in 33 cities. Johnson spoke in St. Louis, Goldberg in Miami Beach, Cohen in San Diego, and so on. This campaign peaked on May 20 when Kennedy appeared before 20,000 senior citizens at Madison Square Garden in New York, an event carried by the three major television networks. It may have been the worst speech Kennedy ever made. He disliked the address that had been written for him, tried to make some last-minute notes, and then decided to speak extemporaneously. Forgetting that his important audience was seated at home in front of the tube, he made a stump speech that aroused the people in the Garden and alienated the viewers. Sorensen, Meany, and Cruikshank were dismayed. To make matters worse, the AMA rented the Garden the next night and bought half an hour of prime TV time. Its star speaker, Dr. Edward Annis, a Miami surgeon, delivered a blistering attack on Medicare to the television audience in a dramatically empty house. The AMA won that one easily.

Meantime, the Republicans, worried that the Democrats would capitalize on Medicare in the 1962 congressional elections, brought up three alternatives to King-Anderson. Representative Frank Bow of Ohio proposed an income tax credit or certificate for $125 to buy a private health insurance policy, which won limited conservative support. Representative John Lindsay of New York sponsored Governor Rockefeller's bill: either King-Anderson or the option of buying a private policy with higher Social Security benefits. Senator Javits introduced a voluntary choice bill for himself and six other Republican liberals on the theory that only bipartisanship could work politically. Social Security beneficiaries would either receive a health policy with no deductible or would get cash to buy their own. Those outside the system would have the same options, but they would be financed by general revenues. Javits ran into the insistence of Robert Ball, now the commissioner of Social Security, that the solvency of the system depended upon the principle of mandatory coverage, as FDR had required in 1935.

By June there was a palpable political impasse. The situation in Ways and Means was impossible. Only eleven of the twenty-five members favored King-Anderson. In addition, Mills himself had been redistricted and had to run against an extreme conservative whose main issue was opposition to Medicare. Finally, the Democrats on the committee did not want to go on record before the elections.

Since the liberals demanded a vote, the Democratic leadership decided to chance Mills's strategy of attaching Medicare to the welfare bill that the House was sending over to the Senate. Anderson agreed to give Javits several changes—letting the insurance companies in and providing an option—in order to get the New Yorker's support. The new bill, introduced on June 29, was sponsored by twenty Democrats and five Republicans.

The Senate debate consumed almost three weeks in the first part of July and attracted great public interest. It was notable for behind-the-scenes shady dealing on the part of Kerr, the leader of the opposition. The vote clearly was going to be extremely close. Mike Mansfield, the majority leader, and Anderson believed they could rely on fifty votes, in which case the Vice President would break the tie for the administration. But Mansfield depended upon a nose count by his assistant, Bobby Baker. When Baker had worked for Lyndon Johnson, his forecasts had been extremely accurate. This time Baker faked two votes, leading Mansfield to believe that he was stronger than the facts warranted. Baker had shifted his loyalty to Kerr and, the day after the vote, took out a large personal loan at an Oklahoma bank partly owned by the senator. Further, Kerr turned around Jennings Randolph, who had supported King-Anderson. West Virginia had overspent public welfare funds by $21 million. Kerr wrote a forgiveness kickback into the welfare bill for Randolph's vote on Medicare. Carl Hayden, Arizona's 84-year-old senator, was up for re-election and preferred to vote against King-Anderson in a state rife with Goldwaterism. But he promised the President that he would vote for Medicare if he was needed for the fiftieth vote. When Randolph voted against it, of course, Hayden was released from his pledge.

On July 17, 1962, the Senate defeated King-Anderson 52 to 48. The majority consisted of 31 Republicans, 19 southern Democrats, and 2 northern Democrats; the minority of 39 northern Democrats, 4 southern Democrats, and 5 Republicans. Kerr had broken the back of the Mills strategy.

The defeat in the Senate, of course, was devastating for the sponsors and a joy to the AMA. Senator Anderson, embittered and in ill health, spent the rest of the year in Albuquerque recovering.

There were a few encouraging signs. In the 1962 elections the Democrats picked up four northern liberals in the Senate—George McGovern of South Dakota, Ribicoff of Connecticut, Birch Bayh of Indiana, and Thomas McIntyre of New Hampshire. All were certain or probable Medicare supporters. The historic champion of the old folks, Claude Pepper, easily won a House seat from Dade County, Florida, in the face of massive AMA opposition. Robert Kerr died of a heart attack in January 1963, depriving the opponents of their most effective leader.

But this was not enough to undo the 1962 Senate defeat. While everyone involved in the Medicare fight was busy during 1963, they accomplished virtually nothing. Cohen continued to rework the bill in the hope of putting together a victorious coalition in Congress. On February 21, Kennedy issued a special message on helping senior citizens. There were thirty-six legislative proposals, including what was now called the Hospital Insurance Act, which made many compromises to win over Javits and other Republicans. Ways and Means did not hold hearings until November. The simmering fight between Nestingen and Cohen in HEW broke into the open. Anderson and McNamara had a contest over who would control Medicare in the Senate. The AMA pushed Kerr-Mills vigorously and Mills seemed to go along. The President, deeply concerned about getting his tax program through the Ways and Means Committee, allowed Medicare to fall back in the line. As 1963 drew to a close, Washington observers spoke of it as "the forgotten issue."

There was, however, a hopeful note at the end of the year. The President flew to Arkansas to dedicate a public works project and talked to Mills. There were reports of an understanding on Medicare. In fact, the chairman announced that he was not opposed to the basic philosophy of Medicare. His concern was over its financial soundness and he urged tax flexibility to deal with future uncertainties. Moreover, Mills promptly proceeded to negotiate the financial formula for Medicare with Henry Hall Wilson of Larry O'Brien's staff. They reached agreement on the morning of November 22, 1963, a few hours before the President was assassinated. O'Brien wrote, "With Mills's objections met, the passage of Medicare was assured."

Wilbur Cohen, who had the skin of a rhinoceros, was not discouraged by these delays. His teachers—John R. Commons at Wisconsin and Arthur Altmeyer at Social Security—were gradualists. Eat one meal at a time, they taught, and avoid indigestion. Further, having worked with Wilbur Mills for years, Cohen knew him well and held him in the highest esteem—his incisive mind, his decency, his flexibility, his masterful command of the Social Security system, and his complete control over Ways and Means. Cohen was certain that Mills would come around and that, when he did, Medicare would be enacted easily. But this would take time.[2]

9

The Peace Corps

JOHN KENNEDY WAS 43 when he became President of the United States. He was the youngest man ever to be elected to that office. Many thought it significant that he was the first President to have been born in the twentieth century (not especially early—1917), that this somehow symbolized a generational changing of the guard. Moreover, unlike those who look older than they are, Kennedy retained a boyish aspect. He enjoyed making little jokes about his youth, like riding up in a Capitol elevator and being asked by another passenger to let him off at the fourth floor. His age was both a liability and an asset.

Democrats (before the 1960 convention), as well as Republicans, attacked Kennedy on the "youth" issue. Sam Rayburn's remarks were unprintable. Lyndon Johnson called him "young Jack." Harry Truman made youth his main argument in the attempt to stop Kennedy from gaining the nomination. Dean Acheson considered him an "unformed young man." Many secondary Democratic politicians, having spent long years climbing the party ladder, feared that Kennedy would pass them over by picking people of his own generation. A large number of loyal Democrats wondered why he was so impatient. Why not wait another four, eight, or even 12 years before making the run for the big prize?

The Republicans made a major campaign issue out of Kennedy's age. Eisenhower considered him a whippersnapper. Nixon seized on the youth question, using the code word "inexperience." Though himself only a few years older, he contrasted his "experience" with Kennedy's presumed lack of it. He made it the central issue in the first television debate (but Kennedy so outshone him as to bury the criticism).

But youth also had political advantages. Kennedy drew great numbers of enthusiastic young people to his large campaign rallies. He was especially effective with college audiences, no more so than at 2:00 a.m. on October 14, 1960, at the Student Union at the University of Michigan, where he introduced the Peace Corps concept into his campaign. This led to the major address on that question at the Cow Palace in San Francisco on November 2, 1960. There was a chemistry at work: Kennedy's youth and the youth of America touched hands in the Peace Corps.[1]

At a planning session in Hyannis Port immediately after his election, Kennedy asked Sargent Shriver, his brother-in-law, to run a talent search for "the brightest and the best" people to staff the new administration. Shriver recruited Harris Wofford, Adam Yarmolinsky, and Louis Martin for his team. Buoyed by his unflagging example and leadership, they worked furiously at scouring the country for able people. This was the most exhaustive and productive search ever undertaken by a President-elect. If Kennedy had any doubts about Shriver's competence, this performance laid them to rest.

As the day of inauguration approached, Wofford wrote, "Shriver and I joked that we were rapidly working ourselves out of any job in the new administration." Just before the inauguration Wofford and Yarmolinsky in a farewell gesture delivered the search files to the White House.

In his inaugural address the next day the new President declared:

> To those people in the huts and villages of half the globe struggling to break the bonds of mass misery, we pledge our best efforts to help them help themselves, for whatever period is required—not because the communists may be doing it, not because we seek their votes, but because it is right. If a free society cannot help the many who are poor, it cannot save the few who are rich.
>
> . . .
>
> My fellow citizens of the world: ask not what America will do for you, but what together we can do for the freedom of man.

On January 21, evidently, Kennedy asked a bone-tired Shriver to make a study of how a Peace Corps could be organized. Shriver had been active in the Experiment in International Living during the thirties and had submitted an overseas service proposal to the Eisenhower administration twenty years later. He immediately phoned Wofford, now his close friend, to invite him to join up. Wofford had been interested in a "peace force" for two decades. He had traveled widely in the Third World and had been involved in many programs and proposals for an American volunteer overseas service. "For me," Wofford wrote, "it was . . . an old dream coming true."

They rented a suite at the Mayflower Hotel and were soon joined in the planning by Gordon Boyce of the Experiment in International Living and Albert Sims, who had worked with Shriver on a student airlift from Kenya in 1960. Since the idea of a Peace Corps was very much in the air, they were flooded with reports and proposals. Among the more important were those by Samuel P. Hayes of the University of Michigan, *An International Peace Corps: The Promise and Problems,* and the Colorado State University study authorized by Congress at the instigation of Representative Henry Reuss of Wisconsin, "A Youth Corps for Service Abroad." Two others posed a fundamental policy conflict. Kennedy had asked Max Millikan of MIT to work up a proposal, and he urged a small low-key program. The contrary and much more interesting report came from two young officials of the International Cooperation Administration (ICA), Warren W. Wiggins, the deputy director of far eastern operations, and William Josephson, the far eastern counsel.

Wiggins, still in his thirties, had worked on the Marshall Plan in western Europe, had served as economic adviser in the Philippines, and had administered aid in Bolivia. Josephson, though hardly as experienced, was a legal dynamo. They had begun working on a plan before the Shriver group had started. At the outset Josephson insisted on limiting it to Americans teaching English overseas. But, as they worked it over, the idea grew and eventually came to be called "The Towering Task," a phrase taken from Kennedy's State of the Union message. They urged an all-out commitment at the start and ultimately a very large Peace Corps with 30,000 to 100,000 volunteers. "A small cautious Peace Corps may be worse than no Peace Corps at all." Further, such a program might "maximize the chance of failure."

Though Wiggins and Josephson were not dissatisfied with their jobs at ICA, they were eager to work for the new agency. They did not submit "The Towering Task" directly to Shriver, rather, they sent a copy to Wofford at the White House, hoping that he would take it to Shriver. He received the report and read it late at night on Sunday, February 5. Shriver was swept off his feet and wired Wiggins to attend a meeting the next morning. "The midnight ride of Warren Wiggins" became a Peace Corps legend or, as Josephson put it, a "myth." The report was the most influential factor in shaping the Shriver proposal and both Wiggins and Josephson joined the task force, the latter looking after legal matters, the most important being whether the President had authority to create the agency by executive order.

Kennedy was impatient to receive Shriver's report and inquired several times. A task force team wrote it under forced draft, with Wofford doing the final version. It arrived at the White House on February 24, 1961.

As for a Peace Corps, Shriver opened, "I recommend its immediate establishment." After having studied all the reports and consulting with many experts, he was satisfied that "we have sufficient answers to justify your going ahead." He then outlined the organization he thought would work.

Peace Corps volunteers would go abroad not as members of an official U.S. mission; rather, they would "go to teach, or to build, or to work in the communities to which they are sent." They would live with "the people they are helping."

The need was manifest. The Colorado State team had visited nations in Latin America, Africa, and Asia, all of which desperately needed the help the Peace Corps could provide. They wanted programs in teaching, health—especially the control of malaria—rural development, large-scale construction, and public administration.

The corps would engage in a variety of activities and, therefore, "must have great flexibility to experiment with different methods of operation." Shriver did not want to see "a large centralized new bureaucracy." Thus, there must be an emphasis on farming out programs to private agencies, to universities, to other U.S. government agencies, and to the United Nations. But the Peace Corps would have to administer many projects itself, particularly in the teaching and large-scale construction areas.

Shriver visualized a nationwide recruitment program. "A simple announcement that Peace Corps application forms are available at all post offices . . . would probably produce an initial flood." It was critical that volunteers must be "the best available talent." Thus, recruitment, including testing, must be rigorous. "Widespread competition for Peace Corps positions with such careful screening is essential if people with the best chance of success are to be sent abroad."

Those selected must receive special training, which could range from six weeks to six months. This would include language instruction, preparation for the work assigned, and the problems of health and survival in the destination country. Existing university facilities would be used for the training.

The term of service must be at least one year, preferably two, and in some cases three or more. Volunteers must be subject to dismissal from the corps at any time during their service. There should be no top age limit for admission and both men and women should be welcomed. "In general corpsmen should be single"; the Peace Corps could not pay for a wife or children. Since Nixon had attacked the Peace Corps idea as "a haven for draft dodgers," Shriver wrote, "there should be no automatic draft exemption because of Peace Corps service." Nor should it be a "special domain of conscientious objectors." The pay should be "just enough to provide a minimum decent standard of living." Volunteers "should live in modest circumstances, and as closely as possible to their host country counterparts."

The Peace Corps should possess "its own identity and spirit and yet receive the necessary assistance from those now responsible for United States foreign policy and our overseas operations." Thus, it should be "semi-autonomous," but located within ICA and, like ICA, reporting to Under Secretary of State for Economic Affairs George W. Ball.

The timing of the launch, Shriver wrote, was critical. It could either await the passage of legislation, which could take many months, or it could be started by executive order within two weeks. If the former, summer 1961 training at universities would be lost and the year would be wasted. If the President acted now and the Peace Corps was financed by the transfer of available funds under the Mutual Security Act, it could have 1000 to 2000 volunteers overseas by the end of the year. This suggested a stress on teaching projects in the first year.

For a week there was a sharp debate between Shriver and Sorensen. The latter was concerned that the Peace Corps might become a political liability and was especially disturbed that an executive order might anger Congress. Wiggins argued that, if Kennedy delayed, there might never be a Peace Corps. Josephson pointed out that FDR had established the Civilian Conservation Corps, a roughly parallel agency, by executive action.

On March 1, 1961, only a week after receiving Shriver's report, Kennedy compromised this disagreement both by issuing Executive Order 10924 establishing the Peace Corps on "a temporary basis" and by sending a special message to Congress asking for legislation for a permanent program.

The executive order was extremely brief. It directed the Secretary of State to establish a Peace Corps with a director, barely described its functions, and authorized the secretary to finance it from funds under the Mutual Security Act. The message to Congress proposed a program that closely followed Shriver's report. Surprisingly, despite an opening statement on the Peace Corps summarizing his actions on March 1, the President received no questions about it at his press conference that day.

In a memorandum that accompanied his report Shriver had suggested several names for director of the Peace Corps—Dean Eugene Rostow of the Yale Law School, Professor Carroll Wilson of MIT, Professor Gilbert White of Chicago, and Clark Kerr, president of the University of California. Kennedy was not interested in these "bookish" types, preferring someone more adventurous. He asked Shriver to take the job. While he was eager to face the challenge of building the program, Shriver was concerned about a charge of nepotism, particularly because the President had recently named his brother as Attorney General. Kennedy brushed this aside. Shriver would later joke that the President had appointed him because everyone expected the Peace Corps to be a fiasco and "it would be easier to fire a relative than a friend."

But there was another serious question: Should the director be confirmed by the Senate? Sorensen and Kenny O'Donnell thought the President should not ask for confirmation in order to sidestep potential political controversy. Shriver, Wiggins, and Josephson strongly disagreed. Josephson sent a memorandum to the White House on March 3, 1961, in which he set forth the reasoning in favor of Senate action: (1) Article II, Section 2, of the Constitution provides that the President shall make appointments "with the advice and consent of the Senate." It was customary in the case of heads of agencies, and, indeed, there was some question whether he could do so without Senate approval. (2) All the important federal officials with whom the director must deal had been confirmed and his stature would be diminished if he were not. (3) Since the Peace Corps was now "widely supported," this was "the best time for confirmation." (4) The decision not to seek confirmation "will not clearly avoid controversy."

Shriver added to Josephson's argument: "I agree with this memorandum" and sent it to the President. In fact, Shriver wrote a "Dear Jack" letter in which he stated: "I don't want to embark on a difficult mission with one arm tied behind my back." If Kennedy decided otherwise, Shriver wrote, "I respectfully suggest that you select another person to head the Corps." On March 4 the President sent his name to the Senate and he appeared before the Foreign Relations Committee on May 21. Approval was a formality.

Robert Sargent Shriver, Jr., was born on November 9, 1915, in Westminster, Maryland. He was descended from Andrew Shriver, a Palatine German who emigrated in 1829 and settled along the Schuylkill River in southeastern Pennsylvania. Over the years the family divided between the Episcopal and Catholic churches. His parents, both Shrivers, were Catholic, his mother born to the faith and his father a convert. His father was a small-town banker who later moved to Baltimore and New York. Sargent Shriver attended quality Catholic preparatory schools—Cathedral in Baltimore and Canterbury in Connecticut when the family lived in New York. Though his father's Wall Street commercial bank had been wiped out by the 1929 crash, Shriver enrolled at Yale in 1934. He concentrated on the *Daily News* and became editor-in-chief in 1937. He graduated the next year and then went to the Yale Law School, finishing in 1941. He served in the Navy during the war as a gunnery officer on the battleship *South Dakota*, which was on duty in the South Pacific, and then on submarines in the Pacific. After the war he tried his hand at corporate law with a Wall Street firm, which he disliked, and then jumped at an offer of an editorial job at *Newsweek*.

Shriver met Eunice Kennedy at a cocktail party in 1946 and was enchanted. Joe, Jr., had been killed flying over Belgium in 1944 and his father was looking for an editor to prepare his son's diary for publication. Eunice suggested Shriver, who looked over the papers and concluded that they were not publishable. Despite this bad news, Joe Kennedy was so taken with him that he invited Shriver to work in his multifarious enterprises. In 1948, Kennedy sent Shriver to Chicago as assistant general manager of the Merchandise Mart, the world's largest building, where he performed outstandingly in attracting rental clients.

On May 23, 1953, Shriver's seven-year pursuit was rewarded: marriage to Eunice. Papa Kennedy put on quite a bash. The wedding took place in St. Patrick's Cathedral in New York with Cardinal Spellman officiating. There were 1700 invited guests and thousands more outside, who held up traffic on Fifth Avenue. The reception was in the Starlight Room of the Waldorf-Astoria. The Shrivers then settled into an apartment on the Gold Coast on Chicago's near North Side.

In 1954, Mayor Richard J. Daley named Shriver to the board of education and the next year he became its president. He threw himself enthusiastically into this job and worked closely with Benjamin C. Willis, Chicago's outstanding superintendent of schools, who would later serve on the Hovde task force on education.

Shriver was deeply religious and was a student of Catholic thought. When they shared a hotel room, Wofford wrote, "Shriver would take me along to mass, his common routine (which he thought good for my Episcopalian soul)." He believed in living his religion, which included treating black people exactly as one would treat whites. He became president of the Catholic Interracial Council and worked to bring Negro students into the better parochial schools, to stimulate discussion of racial problems among young people, and to admit blacks to Catholic hospitals. Shriver's idealism and humanitarianism led the Kennedys to call him "the house Communist."

By 1959 there was talk among Democrats of running Shriver for governor of Illinois the next year. But since Daley, the mayor of Chicago, was Catholic, and so too was John Kennedy, the presidential candidate, it would be too much to have another Catholic as the candidate for governor. Thus, in 1960 Shriver worked on Kennedy's campaign staff and as head of his talent search.

When he became director of the Peace Corps, Shriver was forty-five, athletic, trim, and brimming with energy. He was extremely handsome, exceptionally personable, and unusually good at dealing with people. Brent Ashabranner called him a "glittering man." He had a wide range of friends and acquaintances. The Peace Corps mission appealed to all his impulses. "If the Peace Corps was to be a vehicle for American idealism," Arthur M. Schlesinger, Jr., wrote, "Shriver was an authentic and energetic idealist, well qualified to inspire both staff and volunteers with a sense of purpose and opportunity."

Shriver set to work furiously organizing the new agency. He took office space on the sixth floor of the ICA's Maiataco Building on Connecticut Avenue and at the nearby Rochambeau Hotel. NASA lent him John D. Young, its deputy director, who was experienced in setting up agencies. Wiggins assumed responsibility for the central function of planning and developing the overseas programs. Josephson and Morris Abram, the Atlanta attorney who had been active in civil rights, worked on the proposed legislation and routine administrative problems. Josephson soon became Shriver's special assistant and later general counsel. Gordon Boyce dealt with private voluntary agencies and Albert Sims did the same with the universities. Nicholas Hobbs, an authority on testing and selection, was put in charge of choosing the volunteers. Wofford, already in the White House handling civil rights, followed the establishment of the Peace Corps with avid interest and some involvement.

Wofford was instrumental in recruiting Bill Moyers, the 26-year-old star of Vice President Johnson's staff. Moyers had told Johnson during the campaign that, if Kennedy established a Peace Corps, he wanted to work there. Wofford arranged for Moyers to see Shriver. Kenny O'Donnell was annoyed because Moyers was the only member of Johnson's staff the White House trusted and could deal with. Both Kennedy and Johnson urged Moyers to stay, but he insisted on leaving. As the Vice President said, "That boy cajoled and begged and pleaded and connived and threatened and politicked to leave me to go to work for the Peace Corps." He went.

Moyers had the right idea. As Wofford would write later:

> Participation in the making of the Peace Corps was an intoxicating and illuminating experience. Shriver was not a tidy administrator, but he was a great executive. He did not delegate power through an orderly chain of command, but he empowered people. He released their energies, backed their efforts, and drew on their insights. Wives of staff men tended to be jealous because Shriver harnessed their husbands' energies and loyalties—and weekends.

Wofford's estimate was widely shared. Kevin Lowther and C. Payne Lucas, who were extremely critical of the Peace Corps, called Shriver an "executive genius" who "radiated a sense of purpose that infused all who were drawn into his embryonic universe of peace and brotherhood."

Shriver imparted a particular style to the Peace Corps which made it unique among federal agencies. Robert G. Carey summed it up this way:

> There is little about the Peace Corps that is either standardized or scientific. It is an agency nearly devoid of artificial and calculated orthodoxy. The Corps has a litany, to be sure, but it is the litany of the explorer and frontiersman, not the organization man.

In the spring of 1961 the fledgling Peace Corps faced a grave crisis. In March, Kennedy, in order to coordinate U.S. economic and social development programs, proposed that ICA, the Development Loan Fund, and Food for Peace be consolidated into a new umbrella organization to be called the Agency for International Development (AID). From the viewpoint of bureaucratic tidiness it was logical to put the Peace Corps into AID. Secretary of State Dean Rusk and Under Secretary Chester Bowles, both enthusiastic supporters of Shriver's program, favored its independence. But Henry Labouisse, the head of ICA; the bureaucracies at State and ICA; Budget Director Bell; and Ralph Dungan at the White House backed consolidation. The President in his foreign aid message of March 22 seemed to go along with them.

Shriver and his staff were horrified because they thought the Peace Corps had a unique role to play and must be separated from the usual channels of American foreign policy. Shriver wrote a strong protest, but on March 30 Kennedy told him that Labouisse was heading a task force and he expected Shriver's full cooperation. The final meeting was scheduled for April 26. The timing was terrible because Shriver was going abroad four days earlier to persuade potential host countries to accept Peace Corps missions. In the meantime he lobbied Labouisse, Sorensen, and Dungan, but to no avail. In fact, Kennedy was unable to attend the April 26 meeting because he was dealing with the Bay of Pigs disaster. Dungan was in the chair and steered the decision to incorporate the Peace Corps into AID. Kennedy later approved.

Shriver was in India when he learned the bad news. He felt "helpless" and feared that "the Peace Corps was about to die a-borning." All he could think of was to cable Wiggins to ask the Vice President to intercede. Wiggins sent Moyers, who played on LBJ's nostalgia over his New Deal experience in running the National Youth Administration in Texas and stressed his early support for the Peace Corps concept. Johnson responded immediately. "You put the Peace Corps into the foreign service and they'll put striped pants on your people when all you want them to have is a knapsack and a tool kit and a lot of imagination." He went to see the President on May 1, and, as Josephson put it, "Johnson collared Kennedy . . . and . . . badgered him so much that Kennedy finally said all right." While this soured Shriver's relationship with the White House staff, it was a crucial decision to preserve the independence of the Peace Corps.

On April 22, 1961, Shriver had left with Wofford and several others for a twenty-six day trip to Ghana, Nigeria, Pakistan, India, Burma, Malaysia, Thailand, and the Philippines. It was a critical mission because no Third World nation had yet asked for Peace Corps assistance. There was widespread suspicion of American foreign policy, fear that the Peace Corps was a cover for CIA infiltration, and skepticism over the ability of rich young Americans to do any good in poverty-stricken villages. The Bay of Pigs misadventure was no help. Galbraith in New Delhi expected trouble with Nehru.

The skeptics underestimated Shriver. He returned with invitations from all eight countries he had visited for a total of 3000 volunteers. This opened the floodgates. A few days later Kennedy announced more than two dozen requests for aid.

Wofford wrote the President his "personal observations" of the trip on May 25, 1961:

> The Peace Corps strikes the note these nations are waiting for. . . .
>
> Shriver is a born diplomat. I have never been witness to so successful an international operation. His meetings with government officials, newsmen and private citizens all produced good results for the Peace Corps and U.S. relations. Our ambassador and other overseas officers in every country expressed to me and others their admiration and appreciation of Shriver, their amazement at how much was accomplished in such a short time, and their increased hopes for the Peace Corps in their respective countries.
>
> There exists a reservoir of goodwill and hope for you and the new Administration in all these countries, particularly in Africa, in India, and in Burma. The Cuban affair lowered the reservoir in these countries—but not as much as would have been the case if U.S. force had gone in and overthrown Castro. The high expectations for a new American approach to the world which you have aroused in Nkrumah, Nehru, and U Nu, to name three important cases, are a great opportunity.

In New Delhi, Galbraith said, "Amen." In two advance meetings with the Shriver party he urged them to ask for little and to confine the requests to agriculture, and "they accepted generously." On May 2, 1961, Shriver and Galbraith called on the Prime Minister. The meeting, Galbraith wrote, was "a marvellous miscarriage of plans." "Sarge took over and made an eloquent and moving plea on behalf of his enterprise. The effect was just right—natural, uncontrived and sincere." Nehru readily accepted and actually urged a broadening of the program. Galbraith, whose confidence in his own omniscience knew few bounds, confided to his journal that he left "with my reputation as a strategist in poor condition."

The Peace Corps bill, which closely followed Shriver's report and the actual program being put into effect, sailed through Congress. Shriver, Kennedy said, was "the most effective lobbyist on the Washington scene." He dominated the hearings before the Senate Foreign Relations Committee on June 22 and 23 and the House Foreign Affairs Committee on August 11 and 18, 1961. There was a good deal of support and virtually no opposition, though, according to Josephson, there was a lack of enthusiasm in the Senate committee. Despite a statement from Eisenhower that the Peace Corps was "a juvenile experiment" and that volunteers should be sent to the moon because it was undeveloped, Republican opposition was confined to niggling amendments. The Senate committee adopted the bill unanimously on August 4 and the Senate, with virtually no debate, passed it by voice vote. The House committee also accepted the bill without opposition on September 5, and the House voted for it 288 to 97 on September 14. Conservative Republicans succeeded in inserting two symbolic amendments. Senator Bourke Hickenlooper of Iowa got one requiring that volunteers be trained in Communist philosophy and tactics. Representative Walter Judd of Minnesota got the other, requiring volunteers to sign an oath that they did not advocate the overthrow of the U.S. government. The conference committee adjusted minor differences and on September 22, 1961, the President signed the Peace Corps Act.[2]

The launching of the Peace Corps was an exceptionally complex operation. Broadly speaking, it involved the following: the creation of a headquarters staff and the writing of rules; the recruitment and training of volunteers; making contracts with host countries, which included not only acceptance of a Peace Corps mission but also determining what volunteers would do and arranging for their housing, travel, health, and a multitude of other matters; and developing, monitoring, and evaluating the programs.

The creation of a Washington staff was the least of Shriver's problems. He started with a core of gifted people, and many others of similar talent were eager to serve provided that they could abide a measure of chaos. In these early days Robert Carey went to the Maiataco Building to apply for a job. He was soon put to work moving "file cablinets, office furniture, and office supplies into place." He emerged two hours later "convinced that the first volunteers would not be overseas for years." He could not have been more wrong.

One of the main tasks at the outset was to lay down policies for the people in the field. Volunteers would live simply and unostentatiously, approximately at the level of the people they served. They must not match the standards of American diplomatic and AID personnel abroad. They would not be paid salaries, merely subsistence. Each would bank $75 for each month of service, payable as a termination allowance. He or she would also earn leave time at the rate of two and one-half days for each month served. But leaves must be taken in the Third not the First World.

The most controversial question, which took a year to resolve, concerned the marital status of volunteers. Some argued that married people should be excluded. General Counsel William Delano urged a case-by-case determination, asking whether the individual married volunteer could be effective. His view prevailed. The exception was an unmarried female volunteer who became pregnant; she would be sent home because the Peace Corps could not assume responsibility for her child. By the end of 1963 there were 400 married couples, 40 of whom had wed during their service. Brent Ashabranner, who administered programs in Africa and Asia, stood in for the bride's father at a number of these weddings. "I am not sure I can remember all of the Peace Corps brides I have given away," he wrote, "but they are a part of my emotive investment in the Peace Corps." Pregnancy of a married volunteer was also handled on an individual basis by the same standard of effectiveness.

The church-state controversy plagued the Peace Corps. Shriver was extremely sensitive to the President's religious problem and Sorensen did not hesitate to remind him of it. He was also well aware of his own vulnerability as a Catholic. Moyers, himself a Baptist minister, argued strongly against making contracts with religious voluntary organizations because that might raise the question of government support for religious proselytizing in violation of the First Amendment. At the outset Shriver tried to avoid the issue with silence.

This did not work. The effect of avoidance was to exclude religious overseas programs from Peace Corps contracts. This cut out the major potential voluntary agencies—Catholic Relief Services, the Church World Service, Lutheran World Relief, the American Friends Service Committee, among others. All of these organizations were upset by the denial of Peace Corps support. The National Catholic Welfare Council, which had counted on government money and personnel, was outraged.

There was a special problem in Nigeria in 1961. Volunteers were training at Harvard for a teaching program. During the colonial period the British had established mission schools, both Protestant and Catholic, and the new Nigerian government had carried them over. Some suggested that Protestant and Catholic volunteers should teach in schools of their own faith. Shriver, on Josephson's legal advice, vetoed a religious test for the selection and assignment of volunteers. They were allowed to teach secular subjects in mission schools, but were forbidden to teach religion.

The Peace Corps adopted a policy of nondiscrimination, including what would later be called affirmative action, in the recruitment of blacks, Latinos, and other minorities. This was based in part on the sensible premise that overseas programs would be overwhelmingly in nonwhite nations and that Americans who were not Caucasians would fit in more easily. Muslim countries, excepting Tunisia and Morocco later on, did not get contracts because they refused to accept Jewish volunteers. Universities in the South were not offered training programs because they declined to admit Negroes. Despite the policy of nondiscrimination, the number of black volunteers between 1961 and 1963 never exceeded 5 percent. This was because there were very few Negro college graduates and most of them preferred paying jobs at home. Women were also welcomed and by 1963 constituted about 40 percent of the 7000 volunteers.

The Peace Corps considered the selection process to be critical, that it was preferable to avoid a mistake early than to suffer from it later. Thus, the weeding-out process was rigorous. The applicant had to answer a comprehensive twelve-page questionnaire, submit six references, take a six-hour written examination, and submit to a thorough medical checkup, followed by psychiatric observation. It was difficult to pass through all these gates. Moritz Thomsen, a forty-eight year old farmer from Red Bluff, California, wrote that "we were studied and appraised like a bunch of fat beeves about to be entered in the state fair."

In March 1961, even before forms had been distributed to post offices, the Peace Corps received 10,000 letters from potential applicants. When the forms went out Shriver expected 15,000 completed applications, but received only half that number. Only 3500 took the first examination on May 27. The assumption that recruitment would take care of itself did not work out. Shriver then threw himself into a high-pressure drive, including TV and radio advertising, and by mid-year had 10,000 applicants. In 1962 there were 20,000, and in 1963 almost 35,000. This raised internal concern over whether quality was being sacrificed for big numbers.

The Peace Corps developed a three-phase training program: eight to ten weeks at a U.S. university (usually in the summer when professors, classrooms, and dormitories were available); two to four weeks at a Peace Corps outdoor camp in Puerto Rico, Hawaii, St. Croix, or St. Thomas; and one to two weeks in the host country.

The universities provided an eight-course core program: skills for jobs to be performed overseas; language, with a stress on conversation; area studies; world affairs, including the danger of Communist subversion; American studies; physical education; health care; and orientation on the Peace Corps.

Those who completed the training thought that they had suffered through boot camp. Thomsen took his at Montana State in Bozeman. "Peace Corps training," he wrote, "is like no other training in the world, having something in common with college life, officer's training, Marine basic training, and a ninety-day jail sentence." The typical university schedule was 7:00 a.m. to 10:00 p.m. six days a week. There were special supplements. Volunteers bound for the slums of Colombian cities did time in Spanish

Harlem, those going to Nepal climbed the Rockies. The outdoor camps were even tougher: calisthenics at 5:00 a.m. and a ten-hour day six days a week. The American public ate it up. As Gerard Rice wrote, "Reporters filed into Peace Corps' field camps to see trainees swinging through trees, scaling sheer cliffs, and being thrown into rivers bound hand and foot (presumably the sort of tests that would await them in the jungles of the Third World!)."

As this suggests, the training was often inappropriate. University courses tended to be more cerebral and theoretical than needed. Volunteers were not taught to deal with boredom and frustration, two of their most common complaints. The language problem was almost insurmountable. Nigeria alone had 250 dialects. Teachers were unavailable in many of the exotic languages. The Peace Corps had to write its own textbooks and dictionaries for Somali, Tshi, Malay, Twi, and Hausa. By 1963 it was giving instruction in forty-seven languages, half of them never taught before in the U.S. And by that time language had become the focus of training—300 hours of every course. Interestingly, there was more difficulty with Spanish in Latin American and French in West Africa than with less familiar languages. Thomsen, who was bound for an Ecuadorian village, wrote of his Spanish: "We twisted and mauled that beautiful language into a million distorted shapes and watched our instructors, sensitive and dedicated people all, wither and age before our eyes."

"Even in World War II," Shriver wrote, "our troops were generally in organized units where safe food and water could be provided and medical care was at hand. This would not be the case with the Peace Corps." Health was a matter of concern from the outset. Shriver asked the Surgeon General to study the problem. The system adopted was for public health doctors assigned to the Peace Corps to provide preventive measures, and for physicians in the host countries, if available, to provide much of the day-to-day care. The system worked very well. By 1970 only two volunteers had died of disease. One of the most serious problems anticipated by the Public Health Service was hepatitis. Volunteers were injected with large doses of gamma globulin in advance, which prevented the disease.

The "typical" Peace Corps volunteer was a recent college graduate in liberal arts, unmarried, and between twenty-two and twenty-eight. Many had just finished college and were undecided between a job and graduate school. The Peace Corps created what Morris Stein called a "psychosocial moratorium," a pause on the journey to adulthood. While a degree was not required, 86 percent of the volunteers had at least one. There were very few older people and very few manual workers, who preferred to earn money at home. The leading states in supplying volunteers in descending order were California, New York, Illinois, Pennsylvania, Ohio, Massachusetts, and Michigan. The South provided very few. There were exceptions. Lillian

Carter, whose son would become President of the U.S., celebrated her seventieth birthday in the Peace Corps in India and was from Plains, Georgia.

The quality of the volunteers was impressive. Robert B. Textor, a cultural anthropologist who specialized in southeast Asia, helped select the first group to be sent to Thailand. He knew their backgrounds "practically by heart." "The greatest moment in my Peace Corps career came in October 1961 when I went to Ann Arbor to lecture to the Thailand One contingent." He found them on the whole "fine and appropriately motivated." They were learning the Thai language with "gusto." Their questions were "intelligent, sensitive, and above all *relevant.*" "These young Americans," Textor wrote, "were ample reward, ten times over, for all the *Sturm und Drang* of Washington."

At a staff meeting in Washington in March 1961, Lee St. Lawrence, formerly with ICA and having the manner of a soldier of fortune, suggested Tanganyika, soon to be called Tanzania, for the first overseas contract. Shriver told him to look into it. He immediately caught a flight to Dar es Salaam and vanished into the bush. He covered that large country from end to end and learned that there were only 800 miles of paved roads. In an overwhelmingly agricultural society the farmers could not get their crops to market during the rainy season. Prime Minister Julius Nyerere much admired Kennedy and had a warm relationship with Chester Bowles. He readily accepted when St. Lawrence proposed a contract to lay out a road system.

A team of thirty-five engineers, surveyors, and geologists was quickly recruited, trained, showcased at the United Nations and the White House, and dispatched to East Africa. They designed a network of farm-to-market roads and also had plans for harbors, water lines, airports, and towns. But, as road building was about to begin, the Tanzanian government had a budget crisis and there was little money for the program. Nevertheless, according to Josephson, the mission was "very successful" because the Americans were "an extraordinary bunch of guys." NBC made an excellent documentary on the program.

The more natural Peace Corps activity was teaching. The rate of illiteracy in the Third World was staggeringly high, and most of the volunteers, neither professionals nor skilled blue-collar workers, could become school teachers. This was the pattern in Africa, much of South Asia, and the Philippines. In Latin America, however, the programs were oriented to community development, *Acción Comunal.* Frank Mankiewicz, who was in charge in Peru and later for all of Latin America, developed this idea. He hoped to turn apathetic residents of urban barrios and rural campos into citizens who participated in public affairs in order to improve their own conditions and to democratize their societies. This was exceptionally difficult to achieve and became a source of sharp controversy.

The program for the Philippines was particularly interesting. It was the largest, with a third of all volunteers in the first year and a total of 1,014 by 1964, and was typical in placing the main stress on teaching. Lawrence H. Fuchs, professor of American civilization and dean of the faculty at Brandeis, was the director. On one of Shriver's inspection trips they were flying from Manila to the Visayan Islands when Shriver asked, "Hey, Fuchs, why don't you write a book about the Peace Corps when you go back to being a professor?" His initial reaction was negative, but Fuchs later recanted and in 1967 published "*Those Peculiar Americans*," which provides an exceptional insight into the Peace Corps overseas.

When the U.S. took over the Philippines from Spain in 1900, one of the first undertakings was to establish a public school system based on the English language. The next year two American troopships, one named the U.S.S. *Thomas*, brought 1400 school teachers to the islands, who came to be known as Thomasites. The program they launched enjoyed considerable success until World War II, when the occupying Japanese ruthlessly destroyed the schools. After the war the newly independent Filipino government with American assistance sought to restore the system, but with indifferent success.

Language was the key. The Philippine archipelago, 7100 islands of which 466 were settled, spreads across a large swath of the western Pacific. Luzon in the north and Mindanao in the south are the largest islands, but there are nine other major islands. There are more than a dozen important language groups, none of which is dominant. Spanish rule had failed to establish the Spanish language. English, therefore, was the logical language of government, business, and culture. But a study showed that the state of the language in 1960 was far below the standard of 1925. While English was the language of instruction, many teachers used the local dialect and both they and their pupils avoided English outside of class.

Thus, the Philippine authorities and the Peace Corps quickly agreed on a large program to upgrade the elementary school teaching of English and a smaller one to introduce new techniques for instruction in science and mathematics. When Fuchs left in June 1963, more than 500 of the 630 volunteers were "educational aides," about 40 "co-taught" science in high schools, 30 were utility teachers in normal schools and universities, and 22 worked in community development in the barrios of Mindanao. "The elementary school program dwarfed the others," Fuchs wrote, "and to a considerable extent was seen by Filipinos and Americans as the Peace Corps in the Philippines."

Volunteers were sent to rural barrios throughout the islands, including ten to Muslim areas in the Sulu Archipelago. These communities ranged in size from a dozen families to several hundred. Usually two volunteers lived together in a bamboo frame and nipa house, some with open frame and a nipa roof. Most had no water or electricity. Like their Filipino neighbors, they had to deal with snakes, scorpions, lizards, rats, and mosquitoes. The odor of the nearby garbage pit was inescapable. The diet was mainly rice, fish, and fresh fruit, with occasional local vegetables and meat. Some walked to school, others took the bus. Fuchs wrote:

> Whether in the lowlands, the mountains, or in Muslim areas, the barrios of the Philippines were usually poor. Students often came to school hungry and dressed in torn clothes, or they did not come at all because they lacked the money to buy a pencil or paper. Even in the more prosperous barrios of the Visayas (the central islands) or Luzon, electricity and toilet facilities were rare. Everywhere pigs roamed freely, defecating where they chose. Public health measures were virtually nonexistent or only halfheartedly undertaken. Respiratory infections were rampant, and open sores on the bodies of schoolchildren or stray dogs were common.

There was an inevitable clash of cultures. The Americans arrived with an ingrained sense of liberty, individualism, personal independence—"the right and capacity to make choices in selecting mates, jobs, domiciles, religions, and rulers." In the barrio environment personal liberty had little meaning; the Filipinos enjoyed almost none of these options. The educated middle-class Americans arrived with assumptions about personal hygiene, public health, housing, and diet which were simply not fulfillable in a culture of poverty. As one girl wrote, "When I arrived here nothing appealed to my sense of taste—not sights, nor sounds, smells, food."

This clash inexorably carried over to the job. Volunteers were not allowed to be teachers and were classified as teachers' aides. In Filipino eyes the schoolteacher occupied a respected position in the community, not so much because of what he accomplished but because he had a steady job. The volunteer must not threaten his status. Thus, the volunteer soon found that he had a non-job. The innovative projects he proposed usually ended in failure. As Fuchs wrote, "Filipinos wanted volunteers to eat the local food, speak the dialect, dance the regional dances, give pleasure, and enjoy themselves." Few were interested in educational change or improvement.

The Americans, as Fuchs put it, were "self-selected apostles of their own cultural values" and became deeply frustrated by these conditions. As time went by this turned into "cultural fatigue," which took the form of an intense loneliness. As one girl wrote, "I simply and truthfully hated it all, wanted to go home, and the only thing stopping me was pride."

Mrs. Carter wrote from the outskirts of Bombay:

> I have had four days of complete inertia caused by homesickness and no mail . . . nobody loves me, I am forgotten, I hate [roommate] Mabel's guts, they push me too hard here, no clothes, no food, no nothing, I wish I were dead!
>
> I ran home to lunch and sat down to cry when I heard footsteps on the stairs—God! The mailman! Letters from home!
>
> I'm pepped up, everybody loves me, I don't have enough to do, Mabel is real sweet, food is even better than I expected, and life is wonderful!

In fact, very few of the volunteers in the Philippines did return to the U.S. Rather, they gradually learned to cope. As the administrative machinery expanded and became more effective, conditions improved. Volunteers came to accept Filipinos and their society on Filipino terms, not as an extension of America. The volunteers went through a process of self-discovery, an acute awareness of their own strengths and weaknesses, which helped them to adjust. They found individuals in the barrios who were eager to learn, who wanted to work with them. They established especially warm and loving relationships with children. They developed a sense of belonging to their people and their barrios.

In Bombay, Mrs. Carter received a large package of gifts from her devoted family in Plains.

> On New Year's my friends here have a custom of exchanging gifts as we do for Christmas. I realized that their kindness to me was the one thing that sustains me—and I gave all my gifts away, one by one. I couldn't help it. I had already had the joy of receiving them, and I just had no other way to pass that joy along to these people who have come to mean so much to me.

Fuchs concluded:

> The Peace Corps is primarily an instrument of human development, not just economic development. Filipinos and others unquestionably change in the direction of Western, and particularly American, values as a result of contact with volunteers, although the changes may be much slower than many people had thought they would be. But Americans change too. . . .
>
> Filipinos and American Peace Corps volunteers became more alike in some ways, even as they became more aware of their differences, each side tending to draw from the culture of the other those aspects of man's humanness which seemed lacking in themselves and yet attractive. . . .
>
> It was a magnificent adventure, and to the query, "Are you glad you were in it, knowing now, as you could not before, its physically bruising and exhausting qualities?" I would answer with most Peace Corps volunteers, "You bet I am!"[3]

From the moment of its inception the Peace Corps became extraordinarily popular. Ashabranner spoke of "America's love affair with the Peace Corps." In part this was due to Shriver and his unusual gift for public relations. For example, on March 28, 1962, Shriver testified before the

Senate Foreign Relations Committee to ask for more funds and changes in the Peace Corps Act, which the committee gave him immediately. Senator Humphrey wrote him a "Dear Sarg" letter the same day "to commend you again on the outstanding manner in which you presented your case." He thought Shriver "deserve[d] unstinting praise for the splendid manner in which you have administered your agency." He had "earned . . . the respect and support which was so evident in the Senate Committee today."

But the popularity was due to more than its director. As David Hapgood pointed out, "millions of Americans . . . want to believe that their country is capable of doing *something* good overseas and have therefore invested the Peace Corps with a degree of magic, self-sacrifice and piety. . . ." This was an exaggeration, but it had a large kernel of truth. The agency suffered from no scandal. It enjoyed a favored position with the media. Congress looked on it with great favor and gave it more money than Shriver could use.

The Peace Corps' popularity also stemmed from its style. It seemed to be everything that Americans did not expect a government agency to be. It was youthful, brash, courageous, tough, innovative, informal, hard-working, and anti-bureaucratic. Josephson prevented the publication of an organization manual because he thought it would be "stultifying." Ashabranner wrote that most of the volunteers he knew were "strong-minded, strong-willed, high-spirited men and women, suspicious of anything that looked even remotely like a restraining bureaucracy."

Further, Textor wrote, "the Peace Corps is in fact an ethical enterprise, a way for an excessively fortunate country to share its optimism and generosity with parts of the world that, at a moment in time, are in need of what the Volunteers can best offer." This made Americans feel good about themselves and about their country.

Ultimately, of course, the performance of the agency depended upon its volunteers, who were shipped, alone or in small groups, into mainly tiny impoverished and forgotten communities in many of the world's poorest nations. One can hardly avoid thinking of the widowed Lillian Carter, approaching 70 and deeply devoted to her family, church, and home in Georgia, toddling off to work in a birth control clinic in the outskirts of Bombay. Like Mrs. Carter, the volunteers were hardly typical Americans. Given the selection process, it could not have been otherwise. They were far above average in intelligence, education, adventurousness, toughness, and idealism. Peggy Anderson, who was a teacher in Togo in Africa, put it this way:

> Most people who join the Peace Corps have no idea what it will really be like, but we who become volunteers do have some general notions. One is involvement—we expect to learn from another culture from its root hairs on up; and we hope,

through that knowledge, to gain intimacy with a new environment. Another is help—we expect to be useful to the host country by doing work that needs doing. A third is survival, both physical and spiritual—we expect, you should forgive the term, hardship, or challenges of various kinds, that will test or temper us as individuals. I believe that these elements—in proportions that vary from person to person—constitute the basis of what most applicants understand as the "volunteer experience."

The effectiveness of the Peace Corps to the nations of the Third World must be viewed in the proper focus. This was a micro- rather than a macro-program. Thus, it did not bring peace to the world or to its regions. Nor did it stimulate economic development on more than a minuscule scale. Nor did it become an instrument of American foreign policy. Kennedy and Shriver insisted that it should not, with the full approval of Dean Rusk, though some who handled foreign policy thought otherwise. While it was often accused of being a cover for the CIA, Shriver, with firm presidential approval, erected a wall against penetration by that agency. Josephson said, "To my knowledge, we never used the CIA, ever." When Mrs. Carter was told that many Indians thought the Peace Corps worked with U.S. intelligence, her reaction was, "Good Grief!"

The accomplishments were on a small scale. The Peace Corps taught mothers how to keep themselves and their children clean in order to avoid disease. It helped provide public sanitation with water systems and sewage disposal. It taught people to read and write and do sums. It showed farmers how to produce healthy crops, poultry, and farm animals. It planned and helped build roads to give remote villages access to the outside world. It assisted in administering birth control programs. Perhaps most important, it gave extremely poor and forgotten people the sense that they mattered.

Nehru told Shriver in 1961 that "young Americans would learn a good deal in this country and it could be an important experience for them." Probably the most significant Peace Corps accomplishment was the education of Americans. They came to understand the peoples and cultures of the Third World. Paul Tsongas, who later was elected to the Senate from Massachusetts, spent three years teaching math and science in Ethiopia. He said later:

> I ended up in a village in Ethiopia with five other Peace Corps Volunteers, and I didn't go anywhere on vacations, just stayed in the village. I broke away from the others and set up house by myself, with my students. I took the ten best kids in the school and I lived with them, just a total immersion in their culture. And, you know, nothing I've ever done before or since has given me the same feeling.

Many former volunteers suffered from culture shock when they returned to the U.S. But if their overseas experience had taught them anything, it was the ability to adjust. Large numbers were soon off to graduate and professional schools. Others entered the labor market directly, where they found their Peace Corps experience a premium. School systems snapped them up. The State Department, AID, and the U.S. Information Agency were eager to hire them. Multinational corporations and voluntary organizations prized them.

One of the volunteers said, "I'd never done anything political, patriotic or unselfish because nobody ever asked me to. Kennedy asked." No Americans were so deeply and emotionally attached to the President. In the Dominican Republic they called the volunteers "hijos de Kennedy," Kennedy's children. Sorensen wrote, "A special bond grew up between the President and the Peace Corps volunteers. . . . 'Kennedy's children' . . . comes close to describing how he and they felt about each other." Thus, his assassination pierced no Americans outside his family as painfully as it did the men and women of the Peace Corps. "A shudder went through the Peace Corps," Ashabranner wrote. It brought forth "an emotional response for which no one felt embarrassed." Lillian Carter did not go to India until after that tragic event. Her children sent her William Manchester's *Death of a President.* "I'm not going to let ANYBODY else read it," she wrote, "and have it hidden away in my drawer." Shortly she added, "Oh, I looked at 'Death of a President' for days before I could read it—then I read and cried—read and cried—read and cried. I'm not going to let anybody else read it, because there is nobody in this country who loved Kennedy as much as I."[4]

10

If Men Were Angels

IN A FAMOUS PASSAGE in *The Federalist Papers*, No. 51, James Madison justified the carefully crafted separation of powers so central to the new Constitution. He did not believe that man was inherently good.

> It may be a reflection on human nature, that such devices should be necessary to control the abuses of government. But what is government itself, but the greatest of all reflections on human nature? If men were angels, no government would be necessary. If angels were to govern men, neither external nor internal controls on government would be necessary. In framing a government which is to be administered by men over men, the great difficulty lies in this: you must first enable the government to control the governed; and in the next place oblige it to control itself.

To oblige government to control itself the framers devised "inventions of prudence" in the distribution of powers by establishing "opposite and rival interests" between the branches of government, particularly the President as opposed to the Congress. Every President has had trouble with Congress and every Congress has had trouble with the President. It is in the nature of the system. Thus, American political history is in large part a ceaseless struggle between the executive and legislative branches.

In the brief Kennedy era this was the decisive fact of political life. While it was hardly the most acrimonious conflict between the President and the Congress in the history of the republic, it ranked high on that scale. This was because Kennedy insisted on moving the country forward, and Congress, particularly the House, was equally determined to stand pat. As Richard E. Neustadt put it, he was "an innovative President confronting a reluctant

Congress." "In domestic affairs," Arthur M. Schlesinger, Jr., wrote, "Kennedy was a somewhat beleaguered President."

Kennedy both subscribed to the "theory" of the strong and active presidency as it had evolved under Republican Theodore Roosevelt and his twentieth century Democratic predecessors—Woodrow Wilson, Franklin Roosevelt, and Harry Truman—and also brought to the office supportive personal qualities—great ambition and enormous energy. That notable counselor of Presidents, Clark M. Clifford, put the theory this way:

> [A] quality which is vital to the success of a Presidency is an awareness of the potential of the office. Some of our Presidents have been bold and venturesome in this regard, and, in my opinion, those have been the successful periods in our country's history. Other Presidents have felt restrained and constricted and have refrained from innovation and the exploration of new thinking and new concepts. Some men have been content to accept the status quo. Others have been impatient with conditions as they found them, and they have offered new ideas, new solutions, and new experiments. The opportunities are boundless, and the most fascinating experience associated with our government is to watch a strong President expanding the limits of the office to gain the goals he seeks for our country.

Kennedy was bored when he sat in the House of Representatives. "You sensed without his telling you," Larry O'Brien wrote, "that he would not be making a career of the House." In 1952, Kennedy made up his mind to take the plunge. But he had to wait until incumbent Massachusetts Governor Paul Dever decided whether he wanted to stand for re-election or to challenge Republican Senator Henry Cabot Lodge. Kennedy would take the leavings. O'Brien was with him in his Boston apartment on Bowdoin Street when, looking out the window at the State House, he said, "Larry, I don't look forward to sitting over there in the governor's office and dealing out sewer contracts." He was lucky; Dever ran for re-election. When Kennedy was in the Senate he remarked to Sorensen, "It seems to make so little difference sometimes what we do down here. Only the Executive Branch can really move things." He had been bitten by the presidential bug.

On January 14, 1960, when he was on his way, Kennedy addressed the National Press Club on his image of the presidency. He excoriated President Eisenhower's vision of his office. While hopes had been "eloquently stated," he had not followed through. Eisenhower had failed "to override objections from within his own party, in the Congress or even in his Cabinet." Perhaps "this detached, limited concept of the Presidency" was appropriate in the fifties, when the nation needed "a time to draw breath." But now the country was entering a new decade, "the challenging revolutionary sixties," which will demand more than "ringing manifestoes issued from the rear of the battle." It will be vital for the President to "place himself in the thick of the fight, that he care passionately."

> He must above all be the Chief Executive in every sense of the word. He must be prepared to exercise the fullest powers of his Office—all that are specified and some that are not. He must master complex problems as well as receive one-page memorandums. He must originate action as well as study groups. He must reopen the channels of communication between the world of thought and the seat of power. . . .
>
> We will need . . . what the Constitution envisaged: a Chief Executive who is the vital center of action in our whole scheme of Government. . . .
>
> It is the President alone who must make the major decisions of our foreign policy.
>
> That is what the Constitution wisely commands. And, even domestically, the President must initiate policies and devise laws to meet the needs of the Nation. And he must be prepared to use all the resources of his office to insure the enactment of that legislation—even when conflict is the result.

When Kennedy became President the Republicans and southern Democrats who had firm control of the House and often had control of the Senate, including the key committees, did not view the political system that way at all. They had no interest in moving the country and they had no intention of bowing to presidential leadership. They knew that he had been elected by a very thin margin and had no coattails. They had long experience in delaying and blocking legislation.

Kennedy, of course, was equally aware of his own political weakness (if he forgot, Sorensen was always at his elbow to remind him) as well as of the power of his opponents on the Hill. As Sorensen put it, while Kennedy had no interest in mathematics, he spent much of his time counting. In 1960 he counted convention delegates and electoral votes and the additions worked out. "After November, 1960, he was counting Congressional votes; and this time he could not make the sums come out right." There were about seventy moderate Republicans and southern Democrats who sometimes swung from a conservative to a more liberal position. Kennedy needed to pick off forty to sixty of them on an important bill, an extremely difficult task. As he wryly observed in 1962, "When I was a Congressman, I never realized how important Congress was. But now I do."

The conservative powerhouse at the other end of Pennsylvania Avenue compelled him to adopt two unusual policies. The first was heavy, perhaps too heavy, reliance on executive action. His *Public Papers* for 1961 alone list 67 executive orders. A number of important programs were launched by these orders—food distribution to needy families, the President's Advisory Commission on Labor-Management Policy, the Peace Corps, the President's Committee on Equal Employment, the President's Commission on the

Status of Women, collective bargaining in the federal service, public works acceleration, and equal opportunity in housing. Kennedy was so angered by the refusal of Congress to cover laundries under the minimum wage that he ordered agency heads to pay all federal employees, including laundry workers, no less than $1.25. Sorensen wrote: "He vetoed minor bills that he did not like, impounded appropriated funds that he did not need, ignored restrictive amendments that he found unconstitutional and improvised executive action for bills that would not pass."

But for important programs the only route was the main highway through the Congress. Here Kennedy devised a second unusual policy, an effective congressional liaison office under O'Brien. Despite all the plans and task forces of the interregnum, this came about by accident.

At a meeting in Palm Beach in December 1960, Kennedy and Rayburn agreed that it was crucial to open the Rules Committee bottleneck and that the Speaker would ask the House to allow him to add three new members to the committee. But at a breakfast legislative meeting on January 24, 1961, only four days after the inauguration, Rayburn dropped what O'Brien called "a bomb into the new President's lap." "Mr. President," he said, "I don't believe we have the votes to expand the Rules Committee." Kennedy and O'Brien were "stunned."

"The loss," O'Brien wrote, "would be devastating. If the Rules Committee was not expanded, little if any of the Kennedy legislative program was likely to be passed. Beyond that, the defeat would be a stunning blow to his prestige, both nationally and internationally, at the very start of his Administration."

Kennedy and O'Brien were especially concerned because Rayburn did not have a head count; he did not know exactly how many congressmen would vote aye or nay. The President arranged a delay in the showdown to allow time for a turnaround. "We can't lose this one, Larry," he said. "The ball game is over if we do. Let's give it everything we've got."

Though O'Brien had never worked with Congress before, politics was in his blood and he was a quick study. He immediately made a head count, which confirmed Rayburn's judgment. They would lose by seven votes. The 108 southern Democrats held the key—a third would have to support expansion of the committee. They were lobbied by Rayburn, Georgia's Carl Vinson, Governor Terry Sanford of North Carolina, Governor Orval Faubus of Arkansas, and the President himself. It worked. A third of the southerners, 34, gave the Speaker his five-vote margin.

Kennedy and O'Brien had no illusions about this "victory." They were "haunted" by the narrowness of their edge. Robert Kennedy said "that here we had Sam Rayburn and, therefore, the Texas delegation, that we had the maximum strength—the Democrats—and yet we won by only a couple of votes. And how would we do on something that was far more controversial [for example, civil rights], where we didn't have a Sam Rayburn, couldn't bring along a lot of these southerners that he could in this kind of a fight?" The President said of his own experience in Congress: "I was up there for fourteen years and I don't recall that Truman or Eisenhower or anyone on their staffs ever said one word to me about legislation." He would need a well-organized and aggressive congressional relations program.

O'Brien recruited a competent staff and he also coopted forty congressional relations offices in the departments and agencies. He instituted regular weekly reporting with a summary for the President. Kennedy himself launched an unprecedented program of wooing Congress. During 1961, for example, he met with the leadership for breakfast on thirty-two Tuesdays and engaged in private conversations with individual leaders about ninety times. Around 500 members of the House and Senate attended White House coffee hours and an equal number came in for bill-signing ceremonies. Democrats who helped push a measure through got a personal note of gratitude that they could use in their campaigns. The occasional Republican who assisted was thanked by telephone.

Meantime, O'Brien learned his way around the Hill and courted the important people, particularly Rayburn and Vinson. He held Sunday brunches at his Georgetown house for congressmen, cabinet members, and newspapermen—Bloody Marys, scrambled eggs, ham, bacon, "Boston" baked beans, rolls, coffee, and "O'Brien" potatoes. Sam Rayburn would "tear into" the food with "gusto."

The President had full confidence in O'Brien. An example occurred on March 24, 1961, just after the House Education and Labor Committee reported the administration's minimum wage bill at $1.25. The head count showed that the House would vote it down. O'Brien and Goldberg met in the Speaker's office with Rayburn, Vinson, McCormack, Albert, and Powell and carefully reviewed strategy. Vinson said they would pick up some southern votes if they retreated to $1.15. The question was whether to drop the wage or to cut back on coverage of half a million workers, including the laundry workers, who had become a symbol of the underpaid. Since the laundry lobby had put on heavy pressure, O'Brien knew that a concession on coverage would win over four or five votes. Rayburn said this was a decision the administration should make. Goldberg agreed, "This really is a White House decision." O'Brien did not hesitate. "I could have called the President, but I felt that this decision was mine to make, that he would want me to make it because I had all the facts. . . ." O'Brien chose to cut coverage.

Heretofore O'Brien had worked on elections, which he found "an exciting game." He was now in the big time, dealing with "matters of substance, of the utmost national importance—the minimum wage, Medicare, civil rights, education and all the rest of the Democratic legislative agenda." He soon realized that "how well I did my job could affect the lives of millions of my fellow Americans." He took his work seriously. "I gave it all I had."[1]

The American political system has been characterized by what Arthur M. Schlesinger called "tides," or what his son, Arthur, Jr., described as "cycles," an alternation between liberal and conservative periods, between change and preservation. The elder Schlesinger carried the theory back to the eighteenth century, the younger forward into the twentieth.

The latter defined the cycle as "a continuing shift in national involvement between public purpose and private involvement. . . . Each new phase must flow out of the conditions—and contradictions—of the phase before and then itself prepare the way for the next recurrence. A true cycle, in other words, is self-generating." Reform tends to come in bursts, while digestion is stretched out. Thus the country both seems and is more conservative than liberal.

Each phase must be spawned from the internal strains of the last counter-phase. When the private interest is in the saddle, the younger Schlesinger noted, there are "undercurrents of dissatisfaction, criticism, ferment, protest." Many wait for a "trumpet to sound." After a surge of change in the public interest the society becomes "emotionally exhausted." People yearn for "the privacies of life," for "an interlude of rest and recuperation."

In the twentieth century, the younger Schlesinger observed, the cycles have been generational, lasting about thirty years. Dating them by the activist Presidents who led them off, he pointed to Theodore Roosevelt in 1901, Franklin Roosevelt in 1933, and John Kennedy in 1961, each just about thirty years apart. In the case of Kennedy this is only an approximation.

Noon on January 20, 1961, the hour Kennedy became President, was a historical accident produced by the Twentieth Amendment. The nation did not move at that moment, in Schlesinger's words, from the Eisenhower era of "needed respite amidst the storms of the twentieth century" to the "rush of commitment" of the New Frontier and the Great Society. The process was gradual and, politically considered, took six years. The liberal swing began in 1958 with sharp Democratic congressional gains, particularly in the Senate. In 1960, Kennedy won election, but by the narrowest of margins. While the 1962 congressional results appeared to be a standoff, they actually continued the shift. That is, a sitting President's party almost always loses heavily in the midterm elections and the Democrats did not in 1962. The decisive change, of course, occurred in 1964, when Lyndon Johnson won by a landslide and carried with him working liberal majorities in both houses of Congress, a situation that had not existed since the early New Deal.

Politically, therefore, the brief Kennedy presidency was transitional; it took place during the process of movement from the conservative to the liberal phases of the cycle. Progress under Kennedy was possible, but it would be painfully won.

The tendency is to personalize the times and the administration, to pile up the credit, or the blame, on the President. This goes too far. A President almost never invents new ideas; he rarely works out the details of a program; he never writes legislation or executive orders; and his control over Congress is usually marginal. This is not to denigrate his role; despite these limits, it is enormous. "The Presidency," Neustadt wrote, "is not a place for amateurs. That sort of expertise can hardly be acquired without deep experience in political office. The Presidency is a place for men of politics. But by no means is it a place for every politician."

In the American system the President is the supreme political leader. In the case of an activist President this calls for a receptivity to new ideas, for a capacity to recruit able advisers and administrators, for an ability to use his office, in Theodore Roosevelt's phrase, as "a bully pulpit" in order to mobilize public opinion in support of his program, and for skill in persuading a reluctant and often hostile Congress to provide the indispensable majorities.

The Kennedy domestic program, the substance of this book, was not invented by Kennedy. Howard E. Shuman, an acute political observer who worked for Senator Douglas, said, "I couldn't think of any legislation that ever really originated in the executive branch. . . . I think of very little constructive legislation that was not first proposed by a member of the House or Senate." He was, of course, right. Civil rights, the Keynesian tax cut, area redevelopment, manpower training, equal pay for women, federal sector collective bargaining, federal aid for education, Medicare, and the Peace Corps all had their legislative origins in bills introduced earlier in Congress.

As Arthur Schlesinger, Jr., observed, the passing of the conservative phase of the political cycle is the inception of the next liberal phase. During the second Eisenhower term the Democratic Congress, notably the Senate after the election of "the class of '58," fashioned the agenda which became the Democratic platform of 1960.

The Democratic platform dealt with all the issues that became Kennedy's New Frontier program. Some were specific—civil rights, area redevelopment, the minimum wage, equal pay for women, federal aid for education, and Medicare. The others were phrased generally—a tax cut, manpower training, federal sector collective bargaining, and the Peace Corps.

Later Larry O'Brien was mystified. When Congress recessed in the latter part of 1961, "on balance I thought our 1961 [legislative] record was a good one, even an outstanding one." The President had sent up 53 bills and 33 had been enacted, "more than in the final *six* years of the Eisenhower administration." He was "astounded" to read the newspapers and learn that Kennedy had "failed" with Congress. The President was "frustrated" and asked Sorensen to brief the reporters. About two dozen showed up at the home of Carroll Kilpatrick of the *Washington Post.* Despite O'Brien's fact sheets, they were unimpressed and resented being manipulated by an "administration snow job." Their papers then denounced this attempt "to dictate to the press."

It was the same story in 1962. When Congress adjourned on October 13, O'Brien thought the session had been "a successful one, perhaps even more so than 1961." Congress had passed 40 of the 54 bills the administration had proposed. Yet the press continued to report a "deadlock." The year 1963 was a reprise. Of the 58 "must" bills, 35 had been enacted. Again, the newspapers reported "stalemate."

O'Brien remained baffled. He wrote later, "A myth had arisen that he [Kennedy] was uninterested in Congress, or that he 'failed' with Congress. The facts, I believe, are otherwise. Kennedy's legislative record in 1961–63 was the best of any President since Roosevelt's first term."[2]

Richard Neustadt has pointed out that the key years of accomplishment for a two-term President, the years when there are "signs of pattern, clues to conduct," are the third, the fifth, and the sixth. Kennedy had no fifth or sixth and had only ten months of the third year of his presidency. "The intensive learning time," Neustadt wrote, "comes at the start and dominates for the first two years." Thus, the third year is crucial in the development of a President, and this was certainly the case with Kennedy.

James L. Sundquist has stressed the legislative importance of the third year—1963—in his assessment of the Kennedy presidency. In 1961 Congress had not given Kennedy the traditional "honeymoon" accorded most new Presidents. The year 1962 was not much different. The President and his White House team were still feeling their way. Following the death of Rayburn, the new Speaker, John McCormack, and the new majority leader, Carl Albert, took their "shakedown cruise" in 1962. The 87th Congress was assuredly not a Kennedy congress.

But, Sundquist wrote,

> Looking back upon the legislative successes of the middle 1960s, what appears remarkable is not the prodigious output of the post-Goldwater Eighty-ninth Congress (1965–66), but the achievements of the pre-Goldwater Eighty-eighth (1963–64). The Eighty-ninth Congress performed about as expected. . . . But the performance of the Eighty-eighth Congress was not at all what might have been expected upon the basis of the previous record of its members. The Democratic program was unblocked in 1963 and 1964 by a Congress substantially the same in makeup as the one that had blocked that program two years earlier.

Among the many signs of the change in 1963 were the following: The civil rights bill was approved by the House Judiciary Committee and Rules Committee clearance, and House passage seemed likely. The House had already passed the tax cut and a Senate majority was a virtual certainty. The logjam over education had been broken with enactment of the higher education law and a number of lesser bills.

In assessing the 88th Congress one must separate 1963 and 1964 by the assassination. "The considerable progress of the Kennedy program prior to November," Sundquist wrote, "suggests strongly that most of what happened would have happened—more slowly, perhaps, but ultimately—if Kennedy had lived."

Why was Kennedy's legislative record so much stronger in 1963 that it had been in 1961–62? The President persisted. He was convinced that the country needed and wanted his program. When he got licked on a particular issue, he admitted that he had lost a battle but insisted that he would win the war. "The President of the United States," Sundquist pointed out, "is always heard." He slowly wore down congressional resistance.

In addition, Kennedy was much more self-confident and a far stronger President in 1963 than he had been in 1961. This was evident in both his domestic and foreign policies. While this is hardly the place for a detailed account of his diplomacy, a brief review of U.S.-Soviet relations will substantiate this conclusion. Because domestic policy is linked to congressional sessions, they form a neat annual pattern. But foreign policy has no such discipline under the calendar.

Domestically, by 1963 Kennedy's close 1960 victory was history, and the elections of 1962, in which he campaigned vigorously for Democratic candidates and his party did well, were fresh. His relations with business, as evidenced by the tax bill, had improved markedly. His popularity in the opinion polls was extraordinarily high. His congressional relations team, particularly O'Brien and Keppel, was more effective. Congressional Democrats, including southerners, expected him to head the ticket in 1964 and became reluctant to oppose him if for no other reason than that they would be running with him. The McCormack-Albert leadership in the House had established itself and was working. The polls, excepting civil rights in the South, revealed overwhelming support for the Kennedy legislative program. According to Sundquist, they "showed a substantial readiness among Republicans and Democrats to embrace activist solutions to domestic problems." Voters who opposed a Democratic proposal, as with Medicare, preferred a moderate Republican alternative; those who were against everything had become a tiny, though dedicated, band.

In foreign policy Kennedy went through the same progress from early confusion and failure to later self-confidence and success, really triumph. The great issues involved the Soviet Union and its allies, Cuba and East Germany. The major events were the Bay of Pigs, Berlin, the Cuban missile crisis, and the nuclear test ban treaty.

The Bay of Pigs may have been the most harebrained and inept venture in the history of American foreign policy. Following the success of Fidel Castro's revolution in 1959–60 a stream of Cuban exiles of all political shades poured into Miami. On March 17, 1960, Eisenhower directed the CIA to recruit and train a Cuban force capable of waging guerilla warfare. The agency established a "secret" training base on a coffee plantation in the Guatemalan mountains. But the CIA soon changed its mind, and began training for an invasion of Cuba by a pocket army. This assumed that the Cuban people would rise in support of a landing against Castro.

On Novermber 29, Allen Dulles briefed Kennedy and the President-elect told him to go forward with the training. The CIA plan was for an amphibious landing at the thinly populated Bay of Pigs on the south coast and it gained the approval of the Joint Chiefs of Staff.

The President now had to decide whether to go forward with the invasion and he struggled over the decision. The affirmative reasons were the following: Castro, anticipating the landing, was receiving a heavy flow of Soviet arms and the CIA reported that a delay to May would be fatal. If Kennedy scrubbed the attack, 1400 embittered Cubans would be dumped into the boiling Cuban community in Miami, which would accuse him of betrayal. The Republicans would call him soft on Communism. The Guatemalans were impatiently pressing to rid their country of the Cuban brigade. If the plan succeeded, Kennedy would have gotten rid of Castro without using American forces. On the other side, the military soundness of the plan was highly questionable. In early April he approved it reluctantly on condition that U.S. units would not be directly involved.

The brigade sailed from Nicaragua in seven small ships and made a landing at the Bay of Pigs on April 17. Castro's forces wiped them out in three days, seizing more than 1100 prisoners. The military plan had been ridiculous. The force was much too small; the landing site could hardly have been worse; there was no route of escape; the Cuban people did not support the exiles. Kennedy, swearing silently at those who had given him atrocious advice, took full responsibility for the disaster. As is customary, the American people rallied to their President's support in a crisis and his Gallup poll approval rating rose from 72 percent in the week of February 10 to 15 to 83 percent in the week of April 28 to May 5, 1961. But this did not mask the fact that he had suffered a severe blow.

On June 3 and 4, 1961, less than two months later, an overbearing and cocky Khrushchev at the Vienna summit warned Kennedy that East Germany might cut off West Berlin and that, if the West responded with force, the East would do the same. Convinced that the city was the linchpin of the NATO alliance, Kennedy determined to be ready. He studied the Berlin problem intensively, beefed up conventional forces, asked Congress for an extra $3.2 billion for defense, and primed the NATO allies. In fact, East Germany did not close off Berlin, though it did build the Wall on its own territory to prevent the drain of its citizens to the western part of the city.

In June 1963, as an exclamation point on Berlin, Kennedy made a triumphal tour of Europe climaxed by a speech to a stupendous and delirious crowd in Rudolf Wilde Platz in West Berlin. He delivered the memorable words, "Ich bin ein Berliner."

The Cuban missile crisis of October 1962 was the most dangerous confrontation of the Cold War. A routine U-2 overflight of western Cuba on August 29 revealed SAM installations. By mid-October the photographs conclusively proved that the Soviets were installing missiles which could reach targets in an arc from Hudson's Bay to Peru.

On October 16, Kennedy brought his top advisers together as a crisis team that would be in virtually continuous session for 13 days. Since it was vital that the Soviets not learn that he knew, Kennedy continued with his scheduled routine, including campaigning for Democratic candidates.

The basic policy problem was to get the missiles out of Cuba without a nuclear war. Many options were examined and two came to the top: a naval blockade, later called a "quarantine," including searching merchant vessels approaching the island, or a "surgical" air strike to take out the missile sites. On October 20 the President chose the blockade for several reasons: if it failed, he would have the other as the next option; bombing was not a precise instrument; and he wanted Khrushchev to have multiple options and an air strike might force him to select nuclear war. Dean Acheson was sent with the U-2 photographs to inform the British (who had already figured it out), the French, and the Germans. Senegal and Guinea, which had the only airfields in West Africa from which Soviet aircraft could reach Cuba, agreed to deny them landing rights, if asked. American forces mobilized in the Caribbean.

On October 22 the President addressed the world, but specifically the

Soviet leadership and the American people. The Soviet action, he said, is "deliberately provocative" and "cannot be accepted by this country, if our courage and commitments are ever to be trusted again by either friend or foe." A missile launched from Cuba "against any nation in the Western Hemisphere would be considered as an attack by the Soviet Union on the United States, requiring a full retaliatory response upon the Soviet Union." "All ships of any kind bound for Cuba from whatever nation or port will, if found to contain cargoes of offensive weapons, be turned back." He would immediately ask the Organization of American States to invoke the Rio Treaty and the U.N. Security Council to demand that the Soviet Union remove its missiles. Kennedy then turned to Khrushchev, demanding that he "halt and eliminate this clandestine, reckless and provocative threat to world peace."

The OAS quickly voted 19 to 0 to support the blockcade. Despite valiant efforts by U.N. Ambassador Adlai Stevenson, the Security Council became ensnarled in a bitter and prolonged debate.

On October 24 the Second Fleet took up stations in a great arc 500 miles from the eastern tip of Cuba. Twenty-five Soviet ships were en route to the island. That afternoon a dozen changed course or stopped and later turned back to Soviet ports, presumably because they carried incriminating cargoes. The U.S. deliberately chose a neutral vessel as the first ship searched and its cargo proved innocent.

The Kremlin seemed to be taken completely by surprise and Khrushchev's first two letters to the President were extremely confusing. At Robert Kennedy's suggestion, the President treated part of one as lucid and dispatched a three-point reply: (1) The Soviet Union would remove its weapons and halt further missile introduction, both under U.N. supervision. (2) The U.S. would terminate the blockade. (3) The U.S. would pledge not to invade Cuba. On October 28, Khrushchev accepted and began pulling his missiles out of Cuba. This crisis, which brought the world to the brink of nuclear war, ended without the firing of a shot.

The outcome was a triumph for the U.S. and for Kennedy. Prime Minister Harold Macmillan of Great Britain, the most concerned of world leaders about nuclear war, wrote in his diary that "President Kennedy conducted this affair with great skill, energy, resourcefulness, and courage." His approval rating in the Gallup poll shot up from 62 percent in the week of September 20 to 25 to 74 percent in the week of November 16 to 21, 1962.

International concern over nuclear testing emerged with the detonation of the U.S. hydrogen bomb BRAVO at Bikini Atoll on March 1, 1954. Over the next nine years there were long and fruitless talks in Geneva looking toward a test ban treaty. The Eisenhower administration was not interested because it considered a ban by itself a danger to the nation's security.

As a senator, Kennedy had criticized Eisenhower's policy and had advocated a comprehensive ban or, failing that, a limitation to underground testing. When he became President he used the scientists in his administration to take a crash course in nuclear physics, much as he did with his economists and economics.

During the latter part of 1961 and 1962 the U.S. and the USSR engaged in a nuclear testing race, including the explosion of extremely large weapons in the atmosphere. Further, because of Chinese advances, there was a growing concern over nuclear proliferation. If this madness had any virtue, it was to convince Kennedy that it must be stopped. On April 24, 1963, he and Macmillan jointly asked Khrushchev for a comprehensive test ban. While the response was churlish, Khrushchev did agree to receive emissaries on July 15. This was an opening for Kennedy.

In a powerful address at American University on June 10 he called on the Soviets to join in a test ban and pledged that "the United States does not propose to conduct nuclear tests in the atmosphere so long as other states do not do so." While overshadowed in the U.S. by the civil rights crisis and in Britain by the Profumo sex scandal, the speech rang loud in the Kremlin. Further, Khrushchev was delighted by the American emissary, the old Russia hand Averell Harriman.

Kennedy's hope for a comprehensive test ban was soon dashed. The Soviets were firmly opposed. U.S. defense forces, particularly the Air Force, had until now relied on the Soviets to kill any ban; now, with a treaty in the offing, they insisted on a continuation of underground testing. A comprehensive treaty could not win 67 votes for ratification in the Senate. Thus, on July 25, Gromyko, Harriman, and Lord Hailsham of Britain initialed the Treaty Banning Nuclear Weapons Tests in the Atmosphere, Outer Space, and Under Water, one of the few Cold War milestones on the road to nuclear disarmament. Bipartisan support in the Senate yielded ratification by a vote of 80 to 19.

"No other accomplishment . . . ," Sorensen wrote, "gave Kennedy greater satisfaction." Glenn T. Seaborg, who as chairman of the Atomic Energy Commission was deeply involved in the process, stated:

> I believe that the achievement of the treaty can be traced in large part to the deep commitment of President Kennedy, to his persistence in pursuing the goal despite numerous discouragements, to his skilled leadership of the forces involved within his administration, and to his sensitive and patient diplomacy in dealing both with the Soviet Union and with the United States Senate.[3]

How did Kennedy score? Wilbur Cohen was perhaps the most acute observer of Congress of the mid-twentieth century. His experience and his memory stretched back to FDR's first term. In 1979, at a conference on the presidency and Congress at the University of Texas, Cohen set forth what he called "a Richter-type scale to measure the legislative effectiveness of various Democratic Presidents." On a yardstick of 10, he rated these Presidents "about" as follows:

LBJ	9.8
FDR	6.7
JFK	5.1
Harry Truman	5.0
Jimmy Carter (so far)	2.4

While Kennedy comes out far below Johnson and a considerable distance behind Roosevelt, this is to be expected. Both LBJ and FDR enjoyed huge congressional majorities, and JFK had to face a hostile Congress. Nevertheless, he surpassed the other Democratic Presidents, Truman and Carter, who confronted the same dilemma. Cohen, clearly, considered Kennedy a very successful President in domestic policy.

George C. Edwards III made a statistical study of presidential influence in Congress from Eisenhower to Ford or Carter based on data published mainly in *Congressional Quarterly*. Boxscores of this sort have serious shortcomings—determining whether an amended bill is still the President's bill, weighting all bills equally despite sharp differences in their importance, exclusion of the relationship between the President and Congress, among others. A summary of the findings with annual data converted to annual averages for each presidency follows: The two Democratic Presidents of the sixties, both activists, had a much higher annual average of proposals submitted to Congress—351 for JFK and 380 for LBJ—than the Republicans—216 for Eisenhower, 163 for Nixon, and 110 for Ford. The same pattern continues for the average annual number of proposals passed by Congress—138 for Kennedy and 172 for Johnson, compared with 98 for Eisenhower, 56 for Nixon, and 34 for Ford. But, when the average annual percentage of proposals approved is calculated, Eisenhower jumps up to a tie with LBJ for first at 45 percent and JFK drops to third at 39 percent.

More significant are roll-call votes on bills on which the President took a "clear position." Here there are data on Congress as a whole as well as breakdowns for the House and the Senate. Again, the annual figures have been averaged for the presidential term.

Having taken many more positions on issues, the Democratic Presidents—JFK, LBJ, and Carter (two years)—educed many more roll-call votes from both houses. Here Johnson was the leader with 242 as an annual average, Carter was second with 215, and Kennedy was third with 187. The Republicans came out with much smaller numbers. Nixon, the most activist Republican President, had an annual average of 157 roll calls, Ford had 136, and Eisenhower was at the bottom with 115. The Democrats were much more successful in winning approval for the bills they clearly supported. Here Kennedy was the leader, winning congressional support on 84 percent of the bills he backed. Johnson followed closely at 83 and Carter was at 77 percent. Eisenhower led the Republicans with a 72 percent success ratio, with Nixon at 67 and Ford with 58 percent.

In winning roll calls in the House LBJ was at the top with 86 percent, and JFK was a close second at 84 percent. The followers in order were Nixon at 73 percent, Carter 72, Eisenhower 70, and Ford with an abysmal 51 percent. In the Senate, however, Kennedy moved into first place at 86 percent. Carter was at 81, and Johnson trailed with 80 percent. There was a Republican falloff: Eisenhower 74, Ford 64, and Nixon 63 percent.

Another test of a President's relationship with Congress is the number of times his vetoes are overridden. Facing Democratic Congresses, Republican Presidents were far more prone to veto. Eisenhower had a total of 73 during his presidency, Ford had 50, and Nixon 24. The Democratic numbers were much lower—Carter 19, Johnson 16, and Kennedy 12. It is remarkable that Congress overrode no veto issued by the three Democratic Presidents. By contrast, Ford was overridden on 12 bills, Nixon 5, and Eisenhower 2.

Edwards also compared the performance of these Presidents with members of Congress from the North and the South. He expected LBJ, "the master legislative technician," to come out on top. In fact, Kennedy was first, "receiving more support in the House from both Northern and Southern Democrats than did Johnson." LBJ barely surpassed Carter during the latter's first two years. The same pattern emerged in the Senate: "Kennedy did better than Johnson among Northern Democrats, the same as Johnson among Southern Democrats."

The Edwards analysis differs from Wilbur Cohen's in one significant way: "Johnson's mastery of the legislative process seems to have been considerably less significant than conventional wisdom indicates." But he agrees with Cohen in rating Kennedy highly, in fact, more so than Cohen. His data do not suggest in any way that JFK was overly cautious with Congress, or that he ignored Congress, or that he failed with Congress. Quite the contrary.

The most significant yardstick for measuring the President's effectiveness is qualitative, examining carefully his performance on important issues, as I have tried to do in this book. In the case of Kennedy there are degrees of weight which fall into three categories. Four legislative proposals were of paramount significance—civil rights, the tax cut, aid for education, and Medicare. Charles and Barbara Whalen, historians of the Civil Rights Act, wrote that it "was hailed by many as one of the most important pieces of legislation ever enacted by Congress." The adoption of a peacetime Keynesian economic strategy with tax reduction in the sixties marked what even so conservative an economist as Herbert Stein called "the fiscal revolution in America." Massive federal aid was indispensable in helping to resolve the crisis in the nation's immense educational establishment. Medicare would become by far the most important amendment to the Social Security Act and would provide hospital and medical insurance (better health and greater longevity) for tens of millions of the country's elderly citizens. A President who succeeded in persuading Congress to enact even one of these measures would have made a large historical impact.

The second group of programs, while important, were markedly less so than the Big Four. But they comprised a large part of the Kennedy agenda: the Area Redevelopment Act, the Manpower Development and Training Act, the minimum wage and extended coverage, federal sector collective bargaining under Executive Order 10988, several of the particular education statutes, notably the Health Professions Educational Assistance Act and the Mental Health Facilities and Community Mental Health Centers Construction Act, and the Peace Corps, both Executive Order 10924 and the Peace Corps Act.

The final category contained only a single enactment, the Equal Pay Act, which was more symbol than substance.

The fate of the Big Four, narrowly viewed, gives comfort to the revisionists because, except for executive orders and administrative actions on civil rights and the higher education law, none of the important statutes were actually signed by President Kennedy. But to argue this way is both inaccurate and unfair.

In civil rights Kennedy had begun his presidency with a policy of executive action now and legislation later. This was a political judgment that a civil rights bill in 1961 or 1962 would divide the country, shatter the Democratic party, and be rejected by Congress. In my view this reasoning, while morally questionable, was politically unassailable. Thus, the Kennedy administration devoted two years to pushing civil rights without turning to Congress: Federal agencies were directed to hire blacks and to desegregate facilities; Executive Order 10925 established Lyndon Johnson's Committee on Equal Employment Opportunity to promote fair employment in the federal service and among government contractors; the Interstate Commerce Commission, under pressure from the Freedom Riders and the Department of Justice, made Jim Crow illegal in interstate transportation; the civil rights division filed suits to protect the right of Negroes to vote in the Deep South; discrimination in public housing was prohibited; James Meredith was admitted to the University of Mississippi; Birmingham took its first halting steps toward desegregation; despite Governor Wallace standing in the schoolhouse door, the University of Alabama admitted black students.

But on June 11, 1963, convinced by the showdown in Tuscaloosa and the violence in Birmingham, Kennedy reversed himself and came out for passage of a strong omnibus civil rights act. This totally transformed the politics of civil rights because the President automatically lost southern support. Only the Republicans could provide the votes needed to create a majority on cloture in the Senate and on substance in both houses. Kennedy set out to win the opposition over and succeeded handsomely. The bill cleared the House Judiciary Committee on November 20, 1963, two days before the assassination. It still must go to the Rules Committee, to the floor of the House, and to the Senate, including cloture. While there was no certainty that these hurdles would be surmounted, the signs were favorable. But this would take time.

On November 14, 1963, President Kennedy held his sixty-fourth and final press conference. There were several questions about his legislative program, particularly civil rights and tax reduction. He said he expected favorable House action on civil rights "in the next month, maybe sooner." But the Senate was another story. "There may be a very long debate." This could complicate the timing of the vote, and the tax bill might get entangled in the battle over civil rights. While he was sure that Medicare would pass, he expected it to take even longer. His final sentence is interesting: "This is going to be an 18-month delivery!"

He must have been thinking about the upcoming presidential election, now almost exactly a year away. Never one to dally about campaigning, Kennedy was already on the hustings. In fact, he would be in Texas, including Dallas, in the following week trying to heal a nasty fracture in the Democratic party. He was supremely confident of his own re-election. The religious problem, which had cost him four or five percentage points in 1960, now seemed dead. The polls showed him exceptionally popular and easily defeating any possible Republican candidate. Equally important, they indicated that the country was moving quietly but firmly leftward, that the public strongly favored his activist agenda, including, except the South, civil rights. If the Republicans nominated Goldwater, as seemed increasingly likely, Kennedy expected to win in a landslide, sweeping up big Democratic majorities in both houses of Congress. Civil rights might pass in 1964. If not, he would get it within his eighteen-month frame in early 1965, along with primary and secondary education and Medicare.

Because of the assassination and Johnson's skillful leadership, the process may have been speeded up. Johnson signed the Revenue Act of 1964 on

February 26 and the Civil Rights Act on July 2, 1964. There can be no doubt that, if Kennedy had been alive, he would have signed the Civil Rights Act within his 18-month prediction. Nicholas Katzenbach, who helped to steer the civil rights bill through Congress for both Kennedy and Johnson, thought the assassination made almost no difference. He was asked on November 16, 1964, whether the law would have been passed in approximately the same form had Kennedy lived. He responded:

> I think it would have. I think the same pressures would have been present. I think the same things would have happened. I think if there was an effect on this from the President's death it was largely in terms of the debate in the Senate. I think the debate was less bitter than it would have been if President Kennedy had been alive. I think the votes would have come out exactly the same way.

The outlook for the tax bill was much more favorable. The House had passed it on September 25, 1963, by a large majority with heavy southern Democratic support. The Senate Finance Committee hearings began on October 15, 1963. The Senate, much more liberal than the House on economic issues, could be expected to approve the bill in early 1964, which, in fact, it did.

Federal aid for education was in this context, as in all others, complicated. The important Higher Education Facilities Act and the six lesser education bills had for all practical purposes been passed by November 22, 1963, though all but one had been briefly delayed by Senator Morse's legislative strategy. But this was only part of the Kennedy package. Scholarships for college students and assistance for elementary and secondary education were laid over. While part of this outstanding group of bills might have passed in 1964, it is more likely that all of them would have been enacted in 1965, which is actually what took place.

The prospects for Medicare, which Kennedy had been unable to move at all, were improving. The Senate, particularly after Robert Kerr's death, seemed narrowly safe. On July 17, 1962, it had defeated the bill by a very close margin under bizarre circumstances. Four new liberal Democrats joined that body following the elections in November 1962. There was even evidence of movement in the more conservative House, particularly from Wilbur Mills, whose influence was decisive. On the morning of November 22, 1963, he agreed to a financing formula and now, according to Larry O'Brien, the passage of Medicare was "inevitable." It might have been enacted in 1964. If not, as Kennedy stressed at the November 14 press conference, it surely would have been approved in 1965, as, in fact, it was.

The President was successful in gaining, mainly by legislation but also by executive order, his complete second category program—area redevelopment, the Manpower Development and Training Act, a higher minimum wage with extended coverage, federal sector collective bargaining, the lesser education statutes, and the Peace Corps. In the aggregate this was a significant achievement.

Finally, for the third category, the Kennedy administration gave women's rights a boost. The report of the President's Commission on the Status of Women laid out a program for the future. The Equal Pay Act established the principle of equal pay for equal work without regard to sex in federal law, though this would have little substantive impact.

On Capitol Hill they like to say that politics is the art of the possible. Kennedy, with his penchant for history, preferred to quote Jefferson: "Great innovations should not be forced on slender majorities." Kennedy confronted a reluctant and sometimes hostile Congress. His major proposals—civil rights, tax reduction, aid for education, and Medicare, and some of the secondary ones as well, like the minimum wage—were inherently divisive and aroused bitter controversy. In his first two years he needed to grow into his job. He sometimes seemed unsure of himself and made mistakes. But he settled in during 1963. He now knew how to handle himself, was surrounded with competent advisers, and was very popular. This manifested itself in significant legislative breakthroughs in Congress. When he predicted on November 14, 1963, that his entire program would be enacted within eighteen months, he was right; if anything, perhaps a bit on the conservative side. He was emerging as a President of great stature when, eight days later, a mindless assassin in Dallas cut his life short.[4]

Notes

Prologue

1. Dean Acheson's little book, *A Democrat Looks at His Party,* was republished in *Private Thoughts on Public Affairs* (New York: Harcourt, Brace & World, 1955 et seq.), 95, 97, 98. The Schlesinger poll is in Thomas E. Cronin, *The State of the Presidency* (Boston: Little, Brown, 1980), 2d ed., 387; James MacGregor Burns, *Presidential Government* (Boston: Houghton Mifflin, 1965), 81; Kennedy's National Press Club speech is in Arthur M. Schlesinger, Jr., Fred L. Israel, William P. Hansen, *History of American Presidential Elections, 1789–1968* (New York: Chelsea House, 1985), vol. IX, pp. 3536–40; Arthur M. Schlesinger, Jr., *A Thousand Days* (Boston: Houghton Mifflin, 1965); Theodore C. Sorensen, *Kennedy* (New York: Bantam, 1966). An example of contemporary press criticism of Kennedy's relationship with Congress is the first half of *New York Times* correspondent Tom Wicker's *JFK and LBJ: The Influence of Personality upon Politics* (New York: Morrow, 1968). The revisionist works are the following: Lewis J. Paper, *The Promise and the Performance: The Leadership of John F. Kennedy* (New York: Crown, 1975), 365; Grant McConnell, *The Modern Presidency,* (New York: St. Martin's, 1976), 2d ed., 45, 82; Jim F. Heath, *Decade of Disillusionment, The Kennedy–Johnson Years* (Bloomington: Univ. of Indiana Press, 1975), 160, 163; Henry Fairlee, *The Kennedy Promise: The Politics of Expectation* (Garden City: Doubleday, 1973), 1. Fairlee was obsessed by Kennedy's "rhetoric." His denigrating comparison of JFK to Polk does not work because Polk is held in quite high esteem by American historians. Alan Shank, *Presidential Policy Leadership, Kennedy and Social Welfare* (Boston: University Press, 1980), 158, 263; Herbert S. Parmet, *JFK: The Presidency of John F. Kennedy* (New York: Dial, 1983), 353; Gary Orfield, *Congressional Power: Congress and Social Change* (New York: Harcourt Brace Jovanovich, 1975), v, 18, 48; Bruce Miroff, *Pragmatic Illusions: The Presidential Politics of John F. Kennedy* (New York: McKay, 1976), 200, 272, 273.

Chapter 1. America in 1960

1. Louis J. Halle, *The Cold War as History* (London: Chatto & Windus, 1967); André Fontaine, *History of the Cold War* (New York: Pantheon, 1968, 1969), 2 vols.; Richard H. Rovere, *The Eisenhower Years* (New York: Farrar, Straus & Cudahy, 1956), 351, 365; Stephen E. Ambrose, *Eisenhower, the President* (New York: Simon and Schuster, 1984), 424–25.

2. The statistics are from U.S. Bureau of the Census, *Historical Statistics of the United States, Colonial Times to 1957* (Washington, 1960), *Continuation to 1962 and Revisions* (Washington, 1965), and various *Current Population Reports* of the 1960s. See also Donald J. Bogue, *The Population of the United States* (Glencoe: Free Press, 1959); Ben J. Wattenberg, *This U.S.A.* (Garden City: Doubleday, 1965); Landon Y. Jones, *Great Expectations, America and the Baby Boom Generation* (New York: Coward, McCann & Geoghegan, 1980), 38.

3. Max Lerner, *America as a Civilization* (New York: Simon and Schuster, 1957), 172–82; Louis H. Masotti, "Suburbia Reconsidered—Myth and Counter-Myth," in Louis H. Masotti and J. K. Hadden, *The Urbanization of the Suburbs* (Beverly Hills: Sage, 1973), 16; William M. Dobriner, *Class in Suburbia* (Englewood Cliffs: Prentice-Hall, 1963), ch. 4; William H. Whyte, Jr., *The Organization Man* (New York: Simon and Schuster, 1956), pt. VII; Editors of *Fortune, The Changing American Market* (Garden City: Hanover House, 1955), 25, 29, 131, 230; Lewis Mumford, *The City in History* (New York: Harcourt, Brace & World, 1961), ch. 16; David Riesman, "The Suburban Sadness," in William M. Dobriner, *The Suburban Community* (New York: Putnam, 1958), 375–402; A. C. Spectorsky, *The Exurbanites* (Philadelphia: Lippincott, 1955), 4; Bennett M. Berger, *Working-class Suburb* (Berkeley: Univ. of California Press, 1960); Herbert J. Gans, *The Levittowners* (New York: Pantheon, 1967); Robert C. Wood, *Suburbia, Its People and Their Politics* (Boston: Houghton Mifflin, 1958), ch. 5; Frederick M. Wirt, Benjamin Walter, Francine F. Rabinovitz, and D. R. Hensler, *On the City's Rim: Politics and Policy in Suburbia* (Lexington: Heath, 1972); Philip C. Dolce, ed., *Suburbia: The American Dream and Dilemma* (Garden City: Doubleday, 1976).

4. Louis Wirth, *The Ghetto* (Chicago: Univ. of Chicago Press, 1928), ch. 1; J. H. Bracey, Jr., August Meier, Elliott Rudwick, eds., *The Rise of the Ghetto* (Belmont: Wadsworth, 1971), 40–42, 155; Karl E. and Alma F. Taeuber, *Negroes in Cities* (Chicago: Aldine, 1965), 32–36, 57; *Report of the National Advisory Commission on Civil Disorders* (New York: Bantam, 1968), 248, chs. 7 and 8; Davis McEntire, *Residence and Race* (Berkeley: Univ. of California Press, 1960), 36–37, 73–78, 89–91; Saul Bernstein, *Youth on the Streets* (New York: Association, 1964), ch. 2; E. Franklin Frazier, *The Negro Family in the United States* (Chicago: Univ. of Chicago Press, 1939); Nathan Glazer and Daniel Patrick Moynihan, *Beyond the Melting Pot* (Cambridge: MIT Press, 1964), 50–53; Department of Labor, *The Negro Family, the Case for National Action* (1965); Charles E. Silberman, *Crisis in Black and White* (New York: Random House, 1964), 45; Morton Grodzins, *The Metropolitan Area as a Racial Problem* (Pittsburgh: Univ. of Pittsburgh Press, 1958), 10: Lee Rainwater, *Behind Ghetto Walls* (Chicago: Aldine, 1970), 99, 105, 107, 166, 230, 287, 294, 387, 397.

5. W. W. Rostow, *The Stages of Economic Growth* (London: Cambridge Univ. Press, 1960); *Economic Report of the President* (Jan. 1965), 189, 190, 234; Bureau of the Census, Herman P. Miller, *Income Distribution in the United States,* 1960 Census Monograph (Washington, 1960), 7, 9; George Katona, *The Mass Consumption Society* (New York: McGraw-Hill, 1964), 5; *Statistical Abstract of the United States, 1964,* 564; W. W. Rostow, "The Dynamics of American Society," in *Postwar Economic Trends in the United States,* Ralph E. Freeman, ed. (New York: Harper, 1960), 8; Peter Henle, "Recent Growth in Paid Leisure for U.S. Workers," *Monthly Labor Review* (Mar. 1962), 249–57; Helen H. Lamale, "Workers' Wealth and Family Living Standards," and Ida C. Merriam, "Social Expenditures and Worker Welfare," *Monthly Labor Review* (June

1963), 676–86, 687–94; "Supplementary Wage Benefits in Metropolitan Areas, 1959–60," *Monthly Labor Review* (Apr. 1961), 379–87; Helen H. Lamale and Margaret S. Stotz, "The Interim City Worker's Family Budget," *Monthly Labor Review* (Aug. 1960), 785–808; Geoffrey H. Moore, "The 1957–58 Business Contraction: New Deal or Old?," *Readings in Unemployment,* Senate Special Committee on Unemployment Problems, 86th Cong., 2d sess. (1960), 102–16; Bureau of Labor Statistics, Bull. No. 1312, *Employment and Earnings Statistics 1909–60* (1961), 535; Robert Aaron Gordon, *Economic Instability and Growth* (New York: Harper & Row, 1974), 123–29; Harold G. Vatter, *The U.S. Economy in the 1950s* (New York: Norton, 1963), 115–20; Abramovitz to Heller, May 30, 1962, Sorensen Papers, Kennedy Library.

6. V. O. Key, Jr., *The Responsible Electorate* (Cambridge: Belknap, 1966), ch. 4; Samuel Lubell, *The Future of American Politics,* 2d ed. (Garden City: Doubleday, 1956), 212, 241, 251; the party identification figures are calculated from Table 2–1 in Angus Campbell, Philip E. Converse, Warren E. Miller, and Donald E. Stokes, *Elections and the Political Order* (New York: Wiley, 1966), 13; James MacGregor Burns, *Deadlock of Democracy* (Englewood Cliffs: Prentice-Hall, 1963), ch. 10; James L. Sundquist, *Politics and Policy* (Washington: Brookings, 1968), ch. IX; Theodore C. Sorensen, *Kennedy* (New York: Bantam, 1966), 107; Arthur M. Schlesinger, "Tides of American Politics," *Yale Review* (Dec. 1939), 220, 225–26. In 1959, Arthur M. Schlesinger, Jr., wrote a memorandum, "The Shape of National Politics to Come," which was circulated within the liberal Democratic establishment. Using his father's "tides," he expected a new period of "affirmation, progressivism, and forward movement" to begin about 1961–62. This would demand "vigorous public leadership." Kennedy was impressed with the argument and, evidently, had already reached the same conclusions. Arthur M. Schlesinger, Jr., *A Thousand Days* (Boston: Houghton Mifflin, 1965), 17–18. Schlesinger developed this theory more fully in *The Cycles of American History* (Boston: Houghton Mifflin, 1986), ch. 2. Chester Bowles made the same analysis in *The Coming Political Breakthrough* (New York: Harper, 1959).

7. Chester Bowles, *Promises to Keep* (New York: Harper & Row, 1971), 285–92; Harris Wofford, *Of Kennedys and Kings* (New York: Farrar, Straus, Giroux, 1980), 51–52; "How Bowles Did It," *New Republic* (July 25, 1960), 6–7; James L. Sundquist to author, Oct. 31, Nov. 29, 1989. The account of the Compact of Fifth Avenue is based on Theodore H. White, *The Making of the President, 1960* (New York: Atheneum, 1961), ch. 7. Nixon's version differs sharply and makes him look much better. See Richard M. Nixon, *Six Crises* (Garden City: Doubleday, 1962), 313–16. The Democratic and Republican platforms are in Arthur M. Schlesinger, Jr., Fred L. Israel, William P. Hansen, eds., *History of American Presidential Elections, 1789–1968* (New York: Chelsea House, 1971), vol. IV, pp. 3471–3535, the New Frontier speech is at pp. 3451–45.

8. White, *Making of the President, 1960,* in general with quotations at pp. 291, 321; Sorensen, *Kennedy,* ch. VII; Stanley Kelley, Jr., "The Presidential Campaign," in Paul T. David, ed., *The Presidential Election and Transition, 1960–1961* (Washington: Brookings, 1961), 57–87; Nixon, *Six Crises,* 302; Schlesinger, *A Thousand Days* (Boston: Houghton Mifflin, 1965), 75; Wofford, *Kennedys and Kings,* ch. 1, pp. 245–50; for a detailed account of the King arrest and Kennedy phone call from the Kings' point of view, see David J. Garrow, *Bearing the Cross* (New York: Morrow, 1986), 141, 143–49.

9. Angus Campbell, Philip E. Converse, Warren E. Miller, Donald E. Stokes,

Elections and the Political Order (New York: Wiley, 1966), 94, 97, 112; Lucy C. Dawidowicz and Leon J. Goldstein, *Politics in a Pluralist Democracy* (New York: Institute of Human Relations Press, 1963), 11, 19, 24, 29, 43, 47, 53, 56; Sorensen, *Kennedy*, 246; White, *Making of the President, 1960*, 350, 361; David Halberstam, *The Best and the Brightest* (New York: Penguin, 1972), 83; Clark to Kennedy, Dec. 15, 1960, POF Legislative Files, Kennedy Library.

10. Sorensen, *Kennedy*, 258, 265; David, *Presidential Election and Transition*, 215–18; Carl M. Brauer, *Presidential Transitions, Eisenhower through Reagan* (New York: Oxford Univ. Press, 1986), ch. 2; *Robert Kennedy in His Own Words*, Edwin O. Guthman and Jeffrey Shulman, eds. (New York: Bantam, 1988), 42, 57, 74; Wofford, *Kennedys and Kings*, chs. 3, 4; Halberstam, *Best and Brightest*, 126, 334–35; Arthur M. Schlesinger, Jr., *Robert Kennedy and His Times* (Boston: Houghton Mifflin, 1978), chs. 4–10. Robert Kennedy's remarks on blacks are from Douglas Ross, *Robert E. Kennedy: Apostle of Change* (New York: Trident, 1968), 49, and Edwin Gutham, *We Band of Brothers* (New York: Harper & Row, 1971), 181; R. Wallace, "Non-whiz Kid with the Quiet Gun," *Life*, Aug. 9, 1963, 55; "Burke Marshall: Quiet Fighter for Civil Rights," *Ebony*, May 1964, 90.

Chapter 2. Civil Rights: Confrontation

1. C. Vann Woodward, *The Strange Career of Jim Crow*, 2d rev. ed. (New York: Oxford Univ. Press, 1966), quotes at pp. 7, 147, 155, 165, 168–70; Pauli Murray, *States' Laws on Race and Color* (Woman's Division of Christian Service, 1950), 89–117; Neil R. McMillen, *The Citizens' Council* (Urbana: Univ. of Illinois Press, 1971).

2. Harris Wofford, Memorandum to President–elect Kennedy on Civil Rights—1961, Dec. 30, 1960, Pre–Presidential Task Force Reports, Kennedy Library.

3. Roy Wilkins, Oral History Interview, 4, 7, Johnson Library; Foster Rhea Dulles, *The Civil Rights Commission: 1957–1965* (East Lansing: Michigan State Univ. Press, 1968), 99–108; for an insider's view of the Eisenhower commission, see the appendix to Harris Wofford, *Of Kennedys and Kings* (New York: Farrar, Straus, Giroux, 1980), other citations to Wofford at pp. 124, 126–28, 134, 141–50, 166–70; Harris Wofford, Oral History Interview, 48, Kennedy Library; Burke Marshall, Oral History Interview, 54–58, Kennedy Library; *Robert Kennedy in His Own Words*, Edwin O. Guthman and Jeffrey Shulman, eds. (New York: Bantam, 1988), 154–56; Theodore C. Sorensen, *Kennedy* (New York: Bantam, 1965), 531–32, 540–42; Arthur M. Schlesinger, Jr., *A Thousand Days* (Boston: Houghton Mifflin, 1965), 931–33; Summary of Civil Rights Progress for the Nine Months—Jan. 20 through Oct. 1961, POF, Civil Rights, Kennedy Library.

4. Michael I. Sovern, *Legal Restraints on Racial Discrimination in Employment* (New York: Twentieth Century Fund, 1966), ch. 5, quotes at pp. 105, 113, 140; Executive Order 10925, 26 Fed. Reg. 1977 (1961); Jenkins to Johnson, Jan. 2; Siegel to Johnson, Jan. 3; Draft of Proposed Executive Order, Feb. 3; Katzenbach to Moyers, Feb. 20; Reedy to Johnson, Feb. 28; Johnson to Kennedy, Government Committee on Equal Employment Opportunity, n.d.; Papers of LBJ as Vice President, Civil Rights; all, Johnson Library; Weaver to Goldberg, Summary Statement on the Position and Problems of the Federal Government and Non–discriminatory Employment, Dec. 30, 1960, p. 5; Hill to President's Committee on Equal Employment Opportunity, July 12, 1961; Siegel to Kheel, Dec. 28, 1960; Kheel to Johnson, Dec. 29, 1960; Proposed Executive Order, 1/19/61 Draft, Feild to Holleman, Oct. 12,

1961; all, Goldberg Papers, National Archives; James E. Jones, Jr., "The Genesis and Present Status of Affirmative Action in Employment: Economic, Legal, and Political Realities," unpublished paper given to the American Political Science Association (Washington, 1984), 1–2, 5; U.S. Civil Service Commission, *Study of Minority Group Employment in the Federal Government* (annual); Samuel Krislov, *The Negro in Federal Employment* (Minneapolis: Univ. of Minnesota Press, 1967), 127–42; Arthur M. Schlesinger, Jr., *Robert Kennedy and His Times* (Boston: Houghton Mifflin, 1978), 292; "Plan for Equal Job Opportunity at Lockheed Aircraft Corp.," *Monthly Labor Review* (July 1961), 748–49; "Union Program for Eliminating Discrimination," *Monthly Labor Review* (Jan. 1963), 58–59; Harris Wofford, Oral History Interview, 84, 118, 125–33; Robert F. Kennedy, Oral History Interview, vol. 6, pp. 669, 702–10; Burke Marshall, 1964 Oral History Interview, 62–63; 1970 Interview, 23–24; all, Kennedy Library; Roy Wilkins, Oral History Interview, 5, Johnson Library; Report by Theodore W. Kheel to Vice President Johnson, The Structure and Operations of the President's Committee on Equal Employment Opportunity, 44, Aug. 22, 1962; all, National Archives; Donovan to Wirtz, enclosing Employment in establishments subject to Executive Order 10925 of March 6, 1961, Apr. 24, 1963; Wirtz, Memorandum for the President, June 21, 1963, Wirtz Papers, National Archives.

5. Catherine A. Barnes, *Journey from Jim Crow: The Desegregation of Southern Transit* (New York: Columbia Univ. Press, 1983), quotes at pp. ix, 44–48, 86–100, 120–22, 144–49, 157–75; *Morgan v. Virginia,* 328 U.S. 373 (1946); *NAACP v. St. Louis—San Francisco Railway Co.,* 297 ICC 335 (1955); *Keys v. Carolina Coach Co.,* 644 MCC 769 (1955); *Browder v. Gayle,* 142 F. Supp. 707 (M.D. Ala. 1956); *Boynton v. Virginia,* 364 U.S. 454 (1960); *Plessy v. Ferguson,* 163 U.S. 537 (1896); *Brown v. Board of Education,* 347 U.S. 483 (1954); David J. Garrow, *Bearing the Cross* (New York: Morrow, 1986), Farmer quote at p. 156; Jacob Cohen has a sketch of Farmer in the introduction to James Farmer, *Freedom—When?* (New York: Random House, 1965); Wofford, *Kennedys and Kings,* 153; Schlesinger, *Robert Kennedy,* 294–300, Kennedy quote on Eastland at p. 299; August Meier and Elliott Rudwick, *CORE* (New York: Oxford Univ. Press, 1973), ch. 5; Orrick quote in Victor S. Navasky, *Kennedy Justice* (New York: Atheneum, 1977), 124; Carl M. Brauer, *John F. Kennedy and the Second Reconstruction* (New York: Columbia Univ. Press, 1977), 109–10; Howard Zinn, *SNCC, The New Abolitionists* (Boston: Beacon, 1964), ch. 3; Ross Barnett, Oral History Interview, 18, Kennedy Library.

6. U.S. Commission on Civil Rights, Report 1, *Voting,* 1961, 15, 22, 31, 108; Taylor Branch, *Parting the Waters* (New York: Simon and Schuster, 1988), 334; Wofford, Memorandum to President-elect Kennedy on Civil Rights—1961, Dec. 30, 1960, p. 4, Kennedy Library; Burke Marshall, *Federalism and Civil Rights* (New York: Columbia Univ. Press, 1964), 25–27, quotes at pp. 4–5, 11–12, 30–31, 81; Garrow, *Bearing the Cross,* 162–64; Wofford, *Kennedys and Kings,* 158–64; Brauer, *Kennedy and the Second Reconstruction,* 113–21, Doar and Landsberg quote at p. 118; Navasky, *Kennedy Justice,* 194–95, ch. 5, quote at p. 244; Schlesinger, *Robert Kennedy,* 307–10; Robert F. Kennedy, Oral History Interview, vol. 1, p. 202, vol. 5, p. 579, Kennedy Library; Pat Watters and Reese Cleghorn, *Climbing Jacob's Ladder* (New York: Harcourt, Brace & World, 1967), 44–50, 216–24, quote at p. 47; Steven F. Lawson, *Black Ballots: Voting Rights in the South, 1944–1969* (New York: Columbia Univ. Press, 1976), ch. 9; the Doar quote is from Walter Lord *The Past That Would Not Die* (New York: Harper & Row, 1965), 247. The Marshall theory of federalism was sharply criticized. See

Navasky, *Kennedy Justice,* ch. 4, and Zinn, *SNCC,* ch. 4. Marshall recalled that President Kennedy found federalism "frustrating." He could hardly believe that things that needed doing were unconstitutional. Burke Marshall, 1964 Oral History Interview, 70, 89, Kennedy Library. According to Nicholas Katzenbach, the only judicial appointments the President was interested in were those to the Supreme Court. While he wanted to avoid bad civil rights choices to the lower federal courts in the South, the President did not involve himself in the appointment process. Thus, the deputy attorney general and the interested senator pretty much made the appointments. Byron White was deputy when racist selections were made. After White went onto the Supreme Court and Katzenbach succeeded him as deputy no more racists were named in the South. Nicholas Katzenbach, Oral History Interview, 78–87, Kennedy Library.

7. The general works on the Meredith case are: James W. Silver, *Mississippi: The Closed Society* (New York: Harcourt, Brace & World, 1963); Lord, *Past That Would Not Die;* Russell H. Barrett, *Integration at Ole Miss* (Chicago: Quadrangle, 1965); Michael Dorman, *We Shall Overcome* (New York: Delacorte, 1964), ch. 1; Brauer, *Kennedy and the Second Reconstruction,* ch. 7; Schlesinger, *Robert Kennedy,* 317–27; James Meredith, *Three Years in Mississippi* (Bloomington: Univ. of Indiana Press, 1966), which includes many documents. The quotations are: Lord, *Past That Would Not Die,* 32–33, 98–100; Silver, *Mississippi,* 5–6; Meredith, *Three Years,* 21, 51. See Jean Stein and George Plimpton, *American Journey: The Times of Robert Kennedy* (New York: Harcourt Brace Jovanovich, 1970), 104; Neil R. McMillen, *The Citizens' Council* (Urbana: Univ. of Illinois Press, 1971), 216–19, 344; Meredith to Department of Justice, Feb. 7, 1961, Marshall Papers; Schlesinger to R. F. Kennedy, Sept. 27, 1962, POF; Presidential Recordings Transcripts, Integration at the University of Mississippi, Item 4G1, Belt 4C, Audiotapes 26, 26A, POF; Clifton, Memorandum for the President, Oct. 3, 1962, POF; Burke Marshall, Oral History Interview, 72–84; all, Kennedy Library; *Meredith v. Fair,* 305 F. 2d 341 and 306 F. 2d 374 (1962). According to Marshall and Robert Kennedy, the President was outraged by the Army's delay in getting troops to Oxford and by telling him that they were on their way while they were still at Millington. He was deeply concerned that this might recur in an international crisis and later roasted the Joint Chiefs over the incident. Oral History Interview, 81. Marshall found Meredith "a real character." He proposed driving into Oxford in the gold Thunderbird he had bought. "Boy," the Attorney General said, "that's all we needed. . . ." Marshall got the black comedian Dick Gregory to talk Meredith out of doing so. Gregory, Marshall said, is "somewhat nutty, but he's not that nutty." Robert F. Kennedy, Oral History Interview, vol. 7, 827–28, Kennedy Library; Nicholas Katzenbach, Oral History Interview, 93, 97–99, Kennedy Library.

Chapter 3. Civil Rights: Year of Decision, 1963

1. Michael Dorman, *We Shall Overcome* (New York: Delacorte, 1964), ch. IV; quotations from the *New York Times* are from Anthony Lewis, *Portrait of a Decade* (New York: Random House, 1964), 175, 184; Michael Frady, *Wallace* (New York: World, 1968), 127, 133; Charles Morgan, Jr., *A Time to Speak* (New York: Harper & Row, 1964), 147; David J. Garrow, *Bearing the Cross* (New York: Morrow, 1986), 195, 216–19, 225–29, ch. 5; Taylor Branch, *Parting the Waters* (New York: Simon and Schuster, 1988), 789, 808, 991–92; Stephen B. Oates, *Let the Trumpet Sound* (New York: Mentor, 1985), 201–6; Jean Stein and George Plimpton, *American Journey: The Times of Robert*

Kennedy (New York: Harcourt Brace Jovanovich, 1970), 114–15; Martin to Marshall, Equal Employment in Birmingham, Alabama, Marshall Papers, Audiotapes Nos. 86.2, 88.6; Civil Rights—Birmingham, May 12, 1963, and Cabinet Meeting, May 21, 1963; Burke Marshall, Oral History Interview; R. F. Kennedy, Oral History Interview, vol. 6, 776; Lt. Col. Edelen to White House, May 15, 1963; all, Kennedy Library.

2. The best account of the University of Alabama incident, based on his own coverage and Department of Justice sources, is Dorman, *We Shall Overcome,* ch. IX; the Wallace disability story is at pp. 293–94, 331–32; Theodore C. Sorensen, *Kennedy* (New York: Bantam, 1965), 552, 556; the Robert Kennedy-Wallace confrontation is in Frady, *Wallace,* 150–69; R. F. Kennedy, Oral History Interview, vol. 4, 431–32, vol. 7, 799–800, 841–43; Memorandum of a Conversation between President Kennedy and Governor George Wallace, May 18, 1963, POF, Civil Rights, Kennedy Library, Morgan, *A Time to Speak,* 159–60; Burke Marshall, Oral History Interview, 102, 106, 109–10, Kennedy Library; *Public Papers, Kennedy, 1963,* 468–71; Lewis, *Portrait of a Decade,* 226.

3. R. F. Kennedy, Oral History Interview, vol. 1, 134–36, vol. 6, 699–700; Burke Marshall, Oral History Interview, 61, 64–65; Audiotape No. 88.4, Civil Rights, May 20, 1963, Audiotape No. 90.3; Civil Rights Legislation, June 1, 1963; all, Kennedy Library; *Robert Kennedy in His Own Words,* Edwin O. Guthman and Jeffrey Shulman, eds. (New York: Bantam, 1988), 198; Arthur M. Schlesinger, Jr., *Robert Kennedy and His Times* (Boston: Houghton Mifflin, 1978), 347; LBJ-Sorensen, June 3, 1963, Dictaphone Recording, Johnson Library; Memorandum by Senator Mansfield on Conference with Senator Dirksen, June 13, 1963, POF, Legislative File, Kennedy Library; *Civil Rights—The President's Program, 1963,* Hearings before the Committee on the Judiciary, Senate, 88th Cong., 1st sess., 23–25, S. 1731 at pp. 1–12; Mansfield to JFK, Civil Rights Strategy in the Senate, June 18, 1963, Sorensen Papers, Kennedy Library; Neil MacNeil, *Dirksen, Portrait of a Public Man* (New York: World, 1970) ch. 1; the conclusive study of the legislative history is Charles and Barbara Whalen, *The Longest Debate: A Legislative History of the Civil Rights Act of 1964* (Cabin John: Seven Locks, 1985), quotes at pp. 6, 37, 84; Branch, *Parting the Waters,* 516–17, 835–38, 844–45, 851–59; Garrow, *Bearing the Cross,* 270–75; Arthur M. Schlesinger, Jr., *A Thousand Days* (Boston: Houghton Mifflin, 1965), 968–72; Baker to Mansfield, Civil Rights Possibilities, June 27, 1963, POF, Legislative Files, Kennedy Library; Nicholaus Katzenbach, Oral History Interview, 132–33, 137, Kennedy Library. According to Larry O'Brien, the crisis in the Judiciary Committee led him to urge Kennedy to deal directly with Halleck, "a measure of our desperation." Though a confirmed conservative, he was a "professional" who handled O'Brien in "a cordial manner," plying him with martinis and invariably calling him "O'Toole, much as Ev Dirksen always called me Lawrence." To Kennedy's amazement, Halleck favored civil rights and was flattered to be asked to help. Lawrence F. O'Brien, *No Final Victories* (Garden City: Doubleday, 1974), 145–46. *Civil Rights,* Hearings before Subcommittee No. 5, Committee on the Judiciary, House of Representatives, 88th Cong., 1st sess. (1963); *Civil Rights Act of 1963,* Report of the House Judiciary Committee, No. 914, 88th Cong., 1st sess., Nov. 20, 1963.

4. Thomas Gentile, *March on Washington, August 28, 1963* (Washington: New Day, 1983); Jarvis Anderson, *A. Philip Randolph, A Biographical Portrait* (New York: Harcourt Brace Jovanovich, 1972), 323–32; Garrow, *Bearing the Cross,* 276–77, 283–84; R. F. Kennedy, Oral History Interview, vol. 7, p. 917, Kennedy Library; Schlesinger, *Robert Kennedy,* 352.

Chapter 4. Keynesian Turn: The Tax Cut

1. Council of Economic Advisers (Gardner Ackley, Kermit Gordon, Walter Heller, Joseph Pechman, Paul Samuelson, James Tobin), Oral History Interview, 12, 40–41, 47, 60, 78–79, 145, 195–96, 309–11; Tobin memorandum, Seymour E. Harris, Oral History Interview, 9–13; memorandum on Newport meeting; all, Kennedy Library; Theodore C. Sorensen, *Kennedy* (New York: Bantam, 1966), 442, 457–58; Arthur M. Schlesinger, Jr., *A Thousand Days* (Boston: Houghton Mifflin, 1965), 621, 654; Paul H. Douglas, *In the Fullness of Time* (New York: Harcourt Brace Jovanovich, 1971), ch. 36; Walter W. Heller, *New Dimensions in Political Economy* (Cambridge: Harvard Univ. Press, 1966), 29; C. Douglas Dillon, Oral History Interview, 7, 14, Johnson Library.

2. Paul A. Samuelson, Prospects and Policies for the 1961 American Economy, Jan. 6, 1961, Pre–Presidential Task Force Reports, Kennedy Library. *Current Biography* has sketches of Heller (1961) and Tobin (1984) as does *Political Profiles: The Kennedy Years* (New York: Facts on File, 1976). Heller to the Files, Oct. 4, 1960, and Recollections of Early Meetings with Kennedy, Feb. 21, 1964, Heller Papers, Kennedy Library. Council of Economic Advisers, Oral History Interview, 148–372 passim, Kennedy Library, Tobin memorandum, Kennedy Library; Edward S. Flash, Jr., *Economic Advice and Presidential Leadership* (New York: Columbia Univ. Press, 1965), 169, 183–84, 188–89; Joint Economic Committee, *Hearings on January 1961 Economic Report of the President,* 87th Cong., 1st sess., has the text of the Council's statement; Joseph Kraft, "Treasury's Dillon—The Conservative Power Center in Washington," *Harper's* (June 1963), 51–56; Schlesinger, *A Thousand Days,* 127–36; Richard H. Rovere, *The American Establishment* (Harcourt, Brace & World, 1962), 6; *Time,* Jan. 2, 1961, p. 15; May 19, 1961, p. 22; Seymour E. Harris, Oral History Interview, 54–56, Kennedy Library; Hearings before the Senate Committee on Finance, *Nominations,* 87th Cong., 1st sess. (1961), 7–25; for sketches of Dillon, Fowler, Roosa, Surrey, and Caplin, see *Political Profiles: The Kennedy Years* (New York: Facts on File, 1976); for Surrey's views of the politics of the income tax, see "The Congress and the Tax Lobbyist—How Special Tax Provisions Get Enacted," *Harvard Law Review,* 70 (May 1957), 1145–82; Fowler to Kennedy, Nov. 13, 1961, Kennedy Library; Dillon to Kennedy, Use of the Recession Argument in Connection with the Tax Bill, July 12, 1963, Departments and Agencies, Treasury, Kennedy Library.

3. *Public Papers, Kennedy, 1961,* 19–20; Council of Economic Advisers, Oral History Interview, 186–410 passim, Kennedy Library; Tobin for the record, Luncheon Conversation with Chairman William McC. Martin on May 29, 1961, May 30, 1961, David Bell Papers, Johnson Library; Dillon to Kennedy, Appointment of Chairman of Federal Reserve Board, Jan. 16, 1963, Departments and Agencies, Treasury, Kennedy Library; *Economic Report of the President, 1962,* 97–107, the balance of payments problem is discussed in ch. 3 of the Council's report; Paul A. Samuelson, Prospects and Policies for the 1961 American Economy, Jan. 6, 1961, p. 3, Tax Force Reports, Kennedy Library; John Kenneth Galbraith, *Ambassador's Journal* (Boston: Houghton Mifflin, 1969), 21, 22. Heller took inadvertent revenge on Galbraith for their disagreement over tax reduction vs. spending. The March 17, 1961, entry in Galbraith's diary, p. 36 read: "I had dinner at The Occidental with Walter Heller. My secretary had booked in the name of Ambassador Galbraith. As a result, the headwaiter swept us through the waiting crowd with fulsome assurances that 'Your table is waiting, Mr. Ambassador.' My pleasure was diluted by the fact that

he automatically assumed Heller to be the Ambassador." Hobart Rowen, *The Free Enterprisers* (New York: Putnam, 1964), 39 for the Reuss quote. Tobin's views were somewhat similar to Galbraith's. "At the Council," he wrote later, "I, at least, would have preferred more expenditures on 'good things' to the tax cut, and I would have preferred lower interest rates to the tax cut. But neither was in the cards. I thought the tax cut was better than doing nothing. That's where I disagreed with Galbraith." Tobin to author, Nov. 30, 1989.

4. *Economic Report of the President, 1962,* 8, 185–90; Council of Economic Advisers, Oral History Interview, 282–83, Kennedy Library; Sorensen, *Kennedy,* 457–58; Schlesinger, *A Thousand Days,* 654; for Galbraith on inflation, see *The Affluent Society* (Boston: Houghton Mifflin, 1958), chs. XV–XVII and *Ambassador's Journal,* 22; W. W. Rostow, *The Diffusion of Power* (New York: Macmillan, 1972), 121–23, 138–41; Rostow to Kennedy, Memorandum of Conversation with Secretary of Labor Goldberg, Feb. 1, 1961, Departments and Agencies, Labor, Kennedy Library; Tobin to author, May 24, 1988; William J. Barber, "The Kennedy Years: Purposeful Pedagogy," in Crauford D. Goodwin, ed., *Exhortation and Controls: The Search for a Wage–Price Policy, 1945–1971* (Washington: Brookings, 1975), 136–38; Lloyd Ulman, "Unions, Economists, Politicians, and Incomes Policy," in *Economics in the Public Service, Papers in Honor of Walter W. Heller,* Joseph A. Pechman and N. J. Simler, eds. (New York: Norton, 1982), 112; Heller to Kennedy, Meeting of the President's Labor-Management Advisory Committee, July 10, 1961, Heller Papers, Kennedy Library; Jack Stieber, "The President's Committee on Labor–Management Policy," *Industrial Relations* (Feb. 1966), 1–19; Heller, *New Dimensions,* 42–47. The U.S. wage-price guideposts and the concurrent emergence of incomes policies in Europe generated an enormous literature. The American experience is well covered in the Brookings studies—Goodwin, *Exhortation and Controls,* cited above, and John Sheehan, *The Wage–Price Guideposts* (Washington: Brookings, 1967). When George Taylor objected to the guideposts on the President's committee, he, for all practical purposes, spoke for the expert industrial relations community. See the devastating criticisms of two noted labor economists—John T. Dunlop, "Guideposts, Wages, and Collective Bargaining," and Arthur M. Ross, "Guideline Policy—Where We Are and How We Got There," in *Guidelines, Informal Controls and the Marketplace,* George P. Shultz and Robert Z. Aliber, eds. (Chicago: Univ. of Chicago Press, 1966), 81–96, 97–141.

5. Bernard D. Nossiter, *The Mythmakers* (Boston: Houghton Mifflin, 1964), 40; Rowen, *Free Enterprisers,* in general on Kennedy's relations with business, quotes at pp. 62, 91, 93; Department of Labor, *Collective Bargaining in the Basic Steel Industry* (Washington: Jan. 1961), in general, quote at p. 31; Gardner C. Means used the data collected by the Kefauver Committee to write his basic study of steel pricing, *Pricing Power and the Public Interest* (New York: Harper, 1962); Schlesinger, *A Thousand Days,* 637, 641; *Public Papers, Kennedy, 1961,* 592–94, 604–5, *1962,* 315–17; Heller to Kennedy, Aug. 2, 1961, Heller Papers, Kennedy Library. There are several blow–by–blow accounts of the Kennedy–Blough confrontation: Roy Hoopes, *The Steel Crisis* (New York: John Day, 1963); Grant McConnell, *Steel and the Presidency* (New York: Norton, 1963); Rowen, *Free Enterprisers,* ch. VI; *New York Times,* Apr. 23, 1962, pp. 1, 25. Sorensen, *Kennedy,* 526, 519; New York Stock Exchange, *The Stock Market Under Stress* (1963); *Report of Special Study of Securities Markets,* House Doc. No. 95, 88th Cong., 1st sess. (1963), pt. 5, p. 207; John Maynard Keynes, *The General Theory of*

Employment, Interest and Money (New York: Macmillan, 1936), 154; Council of Economic Advisers, Oral History Interview, 216, Kennedy Library. Blough set forth his muddied defense in "My Side of the Steel Price Story," *Look,* Jan. 29, 1963, and at greater length in *The Washington Embrace of Business* (New York: distributed by Columbia Univ. Press, 1974). Heller to Kennedy, "A Kennedy Bear Market?," May 3, 1962, Departments and Agencies, Treasury, Kennedy Library; Galbraith to Kennedy, The Stock Market, May 29, 1962, Galbraith Papers, Kennedy Library.

6. Schlesinger, *A Thousand Days,* 644–48; John Kenneth Galbraith, *The New Industrial State* (Boston: Houghton Mifflin, 1967); Notes on JFK's Yale Commencement Speech in June 1962, Heller Papers, Kennedy Library (while this memorandum has neither author nor date, internal evidence leaves no doubt that Heller wrote it and did so no earlier than 1965); *Public Papers, Kennedy, 1962,* 470–75. Galbraith applauded the President's attack on the balanced budget at Yale. He had written to Dillon and Heller: "You have now promised a balanced budget for next year although there is little chance that in the end it will be balanced. Therefore, though there is a very good chance you will have continued recovery and continued reduction in unemployment, improvement in balance of payments and stable prices, it will still be possible to say that you have failed. You are so bent on your discredit that you plan for it. I am reminded of the courtesan whose conquests have made her the cynosure of all men and the envy of all women and who at any critical moment in the conversation insists on the absolute importance of chastity." Galbraith to Dillon and Heller, Reflections on Economic Policy, Public Posture and the Will to Put the Worst Foot Forward, Galbraith Papers, Kennedy Library.

7. *Public Papers, Kennedy, 1962,* 73–92, 229, 611–17; Council of Economic Advisers, Oral History Interview, 427, 449–52, Kennedy Library; Administrative History of the Department of the Treasury, 47–48, Johnson Library. The Council's analysis of national budgeting is set out in Heller, *New Dimensions,* 62–67, and the Administrative History of the Council of Economic Advisers, ch. 1, p. 16, ch. 2, pp. 1–4, Johnson Library. Galbraith, *The Affluent Society,* 252, 261; Galbraith to Kennedy, July 10, Aug. 20, 1962, Galbraith Papers, Kennedy Library; Schlesinger, *A Thousand Days,* 649; Galbraith, *Ambassador's Journal,* 331; Dillon to Kennedy with attached Harris memorandum, June 7; Dillon to Kennedy, July 12; Dillon to Kennedy, Visit with New York Bankers, July 27, 1962; Schlesinger to Kennedy, Tax Cut, July 17, 1962; all, Departments and Agencies, Treasury, Kennedy Library; Heller to Kennedy, Recap of Issues on Tax Cuts (and the Galbraithian Alternative), Dec. 12, 1962; Sorensen, Tax Cut, July 12, 1962; Samuelson and Solow to Kennedy, That Second Look at Need for a Tax Cut, July 13, 1962; all, Sorensen Papers, Kennedy Library; Samuelson and Solow to Kennedy, The Changed Mid-Year Outlook, June 6, 1962; Report to the President from the Cabinet Committee on Economic Growth, with covering letter, Dec. 1, 1962, Heller Papers, Kennedy Library; Flash, *Economic Advice,* 230–69; Wilbur Mills, Oral History Interview, 9–11, Kennedy Library; John F. Manley, *The Politics of Finance: The House Committee on Ways and Means* (Boston: Little, Brown, 1970), 346; Sorensen, *Kennedy,* 478–79; *State of the Economy and Policies for Full Employment,* Hearings before the Joint Economic Committee, 87th Cong., 2d sess. (1962), 115, 664–65, 683; Bernard D. Nossiter, "The Day Taxes Weren't Cut," *Reporter* (Sept. 13, 1962), 25–28; Presidential Recordings Transcripts, Tax Cut Proposals, vol. I, No. 7; JFK, Wilbur Mills, Aug. 6, 1962, vol. III, Kennedy Library; Fowler to Kennedy, Conference with Chairman Wilbur Mills re Tax Program, Nov.

15, 1962, Departments and Agencies, Treasury, Kennedy Library; Rowen, *Free Enterprisers,* 50, 53, 237; Philip M. Stern, *The Great Treasury Raid* (New York: Random House, 1962), xvii; Graham to Dillon, Vice Presidential Papers of LBJ, Civil Rights (misfiled), Nov. 29, 1962, Johnson Library; Joseph A. Pechman, "The Case for Tax Reform," *Reporter* (June 9, 1963), 20.

8. The legislative history of the Revenue Act of 1963 is set forth in *Congressional Quarterly Almanac* (1963), 470–99, and in the *Annual Report of the Secretary of the Treasury* for fiscal 1963, pp. 40–48, 293–331. Manley, *Politics of Finance,* 111–12; Paul H. Douglas, "The Problem of Tax Loopholes," *American Scholar* (Winter, 1967–68), 40; *President's 1963 Tax Message,* Hearings before the Committee on Ways and Means, H.R., 88th Cong., 1st sess. (1963); *Revenue Act of 1963,* Report of the Committee on Ways and Means, House Rep. No. 749, 88th Cong., 1st sess. (Sept. 13, 1963); Fowler, Organizing Procedure for the Business and Finance Committee for Tax Reduction in 1963, Apr. 1, 1963, Fowler to Kennedy, n.d., Departments and Agencies, Treasury; Heller to Kennedy, June 7, 1963, Heller Papers; Presidential Recordings Transcripts, Tax Cut Proposals, Vol. I, No. 7, Martha W. Griffiths, Aug. 7, 1962; all, Kennedy Library.

Chapter 5. Wrestling with Structural Unemployment

1. William Glazier, "Automation and Joblessness," *Atlantic* (Aug. 1962), 43; Robert Aaron Gordon, *The Goal of Full Employment* (New York: Wiley, 1967), 57–58; James J. Healy, ed., *Creative Collective Bargaining* (Englewood Cliffs: Prentice-Hall, 1965), 91, 135–65, 220, 244–81; Lincoln Fairley, *Facing Mechanization: The West Coast Longshore Plan,* Monograph Series 23 (Institute of Industrial Relations, UCLA, 1979), chs. II, III, V; Otto Hagel and Louis Goldblatt, *Men and Machines* (San Francisco: ILWU and PMA, 1963); *Progress Report Automation Committee,* formed under agreements of Sept. 1, 1959, between Armour and Company and United Packinghouse, Food and Allied Workers and Amalgamated Meat Cutters and Butcher Workmen; George P. Shultz and Arnold R. Weber, *Strategies for the Displaced Worker* (New York: Harper & Row, 1966); Thomas Kennedy, *Automation Funds and Displaced Workers* (Boston: Graduate School of Business Administration, Harvard Univ., 1962), ch. VI.

2. Paul H. Douglas, *In the Fullness of Time* (New York: Harcourt Brace Jovanovich, 1971), chs. 1, 23, 36. The legislative history is set forth at length in Roger H. Davidson, *Coalition-Building for Depressed Areas Bills, 1955–1965,* Inter-University Case Program #103 (Indianapolis: Bobbs-Merrill, 1966), and more briefly in Sar A. Levitan, *Federal Aid to Depressed Areas* (Baltimore: Johns Hopkins Univ. Press, 1964), ch. 1. Kennedy Task Force on Area Redevelopment, Dec. 27, 1960, Pre-Presidential Task Force Reports, Kennedy Library; *Public Papers, Kennedy, 1961,* 7; William L. Batt, Oral History Interview, 24–25, 37–38, 40, 42–43, Kennedy Library; Myer Feldman, Oral History Interview, vol. 13, p. 10, Kennedy Library; Area Redevelopment Act, *Statutes at Large, 1961,* 47–63; "The Kennedy Program of Aid for Depressed Areas," *Congressional Digest* (Apr. 1961), 99 ff.; *Area Redevelopment,* Hearings before the Subcommittee on Labor, 84th Cong., 2d sess. (1956), pt. 2, pp. 889–90; Paul H. Douglas, *Economy in the National Government* (Chicago: Univ. of Chicago Press, 1952). When Hodges tried to persuade Douglas to put ARA in Commerce, the latter told him that when he had been governor of North Carolina, "You stole all the

plants from the West and Northeast, and everywhere else, and brought them to North Carolina." Luther H. Hodges, Oral History Interview, 5, Kennedy Library.

3. The only serious study of ARA, unfortunately early, is Levitan's helpful *Federal Aid to Depressed Areas,* quotes at pp. 52, 64. See also William L. Batt, Oral History Interview, 44–182 passim, Kennedy Library; James L. Sundquist, *Politics and Policy* (Washington: Brookings, 1968), 92–109; *Area Redevelopment,* Hearings, 1074; Luther H. Hodges, Oral History Interview, 6, 48–58, Kennedy Library; ARA Annual Report, 1964, 46–89; *Public Papers, Kennedy, 1962,* 143–44, 267–69, 379–83; ARA Annual and Final Report, 1965, 13; ARA, Appalachia, Opportunity for Progress, May 1963; The ARA Vote on June 12, 1963, Resulted in the Following Breakdown, n.d.; Summary of ARA Training Programs, Nov. 1961–June 30, 1965; Schultze to O'Brien, Effect of No Additional Authorization for ARA, May 20, 1964; all, Batt Papers, Kennedy Library; Department of Commerce, First Report of the National Public Advisory Committee, Area Redevelopment, Three Years in Review (Dec. 1964), 13; Sar A. Levitan, "Area Redevelopment: An Analysis of the Program," *Industrial Relations* (May 1964), 83–84, 95; William H. Miernyk, "Area Redevelopment," in Joseph M. Becker, ed., *In Aid of the Unemployed* (Baltimore: Johns Hopkins Univ. Press, 1965), 170; Sar A. Levitan and Joyce K. Zickler, *Too Little But Not Too Late* (Lexington: Lexington Books, 1976); Douglas, *Fullness of Time,* 521.

4. There are sketches of Joseph S. Clark, Jr., in *Current Biography* (1952), 107–9, and in *Political Profiles: The Kennedy Years* (New York: Facts on File, 1976), 84–85. *Annals of the American Academy of Political and Social Science* (Sept. 1959), x; Joseph S. Clark, Jr., Oral History Interview, 18–19, 66–67, Kennedy Library; James L. Sundquist to author, Oct. 31, 1989. *Current Biography* has profiles of Arthur J. Goldberg (1961), 178–80, and Willard Wirtz (1963), 474–76. Arthur J. Goldberg, *AFL-CIO, Labor United* (New York: McGraw-Hill, 1956); Nicholas Katzenbach, Oral History Interview, 73, Kennedy Library; *Report of the Special Committee on Unemployment Problems,* Sen. Report No. 1206, 86th Cong., 2d sess. (1960), 112–13; Seymour Wolfbein, Oral History Interview, 2–6, Kennedy Library; Garth L. Mangum, *MDTA, Foundation of Federal Manpower Policy* (Baltimore: Johns Hopkins Univ. Press, 1968), 12–17; *Public Papers, Kennedy, 1961,* 110–11; Report of the Panel of Consultants on Vocational Education, *Education for a Changing World of Work* (Washington: HEW, 1963), ch. 9; *Training of the Unemployed,* Hearings before Subcommittee on Employment and Manpower, Sen., 87th Cong., 1st sess. (1961); *Manpower Utilization and Training,* Hearings before Subcommittee on Unemployment and the Impact of Automation, 87th Cong., 1st sess. (1961); *MDTA of 1961,* Sen. Report No. 651, 87th Cong., 1st sess., July 31, 1961; *MDTA of 1961,* House Report No. 879, 87th Cong., 1st sess., Aug. 10, 1961; *Congressional Quarterly Almanac, 1961,* 77; *1962,* 513; Sundquist, *Politics and Policy,* 89–91.

5. Wolfbein, Oral History Interview, 7–9, 21, 23, 26–27, 33, Kennedy Library; Mangum, *MDTA,* 22–26, 44–47, 49–51, 56–64, 81–86, 97–99; *Manpower Report of the President,* March 1963, p. 195, March 1965, p. 251; The Department of Labor during the Administration of President Lyndon B. Johnson, vol. II, pt. I, 116–17, 123–24, Johnson Library; Kennedy to Wirtz and Celebrezze, June 4, 1963; Wirtz to Kennedy, June 10, 20, 1963; Records of the Secretary of Labor, W. W. Wirtz; all, National Archives; *Public Papers, Kennedy, 1963,* 483, 489; James Tobin, Luncheon Conversation with Chairman William McC. Martin on May 29, 30, 1961, David Bell Papers, Kennedy Library; Sar A. Levitan and Garth L. Mangum, *Federal Training and Work*

Programs in the Sixties (Univ. of Michigan—Wayne State, Institute of Labor and Industrial Relations, 1969), 25–26; Gerald G. Somers, "Our Experience with Retraining and Relocation," in R. A. Gordon, ed., *Toward a Manpower Policy* (New York: Wiley, 1967), 217. For an analysis of the many advantages of OJT over institutional training, see Somers, "Our Experience," 227–30. Gerald G. Somers, *Retraining the Unemployed* (Madison: Univ. of Wisconsin Press, 1968), 7–8; Margaret S. Gordon, "Retraining Programs—At Home and Abroad," *Proceedings*, Industrial Relations Research Association (1964), 138.

Chapter 6. Updating the New Deal

1. *Minimum Wage–Hour Legislation*, Hearings before the Subcommittee on Labor Standards, 86th Cong., 2d sess. (1960), 300; *Public Papers, Kennedy, 1961*, 186.

2. On the origins of the Fair Labor Standards Act, see Irving Bernstein, *A Caring Society* (Boston: Houghton Mifflin, 1985), ch. 5. For Kennedy's move in the late fifties to shed his "conservative" image and become a "liberal," see Myer Feldman, Oral History Interview, 23–39, Kennedy Library. *Minimum Wage-Hour Legislation*, 315, 600, 824; Joseph S. Clark, Jr., Oral History Interview, 29–30, Kennedy Library; *Congressional Quarterly Almanac, 1960*, 79, 309–19; *Public Papers, Kennedy, 1961*, 49, 66, 587; Goldberg to Kennedy, Feb. 6, 1961, Summary of Minimum Wage Bill (H.R. 3935) as Agreed to in Conference, n.d., Records of the Secretary of Labor, A. J. Goldberg, National Archives; *To Amend the Fair Labor Standards Act*, Hearings before Subcommittee on Labor, H.R., 87th Cong., 1st sess. (1961), 8, 178–91, 200, 224–27, 277, 412; *Amendments to the Fair Labor Standards Act*, Hearings before the Subcommittee on Labor, Sen., 87th Cong., 1st sess. (1961), 35–38; *Fair Labor Standards Act Amendments of 1961*, Report from the Committee on Labor and Public Welfare, 87th Cong., 1st sess., Rep. No. 145 (Apr. 12, 1961). For a sophisticated analysis of disemployment under the minimum wage, see Richard A. Lester, "Employment Effects of Minimum Wages," *Industrial and Labor Relations Review* (Jan. 1960), 254–64. *Congressional Quarterly Almanac*, 1961, 471–82; Bureau of National Affairs, *The New Wage and Hour Law* (Washington: BNA, 1961).

3. Bernstein, *A Caring Society*, 290–92; Margaret Price, Survey of Major Presidential Appointments of Women to Positions in Government Service, Dec. 8, 1960, Pre-Presidential Task Force Reports, Kennedy Library; sketch of Esther Peterson in *Current Biography, 1961*, 358–60; Goldberg to Kennedy, April 24, 1961; Records of the Secretary of Labor, A. J. Goldberg, National Archives; Esther Peterson, Oral History Interview, 3, 10, 23, 31, 33, 46, 48, 53, 54, 57, Kennedy Library; Patricia G. Zelman, *Women, Work, and National Policy, the Kennedy-Johnson Years* (Ann Arbor: UMI Research Press, 1980), ch. 2; Cynthia E. Harrison, "A New Frontier for Women: The Public Policy of the Kennedy Administration," *Journal of American History* (Dec. 1980), 630–46; *American Women, The Report of the President's Commission on the Status of Women and Other Publications* (New York: Scribner, 1965), quotes at pp. 4–5, 16, Commission's recommendations at pp. 210–13; Cynthia Harrison, *On Account of Sex: The Politics of Women's Issues, 1945–1968* (Berkeley: Univ. of California Press, 1988), xiv, 113, 163; *Equal Pay for Equal Work*, Hearings before the House Select Subcommittee on Labor, 87th Cong., 2d sess. (1962), 44–56, 58, 66; Walter Fogel, *The Equal Pay Act* (New York: Praeger, 1984), 120; *Congressional Quarterly Almanac, 1962*, 519–20; *1963*, 511–13; *Equal Pay for Equal Work* (Washington: BNA, 1963), text of EPA at p. 73.

4. Irving Bernstein, *The New Deal Collective Bargaining Policy* (Berkeley: Univ. of California Press, 1950); Kurt L. Hanslowe, *The Emerging Law of Labor Relations in Public Employment* (Ithaca: New York State School of Industrial and Labor Relations, 1967), 14–15, 91–103; President's Task Force on Employee-Management Relations in the Federal Service, *A Policy for Employee-Management Cooperation in the Federal Service,* reprinted in Harold S. Roberts, *Labor-Management Relations in the Public Service* (Honolulu: Univ. of Hawaii, Industrial Relations Center, 1968), 8–9; Murray B. Nesbitt, *Labor Relations in Federal Government Service* (Washington: BNA, 1976), 307–14, 317; Michael L. Brookshire and Michael D. Rogers, *Collective Bargaining in Public Employment, The TVA Experience* (Lexington: Lexington Books, 1977), chs. 2, 3; Daniel H. Kruger, "Trends in Public Employment," *Proceedings,* Industrial Relations Research Association (1961), 354–66; City of New York, Department of Labor, *The Right of Public Employees to Organize—In Theory and Practice* (1955), 5; Eli Rock, "The Appropriate Unit Question in the Public Service, the Problem of Proliferation," *Michigan Law Review* (Mar. 1969), 1001, 1002, 1005; Lennox L. Moak, "The Philadelphia Experience," *Proceedings,* Academy of Political Science (1976), vol. 30, pt. 2, pp. 124–33; Harriet F. Berger, "The Grievance Process in the Philadelphia Public Service," *Industrial and Labor Relations Review* (July 1960), 568–80; Ida Klaus, "Labor Relations in the Public Service: Exploration and Experiment," *Syracuse Law Review* (Spring 1959), 197; Raymond D. Horton, *Municipal Labor Relations in New York City* (New York: Praeger, 1973), chs. 2–4, quote at p. 32; the text of Mayor Wagner's Executive Order No. 49 is in *Industrial and Labor Relations Review* (July 1959), 618–25; Jean T. McKelvey, "The Role of State Agencies in Public Employee Labor Relations," *Industrial and Labor Relations Review* (Jan. 1967), 183; "Municipal Employment in Wisconsin," *Wisconsin Law Review* (Summer 1965), 652.

5. Frederick G. Dutton, Memorandum for Arthur Goldberg, April 29, 1961, with attached documents, including Department of Defense objections to the Rhodes–Johnston bill, Feldman Papers, Kennedy Library; Nesbitt, *Labor Relations in the Federal Service,* 17–19, 210–12; Memorandum on Employee–Management Relations in the Federal Service, June 22, 1961, *Public Papers, Kennedy, 1961,* 469–70; Wilson R. Hart, "The U.S. Civil Service Learns to Live with Executive Order 10988: An Interim Appraisal," *Industrial and Labor Relations Review* (Jan. 1964), 206, 219 n. 43; President's Task Force, *Employee-Management Cooperation,* 2, 3, 5, 7, 8, 20–21; Roberts, *Labor-Management Relations in the Public Service,* text of Executive Order 10988 at pp. 34–41, text of Standards of Conduct and Code of Fair Labor Practices at pp. 53–60, other citations at pp. 315, 517, 548–49; Michael H. Moskow, J. Joseph Loewenberg, and E. C. Koziara, *Collective Bargaining in Public Employment* (New York: Random House, 1970), 52–56; B. V. H. Schneider, "Public Sector Labor Legislation—An Evolutionary Analysis," in Benjamin Aaron, Joyce M. Najita, and James L. Stern, eds., *Public–Sector Bargaining,* 2d ed. (Washington: Industrial Relations Research Association, 1988), 195–96.

Chapter 7. Federal Aid for Education

1. Hugh Douglas Price, "Race, Religion, and the Rules Committee," in Alan F. Westin, *The Uses of Power* (New York: Harcourt, Brace & World, 1962), 3, 10–11, 13; *Congressional Quarterly Almanac, 1961,* 210–11; Hugh Davis Graham, *The Uncertain Triumph: Federal Education Policy in the Kennedy and Johnson Years* (Chapel Hill: Univ.

of North Carolina Press, 1984), xvii–xviii, 5–10; Frank J. Munger and Richard F. Fenno, Jr., *National Politics and Federal Aid to Education* (Syracuse: Univ. of Syracuse Press, 1962), quotes at pp. 69, 104, Taft quote at 138–39, also 56–61, 66–69, 150–52, ch. VI; Paul G. Kauper, "Church and State: Cooperative Separatism," *Michigan Law Review* (Nov. 1961), 1, 39; *Cantwell v. Connecticut,* 310 U.S. 296 (1940); *Pierce v. Society of Sisters of Holy Names,* 268 U.S. 510 (1925); *Everson v. Board of Education,* 330 U.S. 1 (1947); Edward S. Corwin, ed., *The Constitution of the United States of America* (Washington: Government Printing Office, 1953), 757–68; Robert F. Drinan, S. J., *Religion, the Courts, and Public Policy* (New York: McGraw–Hill, 1963), 133; *Brown v. Board of Education,* 347 U.S. 483 (1954); Theodore C. Sorensen, *Kennedy* (New York: Bantam, 1966), 401; Katzenbach Oral History, 54, Kennedy Library.

2. Report of the Task Force on Education, Jan. 2, 1961, Pre-Presidential Transition Files, Kennedy Library; Graham, *Uncertain Triumph,* 11–18; Price, "Race, Religion, and the Rules Committee," 21–27, 51, 59–70, Spellman quote at p. 23; D. B. Hardeman and Donald C. Bacon, *Rayburn: A Biography* (Austin: Univ. of Texas Monthly Press, 1987), 444–45, 473; Francis Keppel, Oral History Interview, 2–4, Kennedy Library; Wilbur J. Cohen, Oral History Interview, 61, 69, 74, Kennedy Library; Neil MacNeil, *Forge of Democracy: The House of Representatives* (New York: David McKay, 1963), ch. 15; James A. Robinson, *The House Rules Committee* (Indianapolis: Bobbs-Merrill, 1963), 71–80; William R. MacKaye, *A New Coalition Takes Control: The House Rules Committee Fight of 1961* (Eagleton Institute, No. 29, 1963). In his address to the Ministerial Association, Kennedy had said that he would not hesitate as President to attend a Protestant service. When Sam Rayburn died in November 1961, Kennedy flew to Bonham, Texas, for the funeral at the First Baptist Church. Sorensen, *Kennedy,* 408, also 49–50, O'Brien quote at p. 405, Kennedy Gridiron Club quote at pp. 406–7; *Congressional Quarterly Almanac, 1961,* 221, 227; *Public Papers, Kennedy, 1961,* 107–11, 154–56. There is a sketch of Wayne Morse in Lawrence Key Pettit, "The Policy Process in Congress: Passing the Higher Education Facilities Act of 1963" (unpublished Ph. D. dissertation, Univ. of Wisconsin, 1965), 159–64. See also A. Robert Smith, *The Tiger in the Senate: A Biography of Senator Wayne Morse* (Garden City: Doubleday, 1962), quote at p. 445; *Public School Assistance Act of 1961,* Hearings before the Senate Subcommittee on Education, 87th Cong., 1st sess., pt. 1, pp. 97–98, the HEW brief on constitutionality is at pp. 110–38. The administration's NDEA proposal to the NCWC on parochial schools is treated in Sorensen, *Kennedy,* 403–4, and Graham, *Uncertain Triumph,* 21–22. *Federal Aid to Schools,* Hearings before the House Subcommittee on Education, 87th Cong., 1st sess. (1961); Ribicoff to Kennedy, July 20, 1961; Cohen to Sorensen, Aug. 7, 1961; Sorensen to Kennedy, Aug. 14, 1961; all, Sorensen Papers, Kennedy Library.

3. Ribicoff to Kennedy, Oct. 6, 1961, HEW Papers, Kennedy Library; *Public Papers, Kennedy, 1962,* 9, 110–17; Cohen, Oral History Interview, 68, Kennedy Library; *Engel v. Vitale,* 370 U.S. 421 (1962); Pettit, "Policy Process in Congress," 73–94; Graham, *Uncertain Triumph,* 28–44; *Congressional Quarterly Alamanac, 1962,* 231–40; Keppel, Oral History Interview, 6, Kennedy Library.

4. James L. Sundquist, *Politics and Policy* (Washington: Brookings, 1968), 206; Keppel, Oral History Interview, 4–10, 13, 14–15, Kennedy Library; Dungan to Kennedy, Nov. 19, 1962, HEW Papers, Kennedy Library. There are sketches of Francis Keppel in *Current Biography, 1963,* 220–22, and in *Political Profiles: The Johnson Years* (New York: Facts on File, 1976), 327–28. *Public Papers, Kennedy, 1963,*

489–90. The two Oregonians who unerringly steered the higher education bill through Congress, Wayne Morse and Edith Green, were a crusty pair. Mrs. Green needed no women's rights movement to lean on and she, evidently, terrified the Kennedy administration. The President found her very difficult on the question of breaking up the package and avoided dealing with her directly. At one meeting, Keppel recalled, he said, "Well, now, Cohen, you haven't succeeded with her and Sorensen hasn't, Keppel is a new boy in town, and it's his turn." Pettit, "Policy Process in Congress," ch. VI; H.R., Committee on Education and Labor, 88th Cong., 1st sess., Charles A. Quattlebaum, *Federal Legislation Concerning Education and Training* (June 1964), this report by the Legislative Reference Service neatly summarizes all of the 1963 education and training laws; Graham, *Uncertain Triumph,* 39–52; *National Education Improvement Act,* Hearings before the House Committee on Education and Labor, 88th Cong., 1st sess. (1963); *Education Legislation—1963,* Hearings before the Senate Subcommittee on Labor and Public Welfare, 88th Cong., 1st sess., 7 vols. (1963); *Congressional Quarterly Almanac, 1963,* 188–207; Cohen, Oral History Interview, 70, Kennedy Library. The summary treatment of the lesser education legislation enacted in 1963 is dictated solely by concern over space. It in no way suggests that these laws were unimportant. For example: The distinguished psychiatrist Dr. Milton Greenblatt of the UCLA Medical School was the director of mental health for the state of Massachusetts at the time. In the late fifties, experts concluded that the traditional large state mental hospital should be abolished and replaced by small local community facilities. The states were unable to finance this program and federal aid was indispensable. Thus, there was need for the Mental Health Facilities and Community Mental Health Centers Construction Act. According to Dr. Greenblatt, Kennedy was the first President to recognize the importance of mental health and this law helped fundamentally to modernize the nation's mental health system.

5. Keppel, Oral History Interview, 7, 9, 11–12, 18–19, Kennedy Library.

Chapter 8. A Battle Lost: Medicare

1. The early history of Medicare is set forth in Richard Harris, *A Sacred Trust* (New York: New American Library, 1966), chs. 2–20; Sheri I. David, *With Dignity: The Search for Medicare and Medicaid* (Westport: Greenwood, 1985), chs. 1–3; Theodore R Marmor, *The Politics of Medicare* (Chicago: Aldine, 1970), chs. 1–2; Peter A. Corning, *The Evolution of Medicare . . . from Idea to Law* (Washington: Social Security Administration, 1969), chs. 1–4. See also Paul H. Douglas, *In the Fullness of Time* (New York: Harcourt Brace Jovanovich, 1972), 392; Wilbur J. Cohen, Oral History Interview, 34–35, Johnson Library; Martha Derthick, *Policymaking for Social Security* (Washington: Brookings 1979), 326; *Robert Kennedy in His Own Words,* Edwin O. Guthman and Jeffrey Shulman, eds. (New York: Bantam, 1988), 51, 370; Herbert S. Parmet, *JFK: The Presidency of John F. Kennedy* (New York: Dial, 1983), 208.

2. John F. Manley, *The Politics of Finance: The House Committee on Ways and Means* (Boston: Little, Brown, 1970), 27–29, 38, 214–15; Marmor, *Politics of Medicare* 39–49; Clark to Kennedy, Dec. 15, 1960, Legislative Files, Kennedy Library; Harris, *A Sacred Trust,* chs. 22–24; Cohen, Oral History Interview, 40, Kennedy Library; Health and Social Security for the American People, Pre-Presidential Task Force Reports, 2–7, Kennedy Library; *Health Services for the Aged under the Social Security System,* H.R., Hearings before the Committee on Ways and Means, 87th Cong., 1st sess. (1961), 4 vols.; *Public Papers, Kennedy, 1961,* 77–83, 585–86; *1963,* 191–94; Cohen to Feldman,

Discussions with Wilbur Mills, Jan. 4, 1962, Feldman Papers, Kennedy Library. The dismal history of Medicare in 1962–63 is most carefully treated by David, *With Dignity*, ch. 5. She provides the account of the Kerr-Baker manipulation of the Senate vote, pp. 77–83. See also *Congressional Quarterly Almanac, 1962*, 193–94, 670. Lawrence F. O'Brien, *No Final Victories* (Garden City: Doubleday, 1974), 143. The Senate vote in 1962 was extremely interesting. According to James L. Sundquist, "Few votes had been changed as the result of the two years of some of the most intensive lobbying ever carried out on any measure. When the five unrecorded senators of 1960 are allocated according to their announced or probable positions, the division that year was 53–47. Four Republicans—Javits and Kenneth Keating of New York, John Sherman Cooper of Kentucky, and Thomas Kuchel of California—switched to the affirmative in 1962. But one vote was lost when Republican John Tower replaced Lyndon Johnson as a senator from Texas, and Randolph and Hayden accounted for two more losses. The net gain, therefore, was only one." *Politics and Policy* (Washington: Brookings, 1968), 314 n. 81, 315. For Wilbur Cohen on Wilbur Mills, see Cohen, Oral History Interview, 44–48, Johnson Library.

Chapter 9. The Peace Corps

1. The Peace Corps issue in the 1960 election is treated at pp. 36–37.

2. Harris Wofford, *Of Kennedys and Kings* (New York: Farrar, Straus, Giroux, 1980), 95, 96, 250–55, 267, 268, 281; *Public Papers, Kennedy, 1961*, 1, 3, 134–46; Samuel P. Hayes, *An International Peace Corps: The Promise and Problems* (Washington: Public Affairs, 1961); Report to the President on the Peace Corps from Sargent Shriver, Feb. 22, 1961, Peace Corps, Kennedy Library; Gerard T. Rice, *The Bold Experiment: JFK's Peace Corps* (Notre Dame: Notre Dame Univ. Press, 1985), 34–44, 47–48, 51–73; Shriver to Kennedy, Feb. 22, 1961; Josephson to Goodwin, March 3, 1961; Shriver to "Dear Jack," n.d., Wofford to Kennedy, May 25, 1961; all, Departments and Agencies, Peace Corps, Kennedy Library; *Code of Federal Regulations, 1959–1963*, 447–48. The sketch of Shriver is based on Robert A. Liston, *Sargent Shriver: A Candid Portrait* (New York: Farrar, Straus, 1964). Arthur M. Schlesinger, Jr., *A Thousand Days* (Boston: Houghton Mifflin, 1965), 606; Kevin Lowther and C. Payne Lucas, *Keeping Kennedy's Promise* (Boulder: Westview, 1978), 3; Robert G. Carey, *The Peace Corps* (New York: Praeger, 1970), viii–ix; Brent Ashabranner, *A Moment in History* (Garden City: Doubleday, 1971), 23; William Josephson, Oral History Interview, 3, 6, 9, 15, 26, 41–42, 45, 46, 58, Kennedy Library; John Kenneth Galbraith, *Ambassador's Journal* (Boston: Houghton Mifflin, 1969), 83–84; *Congressional Quarterly Almanac, 1961*), 324, 326, 327; *The Peace Corps*, Hearings before the Senate Committee on Foreign Relations, 87th Cong., 1st sess. (1961); *The Peace Corps*, Hearings before the House Committee on Foreign Affairs, 87th Cong., 1st sess. (1961).

3. Carey, *The Peace Corps*, 22, 57; Moritz Thomsen, *Living Poor* (Seattle: Univ. of Washington Press, 1969), 4, 6, 7; Rice, *Bold Experiment*, chs. 7–9, quote at p. 153; Sargent Shriver, *Point of the Lance* (New York: Harper & Row, 1964), 74–75, App. B. The Stein quote is in David Hapgood, *Agents of Change* (Boston: Little, Brown, 1968), 20; Robert B. Textor, *Cultural Frontiers of the Peace Corps* (Cambridge: MIT Press, 1966), xii–xiii; Lillian Carter and Gloria Carter Spann, *Away from Home* (New York: Simon and Schuster, 1977), 51, 119–20; Lawrence H. Fuchs, *"Those Peculiar Americans": The Peace Corps and American National Character* (New York: Meredith, 1967),

quotes at pp. 5, 9, 11, 50, 110, 139–40, 211, 212, 232. For confirmation of the Fuchs's analysis see Textor, *Cultural Frontiers* chs. 2 and 3.

4. Ashabranner, *A Moment in History,* 1, 119, 139, 141; Humphrey to Shriver, March 28, 1962, Departments and Agencies, Peace Corps, Kennedy Library; Hapgood, *Agents of Change,* 3, 5, 12, 226–28, Peggy Anderson quote at 16–17; Textor, *Cultural Frontiers of the Peace Corps,* ix–x; Josephson, Oral History Interview, 37, 56, Kennedy Library; Carter, *Away from Home,* 1, 63, 65, 117; Rice, *Bold Experiment,* Nehru quote at 291, Tsongas quote at 295, volunteer quote at 299; Carey, *The Peace Corps,* 216–25, Theodore C. Sorensen, *Kennedy* (New York: Bantam, 1965), 599.

Chapter 10. If Men Were Angels

1. *The Federalist Papers,* Roy F. Fairchild, ed., 2d ed. (Baltimore: Johns Hopkins Univ. Press, 1966), 106; Richard E. Neustadt, *Presidential Power: The Politics of Leadership* (New York: Wiley, 1960), 205; Arthur M. Schlesinger, Jr., *The Imperial Presidency* (Boston: Houghton Mifflin, 1973), 171; Clark Clifford quote in Emmett John Hughes, *The Living Presidency* (New York: Coward, McCann, Geoghegan, 1972), 316; Kennedy's speech to the National Press Club is in Arthur M. Schlesinger, Jr., Fred L. Israel, and William P. Hansen, eds., *History of American Presidential Elections, 1789–1968* (New York: Chelsea House, 1985), vol. IX, pp. 3536–40; Lawrence F. O'Brien, *No Final Victories* (Garden City: Doubleday, 1974), 17, 26, 104–23, 124–25; *Robert Kennedy in His Own Words,* Edwin O. Guthman and Jeffrey Shulman, eds. (New York: Bantam, 1988), 52; Theodore C. Sorensen, *The Kennedy Legacy* (New York: Macmillan, 1969), 38; *New York Times,* Jan. 15, 1960; Theodore C. Sorensen, *Kennedy* (New York: Bantam, 1966), 381, 388; *Public Papers, Kennedy, 1961,* App. B; *1962,* App. B.

2. Arthur M. Schlesinger, "Tides of American Politics," *Yale Review* (Winter 1940), 217–30; Arthur M. Schlesinger, Jr., *The Cycles of American History* (Boston: Houghton Mifflin, 1986), 27, 28–29, 32, 34; George C. Edwards III, *Presidential Influence in Congress* (San Francisco: Freeman, 1980), 67–69; Neustadt, *Presidential Power,* 181; Howard E. Shuman, Oral History Interview, 247, Senate Historical Office; the 1960 Democratic platform is in *History of American Presidential Elections, 1789–1968,* vol. IX, pp. 3471–510; O'Brien, *No Final Victories,* 104, 128–30, 137–38, 142–43.

3. Neustadt, *Presidential Power,* 198–99; James L. Sundquist, *Politics and Policy* (Washington: Brookings, 1968), 471–89. Both Schlesinger, *A Thousand Days,* and Sorensen, *Kennedy,* treat foreign policy comprehensively. Sorensen quotes at pp. 667, 740. Useful specialized studies are Haynes Johnson, *The Bay of Pigs* (London: Hutchinson, 1965), Elie Abel, *The Missile Crisis* (Philadelphia: Lippincott, 1966), and Glenn T. Seaborg, *Kennedy, Khrushchev and the Test Ban* (Berkeley: Univ. of California Press, 1981), quote at pp. xiii–xiv. See also *The Gallup Poll* (New York: Random House, 1972), v. 3, pp. 1707, 1717, 1786, 1793; *Public Papers, Kennedy, 1961,* 444, 534; *1962,* 806–8; *1963,* 463–64; Harold Macmillan, *At the End of the Day, 1961–1963* (New York: Harper & Row, 1973), 219.

4. William S. Livingston, Lawrence C. Dodd, and Richard L. Schott, eds., *The Presidency and the Congress* (Austin: L. B. Johnson School of Public Affairs, 1979), 303–4; Edwards, *Presidential Influence in Congress,* 14, 19, 24, 190–93, 200. Russell D. Renka made a study similar to Edwards's comparing JFK and LBJ. With *Congressional Quarterly* data on roll–call votes in both houses of Congress, he showed Kennedy's average annual success ratio at 84.5 percent, Johnson's at 82.3. Renka gives little

weight to congressional skill as a determining factor. "The skill hypothesis prescribed higher Johnson scores. Even a quick scan of the data quickly reveals that this is not the case." JFK even got more southern support than LBJ. "Comparing Presidents Kennedy and Johnson as Legislative Leaders," *Presidential Studies Quarterly*, 15 (Fall 1985), 814, 816, 817. Charles and Barbara Whalen, *The Longest Debate: A Legislative History of the 1964 Civil Rights Act* (Cabin John: Seven Locks, 1985), 232; Herbert Stein, *The Fiscal Revolution in America* (Chicago: Univ. of Chicago Press, 1969). For a view similar to Stein's, see James Tobin and Murray Weidenbaum, eds., *Two Revolutions in Economic Policy* (Cambridge: MIT Press, 1988). *Public Papers, Kennedy, 1963*, 845, 849; O'Brien, *No Final Victories*, 143. According to the Gallup poll, Kennedy's approval rating in the interview week of February 7 to 12, 1963, was 70 percent approved, 18 percent disapproved, and 12 percent with no opinion. In the week of June 21 to 26, following introduction of the civil rights bill, the figures changed sharply to 61, 26, and 13. This reflected a significant Kennedy decline in the South to 33 percent. In the week preceding the assassination, November 11 to 16, he was approved by 59 percent, disapproved by 28 percent, and 13 percent had no opinion. A trial heat during that week between Kennedy and Goldwater showed Kennedy winning 58 to 42 percent, an electoral college landslide, but Goldwater would carry the South 55 to 43. *Gallup Poll*, vol. 3, pp. 1807, 1827, 1847, 1850. According to Robert Kennedy, "We had worked with Goldwater—we just knew he was not a very bright man. He's just going to destroy himself." The President's concern was that he would do so before he got the nomination. In fact, in 1963 he had someone call Walter Lippmann, who had written a devastating article about the Arizonan, to say that it was "too early." *Robert Kennedy in His Own Words*, 373, 392. Nicholas Katzenbach, Oral History Interview, 167, Kennedy Library.

Index